The Placenta and Human Developmental Programming

The Placenta and Human Developmental Programming

Edited by:

Graham J. Burton
Centre for Trophoblast Research, University of Cambridge, Cambridge, UK

David J. P. Barker
MRC Lifecourse Epidemiology Unit, University of Southampton, Southampton General Hospital, UK

Ashley Moffett
Department of Pathology and Centre for Trophoblast Research, University of Cambridge, Cambridge, UK

Kent Thornburg
Heart Research Centre, Oregon Health and Science University, Portland, OR, USA

CAMBRIDGE UNIVERSITY PRESS
Cambridge, New York, Melbourne, Madrid, Cape Town,
Singapore, São Paulo, Delhi, Dubai, Tokyo, Mexico City

Cambridge University Press
The Edinburg Building, Cambridge CB2 8RU, UK

Published in the United States of America by Cambridge
University Press, New York

www.cambridge.org
Information on this title: www.cambridge.org/9780521199452

© Cambridge University Press 2011

This publication is in copyright. Subject to statutory
exception and to the provisions of relevant collective
licensing agreements, no reproduction of any part
may take place without the written permission of
Cambridge University Press.

First published 2011

Printed in the United Kingdom at the
University Press, Cambridge

*A catalogue record for this publication is available from the
British Library*

ISBN 978-0-521-19945-2 Hardback

Cambridge University Press has no responsibility for the
persistence or accuracy of URLs for external or third-party
internet websites referred to in this publication, and does
not guarantee that any content on such websites is, or will
remain, accurate or appropriate.

Every effort has been made in preparing this book to
provide accurate and up-to-date information which is in
accord with accepted standards and practice at the time of
publication. Although case histories are drawn from actual
cases, every effort has been made to disguise the identities of
the individuals involved. Nevertheless, the authors, editors
and publishers can make no warranties that the information
contained herein is totally free from error, not least because
clinical standards are constantly changing through research
and regulation. The authors, editors and publishers
therefore disclaim all liability for direct or consequential
damages resulting from the use of material contained in
this book. Readers are strongly advised to pay careful
attention to information provided by the manufacturer of
any drugs or equipment that they plan to use.

Contents

List of contributors   vii
Preface   xi

1. **Introduction**   1
   Graham J. Burton, David J. P. Barker, Ashley Moffett and Kent Thornburg

2. **The maternal and placental origins of chronic disease**   5
   David J. P. Barker, Johan G. Eriksson, Eero Kajantie, Saleh H. Alwasel, Caroline H. D. Fall, Tessa J. Roseboom and Clive Osmond

   Chapter 2 Discussion   13

3. **Pre- and periconceptual health and the HPA axis: nutrition and stress**   17
   Alan A. Jackson, Graham Burdge and Karen Lillycrop

   Chapter 3 Discussion   32

4. **Nutrition and preimplantation development**   35
   Tom P. Fleming

   Chapter 4 Discussion   44

5. **Maternofetal transport pathways during embryogenesis and organogenesis**   47
   Eric Jauniaux and Graham J. Burton

   Chapter 5 Discussion   54

6. **Imprinted genes and placental growth: implications for the developmental origins of health and disease**   57
   Benjamin Tycko and Rosalind John

   Chapter 6 Discussion   71

7. **Genomic imprinting: epigenetic control and potential roles in the developmental origins of postnatal health and disease**   74
   Elizabeth J. Radford and Anne C. Ferguson-Smith

   Chapter 7 Discussion   91

8. **Trophoblast invasion and uterine artery remodelling in primates**   92
   Robert Pijnenborg, Lisbeth Vercruysse and Anthony M. Carter

   Chapter 8 Discussion   98

9. **The role of the maternal immune response in fetal programming**   102
   Ashley Moffett

   Chapter 9 Discussion   111

10. **Clinical causes and aspects of placental insufficiency**   114
    Irene Cetin and Emanuela Taricco

    Chapter 10 Discussion   123

11. **Uterine blood flow as a determinant of fetoplacental development**   126
    Lorna G. Moore

    Chapter 11 Discussion   144

12. **Placental amino acid transporters: the critical link between maternal nutrition and fetal programming?**   147
    Thomas Jansson and Theresa L. Powell

    Chapter 12 Discussion   158

13 **The maternal circulation and placental shape: villus remodelling induced through haemodynamics and oxidative and endoplasmic reticulum stress** 161
Graham J. Burton and Eric Jauniaux

Chapter 13 Discussion 171

14 **Glucocorticoids and placental programming** 175
Owen R. Vaughan, Alison J. Forhead and Abigail L. Fowden

Chapter 14 Discussion 186

15 **Clinical biomarkers of placental development** 188
Gordon C. S. Smith

Chapter 15 Discussion 198

16 **The placental roots of cardiovascular disease** 201
Kent Thornburg, Perrie F. O'Tierney, Terry Morgan and Samantha Louey

Chapter 16 Discussion 214

17 **Placental function and later risk of osteoporosis** 216
Cyrus Cooper, Laura Goodfellow, Nicholas Harvey, Susie Earl, Christopher Holroyd, Zoe Cole and Elaine Dennison

Chapter 17 Discussion 227

18 **Final general discussion** 229

19 **The placenta and developmental programming: some reflections** 233
Robert Boyd and Richard Boyd

Index 236

The colour plates are to be found between pages 148 and 149.

Contributors

Saleh H. Alwasel
Fetal Programming of Disease Research Chair, College of Science, King Saud University, Saudi Arabia

Susan P. Bagby
Division of Nephrology and Hypertension, Department of Medicine, Oregon Health & Science University, Portland, OR, USA

David J. P. Barker
MRC Lifecourse Epidemiology Unit, University of Southampton, Southampton General Hospital, UK and Heart Research Center, Oregon Health and Science University, Portland, OR, USA

Richard Boyd
Brasenose College, Oxford, UK

Robert Boyd
Maternal and Fetal Health Research Centre, School of Biomedicine, University of Manchester, Manchester, UK

Graham Burdge
Institute of Human Nutrition and National Institutes of Health Research, Nutrition Biomedical Research Unit, Southampton Universities NHS Trust, Southampton General Hospital, Southampton, UK

Graham J. Burton
Centre for Trophoblast Research, University of Cambridge, Cambridge, UK

Anthony M. Carter
Department of Cardiovascular and Renal Research, University of Southern Denmark, Odense, Denmark

Irene Cetin
Unit of Obstetrics and Gynecology, Department of Clinical Sciences 'Luigi Sacco', University of Milan, Italy and Centre for Fetal Research Giorgio Pardi, University of Milan, Italy

Zoe Cole
Clinical Research Fellow, MRC Lifecourse Epidemiology Unit, University of Southampton, Southampton General Hospital, Southampton, UK

Cyrus Cooper
Director and Professor of Rheumatology, MRC Lifecourse Epidemiology Unit, University of Southampton, Southampton General Hospital, Southampton, UK

Hilary Critchley
Centre for Reproductive Biology, University of Edinburgh, Edinburgh, UK

Elaine Dennison
Reader in Rheumatology, MRC Lifecourse Epidemiology Unit, University of Southampton, Southampton General Hospital, Southampton, UK

Susie Earl
Academic Clinical Fellow in Rheumatology MRC Lifecourse Epidemiology Unit, University of Southampton, Southampton General Hospital, Southampton, UK

Johan G. Eriksson
National Institute for Health and Welfare, University of Helsinki Department of General Practice and Primary Health Care, Helsinki University Central Hospital Unit of General Practice, Folkhalsan Research Centre, and Vasa Central Hospital, Finland

Caroline H. D. Fall
MRC Lifecourse Epidemiology Unit, University of Southampton, Southampton General Hospital, UK

List of contributors

Anne C. Ferguson-Smith
Department of Physiology Development and Neuroscience and Centre for Trophoblast Research, University of Cambridge, Cambridge, UK

Tom P. Fleming
School of Biological Sciences, University of Southampton, Southampton, UK

Alison J. Forhead
Department of Physiology Development and Neuroscience, University of Cambridge, Cambridge, UK

Abigail L. Fowden
Department of Physiology Development and Neuroscience, University of Cambridge, Cambridge, UK

Dino Giussani
Department of Physiology, Development and Neuroscience and Centre for Trophoblast Research, University of Cambridge, Cambridge, UK

Laura Goodfellow
Academic F2 in Medicine, MRC Lifecourse Epidemiology Unit, University of Southampton, Southampton General Hospital, Southampton, UK

Nicholas Harvey
Lecturer in Rheumatology, MRC Lifecourse Epidemiology Unit, University of Southampton, Southampton General Hospital, Southampton, UK

Christopher Holroyd
Academic Clinical Fellow in Rheumatology, MRC Lifecourse Epidemiology Unit, University of Southampton, Southampton General Hospital, Southampton, UK

Joan Hunt
University of Kansas Medical Center, Kansas City, KS, USA

Alan A. Jackson
Institute of Human Nutrition and National Institutes of Health Research, Nutrition Biomedical Research Unit, Southampton Universities NHS Trust, Southampton General Hospital, Southampton, UK

Thomas Jansson
Associate Professor, Center for Pregnancy and Newborn Research, Department of Obstetrics and Gynecology, University of Texas Health Science Center, San Antonio, TX, USA

Eric Jauniaux
Institute for Women's Health, Royal Free and University College London Medical School, London, UK

Rosalind John
Genetics Division, Cardiff School of Biosciences, Cardiff University, Cardiff, Wales, UK

Eero Kajantie
National Institute for Health and Welfare, Helsinki, Finland

Michelle Lampl
Department of Anthropology, Emory University, Atlanta, GA, USA

Karen Lillycrop
National Institutes of Health Research, Nutrition Biomedical Research Unit, Southampton Universities NHS Trust, Southampton General Hospital, and Developmental and Cell Biology, University of Southampton, Southampton, UK

Charlie Loke
King's College, Cambridge, UK

Samantha Louey
Heart Research Center, Oregon Health and Science University, Portland, OR, USA

Per Magnus
Division of Epidemiology, Norwegian Institute of Public Health, Oslo, Norway

Ashley Moffett
Department of Pathology and Centre for Trophoblast Research, University of Cambridge, Cambridge, UK

Lorna G. Moore
Dean, Graduate School of Arts & Sciences and Professor, Departments of Public Health Sciences, Anthropology, and Obstetrics & Gynecology, Wake Forest University, Winston-Salem, NC, USA

List of contributors

Terry Morgan
Department of Pathology, Oregon Health and Science University, Portland, OR, USA

Clive Osmond
MRC Lifecourse Epidemiology Unit, University of Southampton, Southampton General Hospital, UK

Perrie F. O'Tierney
Heart Research Center, Oregon Health and Science University, Portland, OR, USA

Robert Pijnenborg
Department of Woman and Child, Katholieke Universiteit Leuven, University Hospital, Leuven, Belgium

Lucilla Poston
Maternal and Fetal health Unit, King's College, London, UK

Theresa L. Powell
Associate Professor, Center for Pregnancy and Newborn Research, Department of Obstetrics and Gynecology, University of Texas Health Science Center, San Antonio, TX, USA

Elizabeth J. Radford
Department of Physiology Development and Neuroscience and Centre for Trophoblast Research, University of Cambridge, Cambridge, UK

Tessa J. Roseboom
Department of Clinical Epidemiology, Biostatistics and Bioinformatics, Academic Medical Center, Amsterdam, The Netherlands

Amanda Sferruzzi-Perri
Centre for Trophoblast Research, University of Cambridge, Cambridge, UK

Colin P. Sibley
School of Biomedicine, University of Manchester, Manchester, UK

Gordon C. S. Smith
Department of Obstetrics and Gynaecology, University of Cambridge, Cambridge, UK

Emanuela Taricco
Unit of Obstetrics and Gynecology, Department of Clinical Sciences 'Luigi Sacco', University of Milan, Italy and Centre for Fetal Research Giorgio Pardi, University of Milan, Italy

Kent Thornburg
Heart Research Centre, Oregon Health and Science University, Portland, OR, USA

Benjamin Tycko
Department of Pathology and Institute for Cancer Genetics, Columbia University Medical Center, New York, NY, USA

Owen R. Vaughan
Centre for Trophoblast Research and Department of Physiology Development and Neuroscience, University of Cambridge, Cambridge, UK

Lisbeth Vercruysse
Department of Woman and Child, Katholieke Universiteit Leuven, University Hospital, Leuven, Belgium

Preface

The size, shape and thickness of the placental surface at birth strongly predict cardiovascular disease, osteoporosis and cancer in later life. These findings suggest that variations in normal placental development lead to fetal undernutrition and hypoxia, and subsequently programme chronic disease. The development of the placenta will be dependent on the mother's nutritional state, but will also be influenced by the processes of decidualization, allocation of cells to the trophectoderm lineage in the pre-implantation embryo, implantation of the blastocyst, trophoblast remodelling of the spiral arteries, growth of the placental villous tree and expression of transporter proteins in the villous membrane. In view of these new important epidemiological findings there is now an urgent need for research into the regulation of placental development in humans.

Chapter 1

Introduction

Graham J. Burton, David J. P. Barker, Ashley Moffett and Kent Thornburg

Developmental programming of the fetus is a phenomenon that has profound implications for the health of individuals and societies. The term describes the process by which gene expression in the fetus is influenced by the intrauterine environment, such that the structure of major organs and the homoeostatic points of metabolic and endocrine systems are set for life. Through this mechanism, perturbations in the intrauterine experience that affect development may predispose to a spectrum of adult diseases, depending on the system principally affected. Thus, in a recent review the National Institute of Child Health concluded that 'coronary heart disease, the number one cause of death among adult men and women, is more closely related to low birth weight than to known behavioural risk factors'. The list of diseases continues to grow, and now also includes diverse metabolic, neoplastic and neurological disorders.

This striking effect first came to light through epidemiological studies of men in Hertfordshire, UK, who had died from cardiovascular disease. Through the records maintained by the midwives attending the births of these men it was possible to show that death rates from the disease fell across the normal range of birth weight. In a later study of blood pressure levels among men and women in Preston, UK, it was possible to relate birth weight to the weight of the placenta. People with small placentas and people with large placentas in relation to their birth weights had the highest blood pressure levels. Although these data highlighted the importance of the fetoplacental relationship to developmental programming, the mechanistic role of the placenta in the phenomenon has received little attention to date. This is a remarkable oversight, given that the placenta evolved to support the fetus *in utero*, and must therefore reasonably be expected to be a key determinant of fetal growth.

This book records the proceedings of a scientific meeting convened to rectify this situation. The meeting, held over two days in Cambridge in December 2009, brought together invited experts from a wide variety of disciplines to present and discuss their data in depth under the skilful and diplomatic guidance of the Chair, Professor Sir Robert Boyd. The contributions presented here explore our current knowledge of the ways in which various aspects of placental development and function may influence fetal programming, and aim to promote further scientific research in their respective fields. When structuring the meeting, we aimed to consider the development of the placenta from the level of molecular genetics through to epidemiological biomarkers, and from pre-conception through to maturity, with the emphasis being placed on human data.

The starting point of the meeting was that changes in the shape of the placental surface, whether it is round or oval, are related to developmental programming. Little attention has been paid to the importance of placental shape in recent years, or to its determinants. What is so striking, however, is that placental shape changed radically over the 20-year period of the Helsinki study, indicating that it is surprisingly plastic and responsive to intrauterine cues. Equally, it is not known whether the tissues along the major and minor axes of the placenta perform the same functional roles. The placenta has a wide variety of functions, including passive and active transport, hormone synthesis, metabolic regulation, acting as a selective barrier to maternal hormones and as an immunological shield. We must therefore be precise when we talk in terms of placental efficiency and insufficiency, and refer to the specific function being monitored.

Morphologically, placental development starts at the time of implantation, but can of course be

---

The Placenta and Human Developmental Programming, ed. Graham J. Burton, David J. P. Barker, Ashley Moffett and Kent Thornburg. Published by Cambridge University Press. © Cambridge University Press 2011.

Chapter 1. Introduction

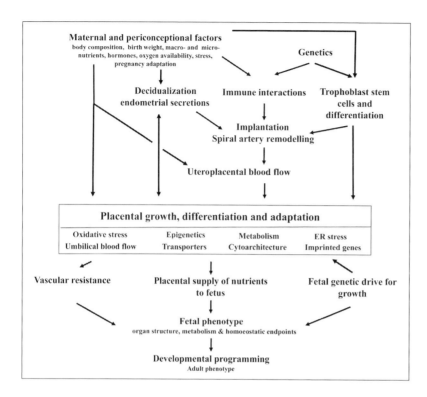

Figure 1.1 Diagrammatic representation of the major factors affecting placental development and function, and the ways in which these may affect the developmental programming of the fetus.

influenced by a variety of preceding events. For example, micronutrients influence the antioxidant composition of the antral fluid during folliculogenesis and the quality of the ovum released. Equally, in animal models maternal diet alters the allocation of cell lineage during formation of the blastocyst, affecting the development of the extraembryonic membranes. Similar perturbations have been observed following culture during *in vitro* fertilization procedures, particularly when the conditions are suboptimal and operate through epigenetic modifications. Thus, greater attention should be given to the pre- and periconceptional care of women in terms of their diet, body mass index and general well-being.

The development of the placenta is not autonomous, but is clearly heavily influenced by the uterine mucosa with which the trophoblast interacts. Appropriate preparation of the endometrium in each menstrual cycle is of central importance to successful implantation, yet surprisingly little is known about either the process of decidualization or, conversely, the triggers for endometrial breakdown at menstruation. Maternal nutrition has a profound impact on the hypothalamopituitary axis, and the effects of anorexia on ovarian function are well documented. In contrast, nothing is known regarding its possible impact on the endometrium and decidualization. It is now appreciated that the influence of the endometrium and its glandular secretions extends well beyond the time of implantation, and that the placenta and fetus are supported by the secretions until the end of the first trimester. Evidence from animal species indicates that the conceptus signals to the glands to upregulate the supply of nutrients and growth factors to meet its needs. Whether a similar mechanism operates in the human remains to be elucidated, but this is of key importance given that the major organ systems are differentiating at this time.

Towards the end of the first trimester there is a need for a greater oxygen supply, which can only be provided by the maternal arterial system. Establishing the intraplacental circulation in the highly invasive haemochorial form of placentation displayed by the human is fraught with immunological and haemodynamic challenges. During implantation, trophoblast cells migrate from the conceptus deep into the decidua and inner myometrium, where their function is to transform the maternal spiral arteries (Figure 1.1). This invasion is far more extensive in humans and closely related great apes than in other primates, raising questions about the evolutionary drive for such a complex and potentially hazardous process. The molecular mechanisms underpinning spiral arterial conversion are not fully understood, but after being surrounded by trophoblast, the smooth muscle in the arterial walls is replaced by an inert fibrinoid material so that the arteries are no longer able to contract and potentially compromise placental blood flow. To achieve this end, the semi-allogeneic trophoblast cells must negotiate components of the maternal uterine immune system. It is now appreciated that interactions between parental MHC ligands expressed by the invading trophoblast cells (HLA-C molecules) and the NK receptors on the maternal decidual natural killer cells (KIR) are an important determinant of pregnancy outcome. This interaction is thought to have a physiological function in terms of regulating trophoblast invasion, necessitating a paradigm shift in our understanding of the roles of these immune cells. Macrophages, dendritic cells and T cells – including T regulatory cells – are also present in the decidua and myometrium, and their role during early placentation requires further research.

The importance of an adequate uterine blood flow bringing a supply of nutrients and oxygen to support fetal growth is highlighted by studies of populations living at high altitude, which can be considered experiments of nature. Blood flow is maintained or even increased in indigenous populations, but reduced in recent migrants, in whom it correlates with a reduction in birth weight. Understanding the mechanisms underlying these adaptations may throw light on important placental homoeostatic pathways.

Maternal factors such as stress and cigarette smoking have a profound but transient impact on uteroplacental blood flow. Stress may also cause elevated levels of glucocorticoids, and recent research on laboratory rodents has shown that these can adversely affect placental differentiation. Glucocorticoids are administered routinely in cases of threatened preterm deliveries to accelerate maturation of the lungs and other organ systems. Although this may provide immediate clinical benefits postnatally, there may be later consequences in terms of fetal programming.

The placenta is not passive to these insults but highly adaptive, responding to various challenges with changes in cellular architecture, vascularity and transporter expression (Figure 1.1). Hence, placental size is not an accurate estimate of placental capacity, and thought should be given to whether the efficiency is best expressed per unit exchange surface area or in some other form. Many of these structural adaptations and changes in transporter expression are regulated through imprinted genes, which allow gene dosage to be strictly controlled and are thought to play crucial roles in the partitioning of nutrients between the mother and her fetus. There are clearly important signals that pass from the fetus back to the placenta regulating expression of these genes: members of the insulin-like growth factor family appear to be strong candidates for these signalling functions. Further research is required to elucidate the components of this fascinating dialogue.

A central question in this respect is whether the placenta, the fetus or both act as the nutrient sensor. The mTOR signalling pathway performs this function at the cellular level, regulating proliferation and differentiation in response to amino acid availability. Activity in this pathway is, however, perturbed by the closely interacting conditions of oxidative and endoplasmic reticulum stress. Thus, although adaptations in placental function in response to fluctuations in nutrient supply may occur in the normal placenta, this may not be possible in cases of pathology secondary to malperfusion, when these stresses are present.

Assessment of placental function *in vivo* is obviously important for clinical diagnosis and monitoring. Most of the biochemical tests used to predict poor placentation and fetal growth originate from other screening tests, for example for trisomy 21, and are therefore not specific for fetal growth retardation. However, in these cases placental products such as pregnancy-associated placental protein-A (PAPP-A) are reduced during the first trimester. Whether this reflects a small placental mass or impaired placental secretory activity due, for example, to endoplasmic reticulum stress, is not known. The difficulty of obtaining placental samples from ongoing pregnancies represents a major problem for research in this area. Ultrasonography of the placenta and uterine artery waveforms combined with a wider range of placental biomarkers may provide a way forward.

The capacity of the placenta to supply adequate nutrients to the fetus is obviously of central importance to the role of the organ in developmental programming, but other aspects of placental function may also operate. The placenta represents a major component of the fetal circulation, and changes in umbilical vascular resistance, through, for example, a reduction in villous vascularization or an imbalance in vasoactive agents, will affect the mechanical loadings on the developing heart (Figure 1.1). Proliferation and differentiation of cardiomyocytes is sensitive to such factors, and so cardiac development may be impaired as a result.

Ultimately, developmental programming arises through the tension between the genetic drive for growth of the fetus and the intrauterine environment to which the fetus is exposed. Changes in the supply of macro- and micronutrients can heavily influence the differentiation of fetal organs and tissues, and alter the homoeostatic set points of endocrine axes. It is important for public health to stress that these effects operate across the birth weight range and are not confined to extreme low-birth weight individuals.

We thank all our participants for their stimulating contributions, the Centre for Trophoblast Research (www.trophoblast.cam.ac.uk) for sponsoring the meeting, and Cambridge University Press for publishing this book, which we hope will provoke interest into the role of a fascinating but often neglected organ, the placenta, in developmental programming.

Chapter 2

The maternal and placental origins of chronic disease

David J. P. Barker, Johan G. Eriksson, Eero Kajantie, Saleh H. Alwasel, Caroline H. D. Fall, Tessa J. Roseboom and Clive Osmond

Introduction

Fetal programming is the process whereby environmental influences acting during development alter gene expression and program the body's structures and functions for life. There is now a body of evidence showing that chronic diseases of adult life, including cardiovascular disease, type 2 diabetes and certain cancers, originate through this process. These diseases are initiated by adverse influences during development [1]. These adverse influences may also slow the growth of the fetus and reduce its body size at birth and during infancy. Compared with people whose birth weights were towards the upper end of the normal range, those whose birth weights were towards the lower end have (a) reduced functional capacity, including fewer nephrons in the kidney; (b) altered metabolic settings, including insulin resistance; and (c) altered production of hormones, including stress and sex hormones. In a recent review the National Institute of Child Health concluded that 'coronary heart disease, the number one cause of death among adult men and women, is more closely related to low birth weight than to known behavioural risk factors' [2]. To date, most epidemiological observations that have demonstrated this link between prenatal life and later disease have used birth weight as the marker of early life. A baby's birth weight reflects its success in obtaining nutrients from its mother [3]. The source of these nutrients is not only the mother's current diet, but her nutritional stores and metabolism, which are the product of her lifetime's nutrition [4]. A baby's birth weight also depends on the placenta's ability to transport nutrients to it from its mother [3]. The placenta seems to act as a nutrient sensor, regulating the transfer of nutrients to the fetus according to the mother's ability to deliver them and the fetus's demands for them [5]. At the outset it is reasonable to assume that how the placenta programs the fetus will depend on the mother's lifetime nutrition.

This review uses new epidemiological data to examine the role of materno-placental interactions in initiating chronic disease in the offspring. It has been possible to pursue this issue epidemiologically because placental function and maternal nutrition are reflected in placental size at birth and in maternal body size, and these can be related to the later occurrence of disease.

Measurements of placental size and shape that reflect its function

The size, weight and shape of the placenta are all subject to wide variations [6]. Its size reflects its ability to transfer nutrients [7]. Small babies generally have small placentas but, in some circumstances, an undernourished baby can expand its placental surface to extract more nutrients from the mother [8]. Low placental weight at birth has been shown to predict hypertension and coronary heart disease in later life [9,10]. The weight of the placenta, however, does not distinguish its surface area from its thickness. In order to increase the surface for nutrient and oxygen exchange, the placenta can increase the surface of its villi, expand its invasion across the surface of the uterine lining or invade the maternal spiral arteries more deeply. The long-term consequences of these three responses may be different.

Placental shape

In the last century the surface of the placenta was described as being either 'oval' or 'round' [11–13]. In order to describe the extent to which the surface was more oval than round, two so-called 'diameters' of the surface were routinely recorded in some hospitals, a maximal diameter (the length of the

---

The Placenta and Human Developmental Programming, ed. Graham J. Burton, David J. P. Barker, Ashley Moffett and Kent Thornburg. Published by Cambridge University Press. © Cambridge University Press 2011.

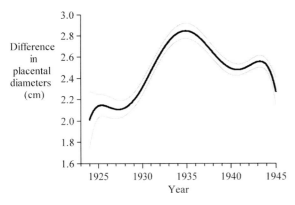

Figure 2.1 Trend line (95% confidence interval) for the difference between the two placental diameters over 21 years in the Helsinki Birth Cohort.

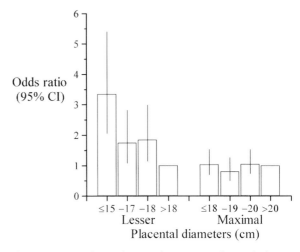

Figure 2.2 Hazard ratios for pre-eclampsia according to the lesser and maximal placental diameters.

surface), and a lesser one bisecting it at right-angles (the breadth) [14]. These measurements are available in the Helsinki Birth Cohort Study. The cohort comprised 20 431 men and women born in the city during 1924–44 [15]. The mean lengths of the two diameters, which were measured to the nearest centimeter, were 19.5 cm and 16.9 cm, and their lengths were highly correlated (correlation coefficient = 0.63). Figure 2.1 shows that the difference between the lengths of the two diameters fluctuated during the 21 years of births, the general trend being for the difference to increase so that the placental surfaces became more oval. The diameters were used to estimate the surface area ($\pi \times$ lesser $\times$ maximal diameter $\times$ 4). How closely this area reflects the total surface area for materno-fetal exchange through the gestational period is not known, but it is reasonable to suppose that they correlate. Combined with placental weight, the two diameters gave an estimate of placental thickness (weight/area).

Pre-eclampsia is associated with reduced placental size. The abnormal placentation is a result of impaired invasion of the maternal spiral arteries by the trophoblast at implantation [16]. Of the mothers of the younger part of the Helsinki birth cohort, born 1934–44, 6410 had their blood pressures and the results of urinary protein tests recorded after 20 weeks of pregnancy. Of these pregnancies, 284 were complicated by pre-eclampsia [17]. Compared to those from normotensive pregnancies, the placentas from these pregnancies had a reduced surface area but the thickness was increased. The increase in thickness in pre-eclampsia could be compensatory for restricted expansion of the surface.

Placentas from pregnancies complicated by pre-eclampsia had a more oval surface than those from normotensive pregnancies because of a disproportionate reduction in the breadth. When the two diameters were analysed together, pre-eclampsia was not associated with the size of the maximal diameter but was strongly associated with a short lesser diameter. Figure 2.2 shows that the relation with the lesser diameter was graded: the shorter the lesser diameter, the greater the risk for and severity of pre-eclampsia. This association with a short lesser diameter depended on its absolute length rather than on its length in relation to the surface area or weight. Processes that underlie the disease may therefore be closely linked to the absolute amount of placental tissue on the lesser diameter. This link may be through a structure or function that it does not share with tissue on the maximal diameter. Alternatively, growth along the lesser diameter may be controlled differently from growth along the major axis. Placental growth may be polarized from the time of implantation, so that growth along the major axis, defined by the maximal diameter, is qualitatively different from that along the minor axis [17]. One possibility is that growth along the major axis is aligned with the rostrocaudal axis of the embryo, whereas tissue along the minor axis may be the nutrient sensor. Tissue on the minor axis may be more sensitive to the mother's nutrition than tissue along the major axis, and it may be more important for nutrient transfer to the fetus.

Table 2.1 Mean systolic blood pressure of men and women aged 50, born at term.

Birth weight (pounds)	Placental weight (pounds)			
	≤1.0	−1.25	−1.5	>1.5
≤6.5	149	152	151	167
−7.5	139	148	146	159
>7.5	131	143	148	153

Placental enlargement

Low placental weight is associated with an increased risk of hypertension in later life [9]. However, a study of men and women born in a maternity hospital in Preston, UK, showed that high placental weight in relation to birth weight is also associated with later hypertension [18]. Table 2.1 shows that, as expected, at any placental weight lower birth weight was associated with higher systolic pressure but at any birth weight higher placental weight was associated with higher systolic pressure. The highest systolic pressure was in people who had the lowest birth weights but the highest placental weights. This observation has been replicated, and high placental weight in relation to birth weight has also been shown to predict coronary heart disease [9,19]. Observations in sheep show that in response to undernutrition in mid-gestation the fetus is able to extend the area of the placenta by expanding the individual cotyledons [8,20]. This increases the area available for nutrient and oxygen exchange, and results in a larger lamb than there would otherwise have been. This is profitable for the farmer, and manipulation of placental size by changing the pasture of pregnant ewes is standard practice in sheep farming. The associations with large placental size suggest that cardiovascular disease can be programmed in association with placental overgrowth as well as restricted placental growth.

Placental efficiency

In Table 2.1, at any birth weight people who had lower placental weights had lower systolic pressures. People in the highest birth weight group but the lowest placental weight group had the lowest systolic pressure. A low ratio of placental weight to birth weight may be an index of placental efficiency, as proportionately more resource has been invested in the growth of the baby as opposed to placental growth. In a recent study of 7000 newborn babies born between 2000 and 2004 in Unizah, a small city in Saudi Arabia, the mean birth weight approached the mean for western populations (Alwasel, unpublished); but the mean placental weight, 473 g in boys and 467 g in girls, and the mean placental to birth weight ratios, 14.6 and 14.9%, were well below those in a recent compilation of European reference values [21]. These differences do not seem to be the result of differences in the procedures used to weigh the placentas. In experimental studies in which uterine blood flow is reduced [22], and in clinical studies [23], a relative decrease in placental size is evident before changes in fetal growth. An interpretation of the findings in Saudi Arabia is that placentas respond to mothers' limited ability to deliver nutrients to them. Placental growth slows but efficiency is increased, so that fetal growth is sustained, albeit with a reduced reserve capacity.

Maternal influences on placental size and shape

In humans, placental growth responds to maternal influences. Maternal anemia and high maternal body mass index are associated with a high placental weight to birth weight ratio [24,25]. Maternal smoking reduces both placental weight and birth weight. The diets of mothers during pregnancy, and their physical activity, are also known to be associated with altered placental weight [5,25].

Changing lifestyle during Ramadan

In Islam, Ramadan is an annual period of day-time fasting that lasts for one month. During Ramadan people in Saudi Arabia change their life style. They take no food or water from dawn until sunset, when they break their fast by eating sweet and fried meals. The next meal is 'Sahoor', which is usually eaten before dawn and comprises foods rich in fat. People reduce their activities during the day, but are more active at night. Although pregnant women are allowed to defer fasting until after the pregnancy, they often prefer to share the spiritual and social experiences of Ramadan with their families. In the studies in Unizah the mean birth weight of babies who were *in utero* during Ramadan was the same as that of babies who were not. However, the mean placental weight of those in the second or third trimester of gestation during Ramadan was lower than in those who were not *in utero* during Ramadan (Alwasel, unpublished). The mean ratio of placental weight to birth weight was also lower. Among boys the mean was 14.4% compared to 14.9 in those not *in utero* during Ramadan. The corresponding figures for girls were 14.7 and 15.0. These findings suggest that the Ramadan lifestyle increases placental efficiency.

The Dutch famine

The 'Dutch famine' was a six-month period of severe food restriction in the west of Holland during the last years of the Second World War, 1944–5. The Dutch famine birth cohort consists of 2414 men and women who were born as term singletons in the Wilhelmina Gasthuis in Amsterdam between 1 November 1943 and 28 February 1947 [26]. The birth records do not include placental weight, but the two diameters were recorded and thickness measured using calipers. Babies who were *in utero* during the famine had lower birth weight and placental area than babies born before, or conceived after, the famine, but their placental thickness was increased (van Abeelen, unpublished).

Maternal/placental programming

Blood pressure

Placental weight predicts the risk of hypertension in the offspring in later life [9,18,27]. Associations with both low placental weight and a high ratio of placental weight to birth weight have been reported. Similarly, placental weight has inconsistent associations with blood pressure levels in children. Again, there are reported associations between increased blood pressure levels and low placental weight and a high ratio of placental weight to birth weight [28,29]. Other studies have found no associations [30]. The observations on hypertension, reviewed below, suggest that these different findings may reflect different materno-placental phenotypes that programme hypertension.

Hypertension in adults

Compensatory expansion of the placental surface in the sheep fetus, in response to undernutrition in mid-gestation, has already been described. This response only occurs, however, if the ewe was well nourished before conceiving [8]. There is evidence of a similar constraint in humans: 2003 members of the Helsinki Birth Cohort were randomly selected to attend a clinic,

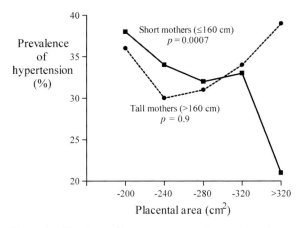

Figure 2.3 Prevalence of hypertension according to placental area in the offspring of short and tall mothers.

644 of them were being treated for hypertension. The prevalence of hypertension fell as placental area increased [14]. The effects of placental area on hypertension, however, varied with the mother's height. Figure 2.3 shows the trends in the prevalence of hypertension in people divided according to whether their mother's height was above or below the median of 160 cm. In people whose mothers were short the prevalence of hypertension fell progressively with increasing placental area. In people with tall mothers the association was U-shaped.

The strong effects of small placental area in people born to short mothers suggest a link between hypertension and the lifetime nutrition of the mother. Short maternal stature is a product of poor fetal or childhood nutrition, or recurrent exposure to infections, though there are also genetic influences [31]. Protein metabolism is established in early life and is related to visceral mass. Short women have less visceral mass than tall women, and have lower rates of amino acid synthesis in pregnancy [32]. An explanation for the combined effect of a short mother and a small placental area on hypertension is that the effects of reduced availability of amino acids in the maternal circulation are exacerbated by restricted placental growth, which limits the transport of amino acids from mother to baby.

Among people with tall mothers there was no overall effect of placental area on hypertension (Figure 2.3). However, when these people were subdivided by their parents' socio-economic status, opposing trends were revealed. In lower social class families, hypertension was associated with small placental area and low placental weight. In middle class families, hypertension was associated with large placental area and high placental weight in relation to birth weight. Tall mothers in middle-class families are likely to have been the best nourished before conceiving. During pregnancy, however, they would have been subject to the food scarcities that occurred in Finland before and during the Second World War [33]. Similarly to sheep, their fetuses may have responded to this by expanding the placental surface. Compensatory placental growth may be beneficial in some circumstances, but if the compensation is inadequate and the fetus continues to be undernourished, the need to share its nutrients with an enlarged placenta may become an added metabolic burden. A long-term cost of this added burden is hypertension, possibly as a result of impaired development of low-priority organs such as the kidney [34].

Of the two diameters of the placental surface the lesser was more closely associated with hypertension [14]. The interaction between the effects of maternal height and placental area (Figure 2.3) was reflected in a similar interaction between the effects of maternal height and the lesser placental diameter; but there was no interaction between mother's height and the maximal diameter. These findings support the idea that growth along the breadth of the placental surface is more nutritionally sensitive than growth along the length. Expansion of the placental surface along its breadth may be one way in which a fetus may compensate for undernutrition.

Blood pressure in children

The first observations on the relationship between placental surface and blood pressure in children were made on 471 children born in the city of Mysore in India [35, Fall unpublished]. Their blood pressures at nine years of age were inversely related to their birth weights. In boys, systolic pressures rose as placental weight and area decreased. The pressures also rose as the breadth of the surface decreased, but there was no similar trend with the length of the surface. This is consistent with the findings among adults in Finland and Holland, in whom hypertension was more closely related to the breadth of the surface than to its length [14, van Abeelen, unpublished]. It is again consistent with the idea that tissue along the breadth of the surface is more closely related to nutrient delivery to the fetus than tissue along the length of the surface, so that shorter breadth is associated with lesser delivery.

The effects of placental size on blood pressure levels among the girls interacted with the mothers' height. Among girls whose mother's height was below the median, systolic pressure rose as the ratio of placental area to birth weight increased. This association depended more on expansion of the surface along its breadth than along its length. The finding can be interpreted as further evidence that raised blood pressure is programmed in association with compensatory placental expansion, though in this instance the compensatory expansion occurred in the offspring of short mothers. Among girls whose mothers' height was above the median, systolic blood pressure rose as the difference between the length and breadth of the surface increased, that is, as the surface became more oval. On its own, small breadth was not related to systolic pressure. It was small breadth in relation to the length that predicted blood pressure. An oval placental surface may be a marker of maternal malnutrition that selectively affects growth in the breadth of the surface, because this is more sensitive to the mother's nutrition than growth in length.

Coronary heart disease

The observations on hypertension established that the relation between placental size/shape and fetal programming depend on the mother. In short mothers a large placental area was beneficial (Figure 2.3); in tall mothers it was deleterious, presumably reflecting the metabolic costs of compensatory placental expansion. There are similar materno-placental interactions in the programming of coronary heart disease. In the Helsinki Birth Cohort coronary heart disease is associated with low birth weight, an association that has been shown in studies around the world [15,36]. Among men, three different materno-placental–fetal phenotypes were associated with coronary heart disease (Erikkson unpublished). The phenotypes were defined by the mother's height and body mass index, and by the placenta having either a small surface, an oval surface, or being heavy in relation to birth weight. Common to each of the phenotypes was that the babies were thin at birth, which suggests that the materno-placental phenotypes led to fetal undernutrition. This may have affected the development of the heart (see Chapter 16).

Chronic heart failure

People in the Helsinki Birth Cohort who developed chronic heart failure were born with a small placental surface area. This association with a small surface was the result of reduced growth along the lesser axis of the surface. The effects of a reduced surface area were confined to people whose mothers were below the median height and whose fathers were manual workers. The reduced rates of protein synthesis in short mothers may have amplified the adverse effects of impaired placentation on the fetus. These effects may have been exacerbated by poor maternal nutrition during pregnancy. There were food shortages in Helsinki before and during the Second World War, the time when our subjects were born [33]. These may have been more severe in families of low socio-economic status. Animal studies support a relationship between fetal malnutrition and heart failure (see Chapter 16). Increased loading of the heart in fetuses with a small placenta is likely to reduce cardiomyocyte numbers for life and hence make the heart vulnerable to developing chronic failure.

Lung cancer

The findings for coronary heart disease suggest that different combinations of maternal and placental size and shape lead to fetal undernutrition and thinness at birth. Findings for lung cancer suggest that different materno-placental combinations also lead to an imbalanced nutrient supply to the fetus. In the Helsinki Birth Cohort men and women who developed lung cancer were short at birth in relation to their weight, so that they had a high ponderal index [37]. One possibility is that this reflected low amino acid – high glucose delivery to the fetus, which impaired the development of its antioxidant systems and made the lungs vulnerable to tobacco smoke and other carcinogens in later life. Three different materno-placental–fetal phenotypes were associated with lung cancer [38]. Common to each was a short mother and a newborn baby that was short in relation to its weight.

The three phenotypes, defined by the mother's height, body mass index (weight/height^2, BMI) were:

(1) *short, low BMI mother: small placental area*. In people whose mothers were short and whose BMI was below the median, lung cancer was associated with a small placental area. The reduced area for amino acid transfer from mother to fetus may have added to the effects of low maternal amino acid synthesis associated with the mother's short stature [32]. Birth weight was less severely affected than length, because it reflects the supply of glucose as well as amino acids.

(2) *Short, high BMI mother: large placental area*. In people whose mothers were short but whose BMI was above the median, lung cancer was associated with a large placental area. This could have led to a high ponderal index through low amino acid availability to the fetus, accompanied by high glucose availability, associated with high maternal BMI. Maternal glucose metabolism is an important determinant of fetoplacental growth [39]. It may be a key environmental cue to which the placenta responds in order to match fetal growth rate with the resources made available by the mother.

(3) *Short, low BMI mother: large placental area*. In people whose mothers were short and whose BMI was below the median, lung cancer was also predicted by large placental area. This contrasts with the second of the three phenotypes and suggests that lung cancer may be associated with large placental area through two different aspects of placental growth. One possibility is that whereas in the second phenotype large placental area depended on high maternal BMI and high circulating maternal glucose, in this third phenotype it was the result of compensatory expansion of the placental surface in late gestation.

Sex differences in placental growth and programming

Sex differences in placental growth

Boys grow faster than girls from an early stage of gestation and this makes them more vulnerable if their nutrition is compromised [31,40]. Among boys and girls in the Helsinki Birth Cohort the small differences in the average measurements of the baby and placenta concealed large differences in body proportions [41]. At any placental weight boys tended to be longer than girls; and boys' placentas were smaller than girls' placentas in relation to the weight of the baby [42]. Similarly, in the Saudi Arabian study the ratio of placental weight to birth weight was similarly lower in boys than, at 14.6% compared to 14.9%, girls

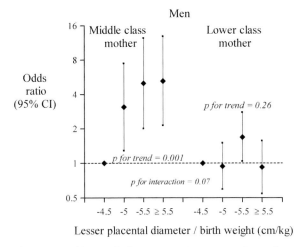

Figure 2.4 Odds ratios for hypertension in men according to the mother's socio-economic status and the ratio of the lesser placental diameter to birth weight.

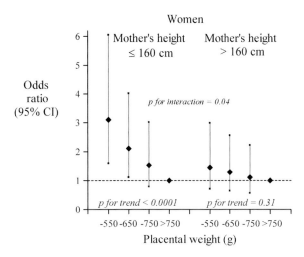

Figure 2.5 Odds ratios for hypertension in women according to mother's height and placental weight.

(Alwasel, unpublished). This suggests that boys' placentas are more efficient but may have less reserve capacity, which therefore increases their vulnerability to undernutrition. There were sex differences in placental responses to the Dutch famine. In boys the ratio of placental area to birth weight was reduced, whereas in girls it was increased (van Abeelen, unpublished).

Boys tend to have larger head circumferences at birth but are thinner than girls [41]. This suggests that they more readily trade off visceral development to protect brain growth. The growth of every human fetus is constrained by the limited capacity of the mother and placenta to deliver nutrients to it [43]. The male fetus, by growing more rapidly and investing in brain growth rather than placental growth, is adopting a more dangerous strategy that puts it at greater risk of becoming undernourished.

Sex differences in placental programming

In the Helsinki Birth Cohort low birth weight predicted hypertension in both sexes, but the relationship between placental size and later hypertension differed between the sexes [42]. In men, hypertension was not related to placental weight or area, but was predicted by a large lesser diameter to birth weight ratio. This points to an association with compensatory expansion of the placental surface along its breadth. In women, hypertension was associated with low placental weight and small placental area. The association with small area depended more on a short breadth than a short length.

Among men the trend in hypertension with a large lesser diameter in relation to birth weight was confined to those whose mothers were middle class (Figure 2.4), but the trends did not differ with the mother's height. Among women the trends in hypertension with low placental weight did not differ with the mother's social class, but were stronger in those whose mothers were short (Figure 2.5). A general interpretation of these findings could be that during development *in utero* boys respond to their mother's diet, as indicated by her socio-economic status, whereas girls respond more to their mother's lifetime nutrition and metabolism, as indicated by her height. The greater effect of the Dutch famine on the number of liveborn boys supports this hypothesis [44], as do the greater effects of experimental maternal malnutrition on male animals [45,46]. Whereas boys' responsiveness to their mothers' current diet enables them to capitalize on an improving food supply and promotes their agenda of rapid growth, it makes them vulnerable to food shortages. The Dutch famine changed the relationship between the lesser diameter and later hypertension in men but not in women (van Abeleen, unpublished). This again suggests that alterations in diet during pregnancy more readily alter the placental programming of chronic disease among men than among women. In men who were born before or conceived after the famine, short breadth of the surface predicted later hypertension. In men who were *in utero* during the famine long breadth predicted hypertension. There were no similar effects of the length of the surface.

Conclusions

This review of the maternal and placental programming of chronic disease leads to five conclusions. (a) Growth of the placental surface is polarized from the time of implantation. Growth along the major axis may align with the rostrocaudal growth of the fetus, whereas tissue along the minor axis may be more important for transfer of nutrients to the fetus. (b) The human fetus may attempt to compensate for undernutriton by expansion of the placental surface along its minor axis. This only occurs if the mother was well nourished before conceiving, and may have long- term costs that include hypertension, coronary heart disease and cancer. (c) Placental efficiency, as indicated by the ratio of placental weight to birth weight, differs between populations. (d) The effects of placental size on long-term health are conditioned by the mother's nutritional state, as indicated by her socio-economic status, height and body mass index. (e) The maternal/placental programming of chronic disease differs in boys and girls. Boys invest less than girls in placental growth but more readily expand the placental surface if they become undernourished in mid-late gestation. In fetal programming boys are more responsive to their mothers' current diets, whereas girls respond more to their mothers' lifetime nutrition and metabolism.

References

1. **Barker DJP**. Fetal origins of coronary heart disease. *Br Med J* 1995; **311**: 171–4.
2. **NICHD**. *Pregnancy and Perinatology Branch: Strategic Plan 2005–2010*. NIH, 2003.
3. **Harding JE**. The nutritional basis of the fetal origins of adult disease. *Int J Epidemiol* 2001; **30**: 15–23.
4. **Jackson AA**. All that glitters. Br Nutr Foundation Annual Lecture. *Nutr Bull* 2000; **25**: 11–24.
5. **Jansson T, Powell TL**. Role of the placenta in fetal programming: underlying mechanisms and potential interventional approaches. *Clin Sci* 2007; **113**: 1–13.
6. **Hamilton WJ, Boyd JD, Mossman HW**. *Human Embryology*. Cambridge: W. Heffer & Sons, 1945.
7. **Sibley CP**. In Case RM, Waterhouse JM, eds. *Human Physiology: Age, Stress, and the Environment*. Oxford: Oxford University Press, 1994; 3–27.
8. **McCrabb GJ, Egan AR, Hosking BJ**. Maternal undernutrition during mid-pregnancy in sheep: variable effects on placental growth. *J Agric Sci* 1992; **118**: 127–32.
9. **Eriksson J, Forsen T, Toumilheto J, Osmond C, Barker D**. Fetal and childhood growth and hypertension in adult life. *Hypertension* 2000; **36**: 790–4.
10. **Forsén T, Eriksson JG, Tuomilehto J** et al. Mother's weight in pregnancy and coronary heart disease in a cohort of Finnish men: follow up study. *Br Med J* 1997; **315**: 837–40.
11. **Hinselmann H**. *Biologie und Pathologie des Weibes*. Berlin: Urban & Schwarzenberg, 1925.
12. **Anderson MC**. *Lessons in Midwifery for Nurses and Midwifes*. London: A & C Black, 1930.
13. **Mays M**. *An Introduction to Midwifery*. London: Faber and Faber 1930.
14. **Barker DJP, Thornburg KL, Osmond C, Kajantie E, Eriksson JG**. The surface area of the placenta and hypertension in the offspring in later life. *Int J Dev Biol* 2010; **54**: 525–30.
15. **Barker DJP, Osmond C, Kajantie E, Eriksson JG**. Growth and chronic disease: findings in the Helsinki Birth Cohort. *Ann Hum Biol* 2009; **36**: 445–58.
16. **Roberts JM, Cooper DW**. Pathogenesis and genetics of preeclampsia. *Lancet* 2001; **357**: 53–6.
17. **Kajantie E, Thornburg K, Eriksson JG, Osmond C, Barker DJP**. In pre-eclampsia the placenta grows slowly along its minor axis. *Int J Dev Biol* 2010; **54**: 469–73.
18. **Barker DJP, Bull AR, Osmond C, Simmonds S**. Fetal and placental size and risk of hypertension in adult life. *Br Med J* 1990; **301**: 259–62.
19. **Martyn CN, Barker DJP, Osmond C**. Mothers pelvic size, fetal growth and death from stroke in men. *Lancet* 1996; **348**: 1264–8.
20. **McCrabb GJ, Egan AR, Hosking BJ**. Maternal undernutrition during mid-pregnancy in sheep. Placental size and its relationship to calcium transfer during late pregnancy. *Br J Nutr* 1991; **65**: 157–68.
21. **Burkhard T, Schaffer L, Schneider S, Zimmerman R, Kurmanavicius J**. Reference values for the weight of freshly delivered term placentas and for placental weight – birth weight ratios. *Eur J Obstet Gynaecol* 2006; **128**: 248–52.
22. **Jansson T, Thordstein M, Kjellmer I**. Placental blood flow and fetal weight following uterine artery ligation. *Biol Neonate* 1986; **49**: 172–80.
23. **Hafner E, Metzenbauer M, Hofinger D** et al. Placental growth from the first to the second trimester of pregnancy in SGA-foetuses and pre-eclamptic

24. **Godfrey KM, Redman CWG, Barker DJP, Osmond C**. The effect of maternal anaemia and iron deficiency on the ratio of fetal weight to placental weight. *Br J Obstet Gynaecol* 1991; **98**: 886–91.
pregnancies compared to normal foetuses. *Placenta* 2003; **24**: 336–42.

25. **Godfrey KM**. The role of the placenta in fetal programming. *Placenta* 2002; **23** (Suppl A): S20–S27.

26. **Ravelli ACJ, Van Der Meulen JHP, Michels RPJ** et al. Glucose tolerance in adults after prenatal exposure to the Dutch famine. *Lancet* 1998; **351**: 173–7.

27. **Campbell DM, Hall MH, Barker DJP** et al. Diet in pregnancy and the offspring's blood pressure 40 years later. *Br J Obstet Gynaecol* 1996; **103**: 273–80.

28. **Moore VM, Miller AG, Boulton TJ** et al. Placental weight, birth measurements, and blood pressure at age 8 years. *Arch Dis Child* 1996; **74**: 538–41.

29. **Taylor SJC, Whincup PH, Cook DG, Papacosta O, Walker M**. Size at birth and blood pressure; cross sectional study in 8–11 year old children. *Br Med J* 1997; **314**: 475–80.

30. **Whincup P, Cook D, Papacosta O, Walker M**. Birth weight and blood pressure: cross sectional and longitudinal relations in childhood. *Br Med J* 1995; **311**: 773–6.

31. **Tanner JM**. *Fetus Into Man*. 2nd ed. Ware: Castlemead, 1989.

32. **Duggleby SL, Jackson AA**. Relationship of maternal protein turnover and lean body mass during pregnancy and birth length. *Clin Sci (Lond)* 2001; **101**: 65–72.

33. **Pesonen AK, Raikkonen K, Heinonen K** et al. Depressive symptoms in adults separated from their parents as children: a natural experiment during World War II. *Am J Epidemiol* 2007; **166**: 1126–33.

34. **Barker DJP, Bagby S, Hanson MA**. Mechanisms of disease: in utero programming in the pathogenesis of hypertension. *Nature Clin Pract Nephrol* 2006; **2**: 700–7.

35. **Hill JC, Krishnaveni GV, Annamma I, Leary SD, Fall CHD**. Glucose tolerance in pregnancy in South India: relationship to neonatal anthropometry. *Acta Obstet Gynaecol Scand* 2005; **84**: 159–65.

36. **Barker DJP, Osmond C, Forsén TJ, Kajantie E, Eriksson JG**. Trajectories of growth among children who have coronary events as adults. *N Engl J Med* 2005; **353**: 1802–9.

37. **Eriksson JG, Thornburg KL, Osmond C, Kajantie E, Barker DJP**, The prenatal origins of lung cancer: I. The fetus. *Am J Hum Biol* 2010; **22**: 508–11.

38. **Barker DJP, Thornburg KL, Osmond C, Kajantie E, Eriksson JG**. The prenatal origins of lung cancer: II. The placenta. *Am J Hum Biol* 2010; **22**: 512–16.

39. **Ericsson A, Saljo K, Sjostrand E** et al. Brief hyperclycaemia in the early pregnant rat increases fetal weight at term by stimulating placental growth and affecting placental nutrient transport. *J Physiol* 2007; **581**: 1323–32.

40. **Pedersen JF**. Ultrasound evidence of sexual difference in fetal size in first trimester. *Br Med J* 1980; **281**: 1253.

41. **Forsén T, Eriksson JG, Tuomilehto J, Osmond C, Barker DJP**. Growth in utero and during childhood among women who develop coronary heart disease: longitudinal study. *Br Med J* 1999; **319**: 1403–7.

42. **Eriksson JG, Kajantie E, Osmond C, Thornburg K, Barker DJP**. Boys live dangerously in the womb. *Am J Hum Biol* 2010; **22**: 330–5.

43. **Ounsted M, Scott A, Ounsted C**. Transmission through the female line of a mechanism constraining human fetal growth. *Ann Hum Biol* 1986; **13**: 143–51.

44. **Ravelli AC, Van Der Meulen JHP, Osmond C, Barker DJP, Bleker OP**. Obesity at the age of 50 y in men and women exposed to famine prenatally. *Am J Clin Nutr* 1999; **70**: 811–16.

45. **Ozaki T, Nishina H, Hanson MA, Poston L**. Dietary restriction in pregnant rats causes gender-related hypertension and vascular dysfunction in offspring. *J Physiol* 2001; **530**: 141–52.

46. **Woods LL, Ingelfinger JR, Rasch R**. Modest maternal protein restriction fails to program adult hypertension in female rats. *Am J Physiol Regul Integr Comp Physiol* 2005; **289**: R1131–6.

Discussion

BAGBY: Is there any information on the histology of the placenta in terms of those diameters?

PIJNENBORG: I would add to this: can the placental diameter be translated to the number of spiral arteries?

BARKER: I have no idea.

ROBERT BOYD: Do you have any idea, Robert?

PIJNENBORG: No. I think we should count the spiral arteries in placental beds in relation to the placental diameters. This is an issue which is understudied. Also, in a normal pregnancy there is definitely a centre-to-periphery gradient in invasion of arteries.

JACKSON: My understanding is that the thickness of the placenta is the other dimension that is critical. Perhaps this relates more to issues around spiral arteries.

BARKER: Thank you for raising this. In pre-eclampsia the placenta is thick. There are also predictions of hormonally related cancers from placental thickness. Astonishing as it may seem, ovarian cancer is predicted by a small, thick placenta, and breast cancer is predicted by a large, thin placenta.

BURTON: Later in this meeting I will discuss how the haemodynamics of the spiral artery interflow may affect placental thickness. The other parameter that may be interesting to look at is the centricity of the insertion of the umbilical cord. This tells the difference between a physiological and a pathological insult to the placenta. I would expect the insertion to be much more eccentric in a pathological situation.

BARKER: So it might be that placental expansion could occur at more than one stage of gestation, and if you have an eccentric cord, you are looking at an event that occurred very early on.

BURTON: Yes. This links back to the timing of the onset of the maternal circulation, around 8–10 weeks, when there is a regression of the villi down to the discoid placenta. If this is symmetrical, then you get a central insertion; if it is asymmetrical you get eccentric insertion.

BARKER: The kind of thing people are talking about in sheep is expansion of the surface in mid-gestation. In this case it is by recruiting more cotyledons. What does it mean in humans?

BURTON: This is where we do get a clear difference in species. The human is unique, in that the environment in which the placenta develops is very different between the first trimester, and the second and third trimesters. In the sheep, it is a much more constant environment. It does not have that potential insult upon opening of the spiral arteries.

BARKER: So what would you guess is going on in this group of humans who are born relatively small, and have a big placenta? For these people it has a huge long-term cost.

BURTON: I don't know at the spiral arterial level, but at the cardiovascular level there is always an issue between the fetus and the placenta. The fetus does not want too big a placenta, and this is what we see at altitude: the placental size does not increase, but it adapts its efficiency through other mechanisms. If the fetus has too big an extracorporeal circulation, this puts extra burden on the developing heart.

THORNBURG: This is an important speculation but I do not know of data to support it.

BARKER: It seems to me that the difference between the situation in humans and sheep is that farmers put the ewes on poor pasture for only 30 days, and then restore them to good pasture. But in the human situation, in the Dutch famine, for example, they weren't restored to a good diet. Then you have an undernourished fetus left with a huge placenta, with which it has to share a nutrient supply.

ROBERT BOYD: Are you sure that the placenta is causative, rather than its shape and position being a visible sign of what has gone on in the first two months. It may be that abnormalities in nutrition in that period could be reflected by the placenta, but that the placenta itself, to some extent, is an irrelevancy.

BARKER: From our data that is the core point in relation to the predisposition to ovarian and breast cancer. Kent and I feel that it is merely a display of the mother's hormonal status. The data on cardiovascular disease suggest that the placenta is more pivotal.

CRITCHLEY: What are you referring to by using this term 'thickness'? Where is the thickness measurement being taken, and what standardizations have been made?

BARKER: In the data I presented there are two measures of thickness. One is an inferred thickness, through knowing the weight along with the breadth and length. In the Dutch famine they measured the thickness with callipers. We have to try to get over the word 'standardization'. These are just clinical observations. Why did they make them? Who knows? They are crude, but the fact that they are highly predictive says that there is a huge bit of biology going on. In general knowing the thickness of the placenta is less helpful than knowing the area.

CRITCHLEY: I was interested in your comment about the three predictors in the male. As we move into this obesity epidemic, which is also a risk factor for coronary heart disease, what do we know about the potential impact of this? Since I trained as an obstetrician we have moved from every placenta being weighed to the placenta never being weighed. Do we have any datasets in those people with high BMI? What is the placental weight of the baby of a mother with raised BMI? These mothers tend to have bigger babies, so what about their placentas, and how is this going to play into risk prediction?

BARKER: Clearly, the overweight mother is relatively new. The point about BMI and coronary heart disease is that what increases the risk of the disease is rapid gain in BMI between the ages of 2 and 11. It is about rates of change. Long ago, D'Arcy Thompson, in his great book, said that to know the size of a child is not to know anything of any great interest, but to know how it achieved that body size is to know a lot. The actual BMI of a mother may not as important as knowing how she got there.

CRITCHLEY: We don't know anything about the placenta ratios, then.

BARKER: One thing we need to know in relation to placental size is the mother's birth weight rather than mother's body mass index. Maternal birth weight is consistently correlated with placental size.

SIBLEY: We are collecting data on obese women, and so far these data back up what David Barker says: there is no relationship between BMI and placental weight, or placental: fetal weight ratio.

RICHARD BOYD: I am worried about this word 'polarized'. When you use this term you are talking about the shape of the placenta, whether it is oval or circular. For some people in the room, polarized refers to the way that transporters are distributed in epithelial cells. I am interested in weights in biology. We need to consider how much water there is per unit dry mass in biological systems. This is a direct reflection of the rate of ATP-generated ion pumping. There are quite subtle differences. If cellular ATP is dropped a bit, you will get more water in a cell than if ATP is slightly higher, and this is a sensitive system. In pathological circumstances, you end up with oedema, but under 'extremes of normal physiology' you can get differences in amount of water per unit dry weight. I think it is quite an important aspect to remember in thinking about how to interpret this fascinating dataset.

BARKER: I am happy to be reprimanded about the use of the word 'polarity'. Perhaps what I didn't say is that from our standpoint as clinicians, we have two measures that were made routinely in wartime Finland, with a ruler, to the nearest centimetre. These measurements are highly correlated (0.65), and only the lesser one tells you about long-term disease. This is remarkable.

THORNBURG: I want to address the point of polarization. We use the term polarization loosely, because we are just referring to the shape of the placenta and it relationship to the embryonic axis. Our interest in this comes from a conversation we had with Alan Enders about implantation. There are many ways that an odd-shaped placenta can be formed, but the fact that the majority of the placentas are oval seems to me to be interesting. This raises the question of how, during implantation, the body of the embryo is related to the formation of the chorioallantoic placenta. Is the placenta polarized in the direction of the embryo itself, and does this set the pattern for growth? If so, we might be seeing a reflection of the axis of the embryo superimposed on the placenta. This is an open question but thinking in this direction might be useful.

CETIN: I have a comment related to obesity. I haven't seen data on obesity and the fetal: placental ratio, but there are many data on maternal diabetes and increased placenta: fetal weight ratio. Now we know that gestational diabetes is mainly related to obesity. What is the problem? Gestational diabetes or obesity? We could infer that it is the maternal body composition that is somehow influencing the placental: fetal weight ratio. In some of the papers in gestational diabetes, fetal weights are increased, but proportionally the babies are also shorter. It is probably a metabolic disease of the fetus that is already having an effect *in utero*, because these babies have a different body composition.

BARKER: Can I say something tangential? In thinking of the three phenotypes associated with lung cancer, what is common to all is that the baby is short in relation to its weight. This could be translated into a baby that is getting more than enough glucose but which is short of protein, so cannot grow. I wonder whether there is any merit in the thought of imbalances between protein and glucose supply.

JACKSON: I think it is one of the fundamental questions that need to be addressed.

JANSSON: I have a question about placental infarcts and how we deal with them. It is reasonable to assume that the placental weight is related to the areal volume of functional placental tissue. But if you have an infarct, which is quite common even in normal pregnancy, and much more common when the mother has vascular disease, this would change the fetal: placental ratio. Could the change in fetal: placental ratio reflect the fact that the mother had vascular disease, and what is seen in terms of hypertension in the offspring is actually reflecting cardiovascular disease in the mother.

BARKER: We know that people who develop toxaemia have high levels of cardiovascular risk factors before they become pregnant. So the placenta could be displaying the mother's cardiovascular incompetence.

ALWASEL: It has been accepted that the timing of an insult is critical in fetal development. But the timing of the insult is also critical in affecting placental size. We found that maternal fasting affects placental size only when it occurs in the second and third trimester, but not the first. So we also need to think about timing of the insult when we talk about the placenta.

BARKER: Clearly the nature of the insult and the timing of it are critical.

ROBERT BOYD: With thalidomide, the day on which it was administered was critical. Timing is everything.

Chapter 3

Pre- and periconceptual health and the HPA axis
Nutrition and stress

Alan A. Jackson, Graham Burdge and Karen Lillycrop

Paper presented at 'The role of the placenta in human development and programming', Clare College, Cambridge, 14–15 December 2009

Introduction

McCance and Widdowson [1] suggested that the critical influence exerted by nutrition during fetal life is on hypothalamic function. Selye introduced the concept of the 'adaptation syndrome' to characterize the body's response to stress, with the fundamental reaction pattern being topical 'inflammation' and systemic 'shock', regulated by hormonal factors [2]. The state of stress may be produced in the body by many causes, but the hypothalamopituitary–adrenal (HPA) axis plays a central coordinating role in the response to both internal and external stressors, substantially mediated through the release of corticotrophin-releasing hormone (CRH) and arginine-vasopressin (AVP) [3,4]. Stressors are a usual feature of all environments and the ability to cope effectively is partly determined by the nutritional status of the host. In turn, stressors modify the handling of nutrients within the body through alterations in appetite, tissue demands for nutrients and their utilization, delivery and partitioning of nutrients to tissues and increased nutrient losses from the body. There is a wide range of potential stressors: biological, such as infection with microorganisms or an unbalanced diet; behavioural, such as use of tobacco or alcohol; or social, such as poverty or deprivation, heavy work, or adolescence [5,6]. Each has its own specific characteristics, but equally there are important commonalities mediated by integrated hormonal and cytokine responses, with or without a systemic inflammatory response. Stresses imposed during pregnancy have the potential to alter the pattern of nutrients available through the maternal circulation to the uterus, thereby modulating the opportunities for early blastocyst, placental and fetal development [7]. Inadequate, excessive or unbalanced availability of energy or nutrients acts as a stressor in its own right, thereby leading to an altered metabolic set in the mother, the placenta or the fetus, potentially reinforcing and exacerbating any other adverse circumstance. The placenta acts to coordinate or balance the relative needs of the mother on the one hand and the fetus on the other.

Growth

Growth, whether for the placenta, the embryo or the fetus, represents a complex pattern of change in structure and function over time [8]. Although during normal health the appearance is of a smooth uninterrupted progression, in practice there is a sequence of distinct phases, each with its own characteristics and drivers [9–11]. Each successive phase builds upon the relative success of earlier phases with the potential for cumulative consequences. The transition from one phase to another represents a period of particular vulnerability. At fertilization the prior nutritional experiences of the female and male gametes come together, and although there is heavy dependence of the resulting zygote on the immediate nutritional support provided by the ovum, the sperm also plays its part [9,10]. Progress through the uterine tube, and the pace of progressive cellular multiplication without substantial increase in overall mass, is dependent upon the ability to draw a suitable mix of nutrients from the immediate environment. Blastocyst expansion, and its further elaboration into the presumptive trophectoderm and inner cell mass, is modulated by the nutritional environment provided by pabulum from the

The Placenta and Human Developmental Programming, ed. Graham J. Burton, David J. P. Barker, Ashley Moffett and Kent Thornburg. Published by Cambridge University Press. © Cambridge University Press 2011.

effective invasion of the uterine wall [7,10,12]. Successful implantation ultimately enables a more refined interaction between the maternal circulation and the evolving embryonic/fetal circulation. As the placenta develops into its mature form the metabolic and nutritional function of the syncytiotrophoblast represents a critical regulatory interface between the vascular endothelia of the maternal and fetal circulations [9].

Delivery of nutrients to the conceptus

At every stage healthy development presumes the ready availability of a suitable mix of nutrients to support the current needs for cellular growth, elaboration, maturation, function and replication [12–15]. As growth is an ordered process with a varying but specific pattern of needs as it progresses, the needs for energy and nutrient vary constantly. The pattern of nutrients to enable cellular replication and growth have either to be present preformed or derived from the immediate external environment, determining the very early developmental choices and setting the platform upon which later development can take place. Established placentation enables a more structured relationship among the availability of nutrients to the placenta from the maternal circulation, the ability of the placenta to effectively extract a suitable pattern of nutrients from maternal blood, and the processing of nutrients to deliver a pattern of nutrients to the fetus that best matches its needs as pregnancy progresses and fetal demands change. As the placenta carries out its own substantial metabolic interchange it places a significant demand on the available energy and nutrients. Under situations of limited energy or nutrient availability a stress is imposed upon the organism, and a delicate balance has to be achieved between the needs of the mother, the placenta and the fetus. The consequence of this fine balance is reflected in the outcome of pregnancy in terms of the size, shape and maturation of the fetus, the structure and function of the placenta, and the mother's nutritional wellbeing and her ability to lactate adequately for the coming months.

The ready availability of an appropriate mix of energy and nutrients may ultimately be derived from the maternal diet, but for the conceptus, the rate at which nutrients are made available is determined by the capacity of the uterine arterial supply, the product of the rate of flow and the concentration of specific nutrients. Thus, the amount and pattern of the available mix is substantially determined by the mother's own reserves and the extent to which her metabolism enables her to effectively draw on those reserves [14–16].

Metabolic anticipation

Later increases in demands for nutrients by the fetus are anticipated from very early in pregnancy through general and local changes in the maternal cardiovascular system. Normal pregnancy is characterized by a decrease in mean arterial pressure accompanied by an increase in cardiac output and a decrease in systemic vascular resistance [17]. There are associated increases in plasma volume [18] and the set point for plasma osmolality is progressively reduced over the first ten weeks of pregnancy [19]. The gestational changes in the osmoregulatory system are matched by increased turnover of vasopressin, despite the maintenance of near normal concentrations [19]. These substantial changes in cardiac output, blood volume and body composition during the first weeks precede the increased quantitative demands imposed by the placenta and fetus and ensure that there is an adequate supply to meet the needs as they increase progressively. Failure to achieve this 'metabolic anticipation' leads to poor fetal growth and potentially a blighted pregnancy [20,21]. Successful trophoblast invasion, adequate placentation and the early growth of the placenta itself are essential prerequisites, dependent upon maternal cardiovascular adaptations, which anticipate the needs for fetal growth. By 12 weeks of pregnancy the size and function of the placenta, itself a manifestation of the quality of development of the preimplantation conceptus, already has a determining effect on the later growth of the fetus. Hence, what the mother brings to her pregnancy in terms of her nutritional state, sets the stage for the effectiveness with which she can make nutrients available to the conceptus, and in turn its ability to access those nutrients most effectively as the stage of development requires. The most evident example of this is the preconceptual need for folic acid to enable effective closure of the neural tube, and the efficacy of dietary supplements of folic acid in preventing neural tube defects [22].

The placenta is the sole source of nutrients and oxygen for the fetus, thereby determining its growth trajectory. The weight, the diameters and thickness of the disc of the placenta at delivery represent a cumulative historical record of its earlier growth and the timing of any potential constraint. The effects on birth size of maternal age, height, weight, parity, social status and race can all be substantially explained as operating directly through placental size and structure [23–25]. Maternal risk factors relate directly to thickness and chorionic plate area. Black race or hypertension beyond 24 weeks is associated with placental growth restriction, whereas anaemia reduces the likelihood of placental restriction, and prepregnancy BMI and pregnancy weight gain are associated with hypertrophy of the placenta [26]. Placentas of different functional dimensions deliver different birth weights for a given placental weight [24]. Differential placental: fetal proportions explain as much as 39% of the variability in birth weight, with important gender differences [27]. Many of these differences in shape can be attributed directly to factors associated with vascularity [28], which in themselves can be accounted for by their relationship to the underlying metabolic demands imposed directly by the growth of the placenta relative to that of the fetus [29].

Imaging of the placenta from early in pregnancy shows that maternal factors relate directly to earlier placental development and the consequent relationship this has upon later fetal growth. Maternal prepregnancy weight and rate of weight gain during early pregnancy were directly related to the size of placenta and the rate of placental growth at mid gestation [30]. Further, the rate of placental growth between 17 and 20 weeks of gestation was significantly related to changes in the shape and size of the fetus at 35 weeks. Placental volume in the second trimester was positively associated with size at birth [31]. Placental volume and abdominal circumference of the fetus in the second trimester related directly to blood pressure in childhood [32]. Taken together, these data indicate that maternal nutrition status and nutrient reserves at the very start of pregnancy determine the pace and extent of placental growth, which in turn determines the opportunity for fetal growth and development, relating directly to the opportunity for health later in life [30].

Stress

Nutrients can act as stressors in a variety of ways, but the general underlying principle is that, because nutrition is a demand-led process, when the available supply does not adequately match the pattern of the demand function is constrained, thereby imposing stress directly, but also exerting influence on the response to other stressors. A sustained mismatch over any extended period is best seen as a failure of growth and development with alteration in body composition [33]. Thus measures of structure, shape and size are a proxy for an inadequate provision of nutrients, a history of exposure to stress and enhanced vulnerability. Hormonal responses are integral to the body's attempts to correct or accommodate any imbalance. Historically, in nutrition great emphasis has been placed on the adequacy of the diet for those nutrients that have to be provided preformed in the diet, such as essential amino acids or fatty acids, vitamins, minerals and trace elements. Much less emphasis has been placed on those nutrients for which the metabolic demand is high and pathways for their formation have been protected developmentally for survival of the species. The groupings of conditionally essential amino acids such as glutamine/glutamate/arginine/proline or serine/glycine/cysteine/taurine are good examples of nutrients for which the pathways for endogenous formation are highly protected. This protection presumably ensures their ready availability to play a continuing, central role in the regulation of intermediary metabolism, growth and development, and responses to stressors [34]. Importantly, although the capacity for the endogenous formation may be considerable, it is finite. The available capacity relates to the maturation of metabolic pathways and the ready availability of sufficient micronutrients to enable appropriate biochemical transformations. The mechanism through which adaptive changes are initiated to maintain physiological stability and metabolic integrity in which the HPA axis is involved has been termed 'allostasis' [35]. These processes play a major part in maintaining the function and integrity of the systems through which the body copes with challenge or stress [36,37], whether intrinsic or extrinsic. In this sense pregnancy should be considered one form of physiologic or metabolic stress. In the state of stress the integrity and *status quo* of the system are challenged, and hence any limitation in the delivery of oxygen, energy or nutrients to the fetus potentially represents a substantial stress. The imposition of a stressor elicits a repertoire of metabolic and behavioural responses that enable a complex interaction among centrally coordinated neuroendocrine, cellular and molecular processes, partly through activation of the HPA axis. The release of CRH and AVP regulates the secretion of adrenocorticotropic hormone (ACTH), leading to systemic secretion of glucocorticoids from the adrenal. Glucocorticoid receptors (GR) are widely but differentially expressed in all cells. Through the action on glucocorticoid response elements, a wide range of genes are either activated or repressed, mediating the effects of glucocorticoids upon metabolic integration and cellular responses. GR are highly expressed in the placenta and mediate metabolic and anti-inflammatory effects. Glucocorticoids exert an effect on placental size and function, depending on the timing and level of exposure. Therefore, a woman's response to stress in the periconceptual period and during pregnancy will be determined by the set and responsiveness of her HPA axis, and also the specific responsiveness of the cells of her tissues: a set which has been influenced or determined by her nutritional and other experiences during her own earlier life. Thus, there is a direct intergenerational communication of responsiveness to a wide range of stressors that affects the function of many organs and the integration of metabolism itself. The evidence indicates that epigenetic processes play a major if not determining role in this process of intergenerational communication of the experience of biological, nutritional, emotional/behavioural and wider social stresses [38–41]. The consequence is most evident in relation to neurocognitive development and behaviour, and the predisposition to altered body composition related to aspects of the cardiometabolic syndrome. Placental function plays a critical determining role in the process of fetal programming and the determination of the fetal phenotype.

Glucocorticoids

Steroid hormones are central to the complex metabolic interplay that sets the opportunity for a successful pregnancy. Glucocorticoids appear to have a special place in determining the pace and pattern of early development and facilitating intergenerational communications. The activity of placental 11β-hydroxysteroid dehydrogenase limits fetal exposure to maternal glucocorticoids. If the barrier is breached the fetus is overexposed to maternal glucocorticoids, with direct effects for fetal growth [38]. The changes in growth are associated with long-term alterations in behaviour, circulating levels of glucocorticoids and the set of the HPA axis in the offspring.

One of the more important features of the early epidemiological studies was the observation that variable risk of ill-health in the longer term was related to differences in size at birth [14,42]. This variable risk was not a feature of the extremes of birth weight, but graded across the usual range for populations generally considered to be of reasonable health. If birth weight is a proxy for nutritional opportunity over the course of the pregnancy, then the clear implication is that variation across the usual range of nutrient exposure leads to structural and functional change in the fetus, which predisposes to later risk of ill health. Our earliest animal studies sought to test this by exploring the least variation in dietary exposure of the mother that could induce structural and functional change in the offspring, and the shortest duration over which this had to act in order to exert a significant effect. In rats receiving a diet in which the amount of protein was varied across the usual range of intake, there was a graded increase in blood pressure as the protein content of the diet was reduced [43,44]. This represents one manifestation of extensive phenotypic modifications which appeared selective but reproducible, involved multiple systems, and were sexually dimorphic. Thus shape and size at birth were but one manifestation of a more general response that included a resetting of endocrine pathways in the hypothalamus and altered responsiveness to stressors [45].

Furthermore, in dams given lower-protein diets during pregnancy the activity of placental 11β-hydroxysteroid dehydrogenase was reduced by about one-third, increasing the likelihood of fetal overexposure to maternal glucocorticoids [46]. The finding that maternal administration of metyrapone abolished the effect of maternal diet on the blood pressure of offspring indicated that maternal glucocorticoids might be directly involved in mediating the effects of maternal diet on the programming of fetal metabolism [47], and provided direct evidence that part of the effect of maternal diet on fetal programming could be attributed to changes in the HPA axis in the mother translated to fixed changes in the HPA axis in the fetus. Further direct evidence comes from observations that maternal low-protein diets induce modulation of the activity of the GR and GR-responsive enzymes in the fetal tissues, including brain and liver [48–50]. Variations in nutritional exposure are one direct stress mechanism through which programmed changes can be elicited; others include alterations in the nurturing environment. Natural variation in maternal grooming of her pups in the first days after birth sets the sensitivity of the HPA axis and GR-related gene expression in the brain of the offspring, leading to life long differences in behaviour, responsiveness to environmental challenge and other stressors [38,50].

It is becoming increasingly clear that the system can be exquisitely sensitive to the consequential effects of changes in maternal diet. A lower level of protein given for as short a period as four days at the start of pregnancy elicited structural changes in the early blastocyst and consequential effects on vascular function and other outcomes [12,13]. As has been noted, for many studies in which different animal or very different dietary or other exposures have been used to explore aspects of programming, there appears to be a commonality of outcome [51]. This invites consideration of the extent to which there might be a final common pathway through which all dietary challenge exerts its effects. If unbalanced nutrient availability is acknowledged as a stressor in its own right, then the question would be whether the stress response itself represents the final common pathway. An alternative consideration would be the extent to which the varied dietary manipulations do lead to a constraint on a critical metabolic interaction or control point. Insofar as an adequate micronutrient status is required for the effective handling of all macronutrients and the effective net deposition and partitioning of energy is dependent upon the effective handling of all macronutrients, it is possible that all the various dietary manipulations effect to a greater or lesser extent different aspects of a limited range of critical regulatory pathways. For animals exposed to a lower-protein diet during pregnancy, the later effects on cardiovascular function can be reversed by supplementing the diet with glycine, but not with alanine or urea [52,53], indicating that the impact of a low-protein diet operates through specific processes. A similar reversal of the cardiovascular effects of the lower-protein diet has been found when the diet is supplemented with folic acid [54], inviting consideration that, as glycine is a one-carbon donor and folic acid is a one-carbon carrier, the common factor is modification in aspects of one-carbon metabolism or methyl group availability, possibly acting through epigenetic processes [55].

Arginine vasopressin

AVP and CRH together regulate the tone and activity of the HPA axis [56]. AVP has a particular role in the maturation of the pituitary–adrenal axis and potentiates the action of CRH under circumstances that require sustained activation. This is important in the epigenetic programming of neuroendocrine and behavioural function, and hypomethylation of the *Avp* enhancer leads to sustained upregulation of *Avp* expression, leading to increased activity of the HPA axis. Quite apart from these central effects, AVP plays a major role in assuring water homoeostasis and is critical to ensuring the nitrogen economy of the body. During pregnancy there is a resetting of the osmotic set point and altered sensitivity to AVP [19].

The regulated movement of water across cell membranes is partly controlled by aquaporins, which are critical for achieving water homoeostasis and regulating the rate of water loss through the kidney [57]. The expression and function of aquaporins is under the direct influence of AVP [58]. Nitrogen balance is achieved through the regulated reabsorption of urea from the collecting duct of the kidney, mediated by a family of urea transporters [59,60]. The urea retained within the system is actively secreted into the colon, where it is available and utilized by the colonic microbiota [59]. Urea nitrogen salvaged in this way is returned to the host in the form of potentially useful nutrients, which includes amino acids in significantly functional amounts [59]. The achievement of nitrogen balance, appropriate intermediary metabolism of amino acids and of water homoeostasis and balance, is a linked regulatory process directly influenced by AVP. In this regard the secretion of AVP is activated by a low-protein diet and/or by water stress, thereby exerting influence on the expression and activity of urea transporters and enabling nitrogen and water balance [61]. The fetus derives up to 35–40% of its energy needs from the oxidation of amino acid. During pregnancy the overall dynamics of urea metabolism, protein turnover and the efficiency with which nitrogen balance is achieved show a close relationship with the rate of fetal growth [15].

AVP acts directly on the placenta, in terms of both aquaporins and urea transporters, although the details of the relationships are far from clear [62]. The liver plays a central regulatory role in integrating the metabolism of the body through a complex array of hormonal, nutritional and neuronal signals. One important effect of AVP on the liver is to modulate cell expansion and cell shrinkage, thereby contributing directly to the cell and metabolic integration [63]. To an extent, the placenta fulfils a similar regulatory function for the fetus and its interaction with maternal metabolism. One critically important difference between the placenta and the liver, however, is the fundamentally different arrangements for their architecture. Whereas the liver has complex zonal cellular differentiation of structure and function, the metabolic activity of the placenta is managed within the syncytiotrophoblast. It is far from clear what, if any, benefit accrues from these highly regulated functions taking place within a syncytium, but one possibility would be that responsiveness in terms of cellular volume regulation is uniquely different, thereby enabling the placenta to have an independence of responsiveness relative to other maternal organs, particularly the liver.

Epigenesis

Variability in risk for chronic disease is graded across the usual range of birth size within the population [42]. For nutritional exposure to be a mediating factor in relation to this risk requires that the important effects are demonstrable across the range of dietary intake usually considered to be compatible with health. If the impact is cumulative during life, it might be explained simply by sensitive periods during development leading to differences in structure and function that constrain the maximal capability of one or other function, or limit the ability to regulate and integrate at later ages [7,8]. However, as these effects can be communicated between generations, and by embryo transfer, acquired genetic mechanisms of retained memory are necessary, considered most likely to be through epigenetic processes such as DNA methylation and covalent modification of histones [39,41,50,64]. This potentially implicates those processes through which one-carbon moieties such as methyl groups are made available to the metabolism, and the mechanisms through which methylation of the promoter region of specific regulatory genes is enhanced or constrained from one situation to another [41].

The induction of changes to the phenotype of the offspring, in response to the prenatal environment that persists throughout the lifespan, implies stable changes to gene transcription resulting in altered activities of metabolic pathways and the set point of homoeostatic control processes, and in differences in the structure of tissues. Studies on gene expression demonstrate stable effects on transcription [41]. Some of the genes that showed altered expression following prenatal undernutrition are transcription factors that affect multiple pathways in development and nutrient homoeostasis: for example PPARα and the GR [39,65]. The methylation of CpG dinucleotides confers stable silencing of transcription. Methylation patterns are largely established during embryogenesis or in early postnatal life [66]. DNA methylation also plays a role in cell differentiation by silencing the expression of specific genes during the development and differentiation of individual tissues, and thus the timing of gene methylation is tissue and gene specific [67]. Covalent modifications of histones influence chromatin structure and the ability of transcriptional machinery to gain access to DNA. DNA methylation can induce transcriptional silencing by blocking the binding of transcription factors and/or through promoting the binding of the methyl CpG binding protein (MeCP)-2. The latter binds to methylated cytosines and, in turn, recruits histone-modifying complexes composed of deacetylases and histone methyl transferases to the DNA, resulting in a closed chromatin structure and transcriptional silencing [68,69].

Perturbations as diverse as lack of maternal grooming, uterine artery ligation and embryo culture have been shown to lead to epigenetic modulation of transcription, structural and functional effects in the short and long term [41,50,70,71]. Varying the maternal intake of nutrients involved in one-carbon metabolism across a wide range can induce graded changes in DNA methylation and gene expression in the offspring, that persist into adulthood [71]. Feeding pregnant dams graded amounts of protein across a range of intakes not associated with any obvious pathology leads to graded increases in blood pressure in the offspring, and has become an established model of phenotype induction [44]. This modest change to maternal macronutrient intake during pregnancy induced hypomethylation of the PPARα and GR promoter and increased expression of PPARα and GR in the liver of the offspring. There was also an increase in the expression of PPARα and GR target genes such acyl-CoA oxidase and phosphoenolpyruvate carboxykinase, respectively, supporting the suggestion that altered epigenetic regulation of transcription factors modifies the activity of important metabolic pathways [64,65]. Sequence analysis of the PPARα promoter showed that the methylation status of only a few CpG dinucleotides was altered by the reduced-protein diet during pregnancy [72]. This suggests that the process of induced epigenetic change is targeted, and that the resulting change in transcription may reflect changes in the interaction of the gene with relatively few transcription factors, thereby inducing specific changes in the regulation of gene function and hence the response to environmental differences. Further, the PPARα promoter was hypomethylated in the whole umbilical cord offspring of rats fed a reduced-protein diet during pregnancy [65], suggesting that hypomethylation of PPARα and GR promoters had already been established very early in pregnancy, before cell lineages had become definitively established. Hypomethylation of the GR promoter was associated with an increase in histone modifications that facilitate transcription, whereas those that suppress gene expression were reduced or unchanged [64].

Induction of the altered phenotype (hypertension and endothelial function) in the offspring of rats fed the reduced-protein diet during pregnancy was prevented by supplementation of this diet with glycine or folic acid [52,53,54]. Hypomethylation of the hepatic PPARα and GR promoters was also prevented by the addition of five times more folic acid than contained in the reduced-protein diet [65]. Thus, one-carbon metabolism plays a central role in the induction of an altered phenotype. There was an important interaction between the metabolism of macronutrients and micronutrients, and these interactions operate through differential methylation of the promoter region of regulatory genes through seemingly epigenetic mechanisms. The regulatory genes themselves play a central role in metabolic integration in terms of responsiveness to stress (GR), macronutrient partitioning and central fat deposition (PPARα). Feeding the reduced-protein diet during pregnancy in the F0-generation results in elevated blood pressure, endothelial dysfunction, insulin resistance and adverse glucose homoeostasis in the F1, F2 and even the F3 generations, despite no further unusual dietary exposure for subsequent generations [73–76]. This implies that transmission of a phenotype induced in the F1 generation to the F2 generation and further into the F3 generation may involve preservation of levels of DNA methylation of specific genes. As the female line appears sufficient for transmission of this epigenetic information between generations, the level of methylation of the PPARα and GR promoters in gametes must be similar to that of somatic cells. Alternatively, impaired adaptations to pregnancy as a consequence of the effect of constraint in the mother's intrauterine experience may induce in her offspring phenotypic and epigenetic changes *de novo*. Hence, epigenetic traits may be transmitted between generations without changes in DNA methylation in germ cells.

Assisted reproductive technologies

The development of assisted reproductive technologies has increased the need for a better understanding of the factors associated with improved selection and survival of the highest-quality sperm, oocyte or zygote, especially over the critical period of preimplantation development. Although factors intrinsic to the putative or developing conceptus are important, the milieu within which development takes place also plays an important determining role [7]. Assisted reproduction can be carried out successfully *in vitro*, but the health and vigour of the sperm, oocyte and zygote are directly influenced by their nutritional status and their ability to withstand the stresses of manipulation and development *in vitro* [77]. For some women with a poor reproductive history oxidative stress might be a cause of poor oocyte quality, and in women with poor trace element status undergoing IVF [78] supplementation with multivitamins and minerals improved the antioxidant status of follicular fluid [79]. The development of the early embryo requires substantial elaboration of membranes, with the attendant need for phospholipids and fatty acids and their associated regulatory metabolic roles. Although linoleic acid is the most abundant fatty acid in bovine follicular fluid, incubation with supplemental linoleic acid *in vitro* has an adverse effect on oocyte development and blastocyst yield [80]. By contrast, supplementation with linolenic acid at lower doses improved early embryo development, although higher doses were detrimental [81].

The unfertilized oocyte and the very early zygote appear exquisitely sensitive to the immediate nutritional environment around the time of conception. Both *in vivo* and *in vitro* studies indicate that mammalian preimplantation development is sensitive to subtle changes in environmental conditions, with effects on placentation and fetal growth in the short term and resulting in long-term changes in physiology and health. Exposure of pregnant rats to a lower-protein diet for the first 4.5 days following conception altered partitioning of cells to presumptive trophectoderm with associated longer-term changes in phenotype [12]. The same lower-protein diet provided exclusively before conception during a period of oocyte maturation resulted in altered behaviour, cardiovascular function and renal development in the offspring [13]. Thus oocyte maturation is sensitive to the nutritional environment provided by the mother.

Before implantation the embryo is bathed in the oviduct and uterine fluids, and the specific amino acid composition of these fluids has an influence on developmental progression [7,82]. The amino acid profiles in the oviduct and uterus are very different from that found in plasma, with most amino acids being at much higher concentrations, indicative of active secretion by the epithelium. The concentrations of individual amino acids differs between oviduct and uterus. Whereas the concentration within the oviduct remains relatively unchanged across the oestrous cycle, in the uterus there is variability in relation to the day of the cycle [82]. The most prevalent amino acid in both locations was glycine, some two to six times higher than alanine or glutamic acid, and 30 times higher than those amino acids present in the lowest concentrations. The authors note possible benefits that can be attributed to these amino acid patterns, such as protecting cellular integrity against pH, low oxygen tension or osmotic stress to enable and promote preimplantation development, or the viability of spermatozoa.

Stressed gametes, zygote and embryo

Animal evidence indicates that maternal diet influences the relative proportions of male and female offspring. Animals deprived of food before and during mating tend to have more females, whereas those that enjoy a better nutritional state tend to have more males [83]. Thus the appearance is that nutritionally stressed females have fewer sons. It is not clear the extent to which the determining factor acts prior to conception leading to a selective advantage for those sperm that will give daughters or to a selective survival of females over male progeny, with some support for both possibilities. There is lower fertility with sex-sorted sperm associated with differences in cleavage and blastocyst development which appears to be due to damage during the sorting procedure, which would suggest differential sex vulnerability to the stresses associated with the sorting process [84]. Attempts to determine the metabolic needs or state of the early embryo by using metabolomics or other approaches to capture the broad pattern of metabolic interactions are at an early stage [85–87], but indicate potentially important shifts in the metabolic profile for macronutrient metabolism with stress. In time, differential patterns will need to be related directly to the extent of maternal stress and maternal diet.

Human zygotes can develop to the blastocyst stage in relatively simple solutions of salts, glucose and amino acids, but the yields are relatively low [88]. The survivors appear to have the lowest rates of metabolism, with relatively low uptakes of glucose and pyruvate. Further, it has been possible to make a non-invasive assessment of embryo health by measuring the turnover of amino acids in the culture medium (the rate of appearance/depletion). On this basis embryos can be classified as either 'active' or 'quiet', with quiet embryos having better survival from transfer at day 2/3 to the blastocyst stage at day 5/6 [88]. It has been suggested that those embryos that are metabolically 'active' might have been developmentally compromised from the oocyte stage, with greater demand for repair/rescue pathways especially related to DNA or RNA damage in the mature oocyte. Thus a key factor in determining progress is the ability to delay activation of the embryonic genome in the blastocyst as it takes over functionally RNA transcripts received from the mature oocyte. The greater the DNA or RNA damage, the earlier the need for activation of the embryonic genome. Premature activation of the embryonic genome therefore indicates that higher levels of DNA damage have been incurred from the maternally derived RNA transcripts. The severity of DNA damage drives the need for an arsenal of repair proteins available as active enzymes or non-degraded RNA, in order to enable effective embryo development. Thus those cells that are metabolically active are more likely to be those that have incurred DNA damage leading to an activated repair pathway and an increased need for oxidative metabolism, and are least likely to progress. By contrast, 'quiet' embryos have developed with a higher degree of fidelity and have to commit fewer resources to repair, thereby being metabolically more efficient and hence 'quiet' [89,90]. Thus morphometric criteria appear less good as predictors of survival than functional measures, and paradoxically, those that survive best at this stage of development appear to have the lowest rates of metabolism, with relatively low uptakes of glucose or pyruvate [91]. At later stages, as the embryonic genome does become activated and differentiation of the trophectoderm becomes dependent upon membrane assembly, a high demand is placed on the embryonic metabolism [92]. Those embryos that were initially 'quiet' appear more efficient at this stage and are most effective in achieving effective junctional biogenesis, showing a higher amino acid turnover associated with the demands for trophectoderm differentiation.

Sufficient numbers of people have been born following assisted reproduction to make it possible to determine whether there are longer-term effects in the children as they grow into adulthood. For gross aspects of physical growth in later childhood and adolescence the results from different centres have been equivocal, and overall there are no consistent marked differences. There is a need for longer-term follow-up and to take into consideration possible intergenerational effects [93]. There may be more subtle differences in terms of body composition and patterns of fat distribution, with possible differences in pubertal development and an advanced bone age in girls [94,95]. However, there is clearer evidence of differences in growth during infancy and early childhood, with those conceived following IVF being shorter, lighter and thinner at three months of age, and having greater gains in length during late infancy. Further, compared to controls, those children born following IVF had higher values for systolic and diastolic blood pressure and higher fasting glucose when at puberty, directly related to the rate of weight gain in early childhood [96,97].

In Britain the nutritional status of young people in the reproductive years is poor [98]. Those with a poor diet are characterized by overconsumption of foods high in sugar, salt and fat and low in complex carbohydrates, leading to excessive intake of energy and a poor pattern of nutrient consumption. The diet of the more vulnerable in society appears to be the worst and contributes directly to an increasing prevalence of overweight and obesity in the parents [99], lower birth size, poor growth and increased vulnerability to ill-health in the children. Thus, there are two groups of women who can be considered particularly vulnerable around the time of conception, and appear increasingly prevalent in many societies: those who are overweight through excessive adiposity, and those who are adolescent and have not yet completed their own physical growth [5,6].

Obesity

Prepregnancy obesity is increasingly common, and despite the reality of obesity-related infertility assisted technologies enable more to become pregnant [100–102]. The increased fat mass is indicative of a persistent positive energy balance over long periods associated with low levels of energy expenditure as physical activity. Albeit usually associated with unusual patterns of macro- or micronutrient consumption, this is not always necessarily the case. The increased prevalence of congenital malformations and the apparent responsiveness to periconceptual supplementation with micronutrients such as folic acid raises the likelihood of serious nutritional impairment, which is not immediately obvious using standard clinical investigations [100,102]. However, the observation that adiposity, especially central adiposity, is associated with enhanced local and systemic inflammatory disease, decreased physical activity, and features of cardiometabolic syndrome, such as abnormal accumulation of lipid in the liver indicative of perturbed metabolic regulation, indicates that in this state all cells exist in an environment which is nutritionally challenging. Separately and together they represent a condition of nutritional metabolic stress, which in itself is liable to predispose to a pregnancy of poorer quality from conception through early development to placentation. The inherent endothelial damage, the associated vascular compromise and impaired placental growth and function from the first trimester [103] may be partly accounted for by an unusual pattern of nitrative and oxidative stress [104]. In obese pregnant baboons there was altered function of the HPA axis and an inflammatory state with marked structural and functional changes in the placenta [105]. In the mouse a high-fat diet before and during pregnancy led to increased adiposity, with activation of the HPA and other hormonal axes and reduced placental and fetal weight [106]. Increased adiposity or obesity is not a single state but embraces a range of possible responses during pregnancy, with consequential effects on the structure and function of the placenta and a wide range of phenotypic outcomes for the fetus. Obesity itself represents impaired utilization and partitioning of the available energy in the mother. Pregnancy extends this impairment to the partitioning among the mother, the placenta and the fetus. There is a need for much more detailed information on the underlying processes that determine the relative partitioning of energy and nutrients.

Adolescence

Mothers who carry a pregnancy while they are still adolescent face a significant biological and emotional challenge. Adolescent pregnancies are relatively high risk, partly because of the competing demands for nutrients for ongoing maternal growth and maturation, interacting with the needs imposed by the pregnancy itself. Thus the frequency of adverse outcomes is increased, with a greater likelihood of low birth weight, being premature or small for gestational age [107,108]. The mothers tend to be shorter and lighter at the start of pregnancy, and even if they achieve similar weight gain during pregnancy their babies are less well grown, being significantly shorter, lighter, and with a smaller head circumference [109]. However, not only are mothers shorter and lighter at the start of pregnancy, but the quality of the weight gained tends to be poor, with a substantial proportion smoking habitually and having a poor folate status, related directly to the poorer outcomes [108]. There was an important interaction between smoking and micronutrient supplementation [110]. For women who smoked during pregnancy the risk of fetal death was progressively greater with the number of cigarettes smoked, with a 70% increase in women smoking more than 20 per day. Supplementation with micronutrients once pregnancy had been established did not significantly alter this risk, but for those women who were supplemented before conception the risk was apparently removed completely. Results from the Camden study indicate that adolescent pregnant mothers who are still growing in length appear to have particular difficulty in partitioning nutrients between mother and fetus, especially during the last trimester, when adipose tissue is preferentially acquired as excessive maternal subcutaneous fat rather than being partitioned towards increasing fetal fat mass and birth weight [111,112].

The increased metabolic demands placed on the pregnant adolescent are illustrated by their inability to enhance gluconeogenesis following a short period of food deprivation, and their greater dependence on glycogenolysis [113], thus the metabolic profile of accelerated fasting of pregnancy is even more pronounced. This altered glucose handling reflects a wider metabolic challenge, and they are, for example, less able to maintain the production of non-essential amino acids, such as glycine, than are more mature women [114]. The effective utilization of amino acids for protein synthesis, the growth of lean tissues and gluconeogenesis is a critical measure of the placenta's ability to provide an adequate profile of nutrients to achieve normal growth and development [15].

The model of overnourishing pregnant adolescent sheep to promote rapid maternal growth has enabled detailed exploration of the mechanisms that enable placental function to support fetal growth. Overnourishment during pregnancy results in major placental growth restriction, which becomes the primary limitation on fetal growth and is most marked in very young adolescent animals. During mid gestation, before alterations in placental mass become obvious, there is attenuated placental blood flow associated with impaired angiogenesis and reduced proliferation of the trophectoderm, secondary to poor uterine blood flow. Nutritionally mediated placental insufficiency and altered placental hormone secretion lead to fetal hypoxia and hypoglycaemia, and accelerated maturation of the fetal HPA axis. In spite of maternal overnutrition and the ready availability of nutrients in the maternal circulation, the fetus is growth restricted by the small size of the placenta *per se*, rather than impaired function appearing to be the major limitation [115,116].

Conditionally essential amino acids

Nutrition is a demand-led process, the demand being met from the dietary intake, metabolic transformation and potential reserves [33]. For some nutrients, the amounts provided preformed in the diet are often marginal in relation to the needs. A greater dietary intake that might be required by those who are especially vulnerable, especially later in pregnancy as the needs increase; choline and some of the long-chain polyunsaturated fatty acids are good examples. For other nutrients the increased demand during pregnancy cannot readily be met simply by increased dietary intake, and there is a need to ensure adequate endogenous formation, sometimes through complex or poorly characterized pathways. Two amino acids, arginine and glycine, which are conditionally essential, appear to have the potential to be critically limiting for normal pregnancy, as the ability to provide sufficient through endogenous pathways appears marginal for the demands [7]. These two amino acids have especially important metabolic functions in cellular replication, the formation of extracellular proteins and wider regulatory processes. The pathways through which they are formed are highly dependent upon intermediary metabolism and adequate micronutrient status, thereby representing a potential secondary vulnerability to a range of dietary limitations.

Arginine

Although there may appear to be generous amounts of arginine in the diet, because of first-pass clearance and metabolism within the gastrointestinal tract, the dietary content is not always available to the general metabolism [117]. Arginine readily interchanges with glutamic acid, glutamine and proline, but the pathway through which it is formed requires complex interorgan cooperation involving muscle, the gastrointestinal tract, the liver and ultimately the kidney. It is the immediate precursor of urea, placing it at the heart of nitrogen metabolism; it also gives rise to nitric oxide, giving it a central role in physiological regulation and the response to stress. An adequate, regulated provision of arginine plays a critical role in cardiovascular adaptations to pregnancy, normal placentation, and the ability of the mother to delivery adequate nutrients to the fetus. Our understanding of the kinetics and stoichiometry of arginine metabolism during pregnancy requires much further detailed study.

Glycine

Glycine is the simplest amino acid, which interchanges immediately with serine and hence cysteine, and taurine [34]. It is directly involved with the metabolic interchange of methionine and homocysteine, and is thereby close to the heart of one-carbon transfer and the regulation of methylation reactions. It is a primary building block for many fundamental cellular components, such as DNA, and haemoglobin. Glycine plays a fundamental role in cellular protection as a component of glutathione, as an osmolyte, and through cellular protection under anoxic conditions. The pathways through which glycine is formed are poorly characterized, and the stoichiometry through which the demands are adequately met is not understood. Simple dietary supplementation cannot meet the demands, which are quantitatively substantial at all times, but especially during pregnancy and other situations where growth or net tissue deposition dominate the metabolic imperative. The delivery of glycine in adequate amounts through the placenta is not clearly identified in humans. The extent to which glycine might be limiting for metabolic interaction can be assessed by measuring the urinary excretion of 5-L-oxoproline [118]. 5-L-Oxoprolinuria is unusually high during pregnancy, in the preterm and term infant and during infancy, but is also raised in chronic diseases, such as diabetes, hypertension and asthma, or when the body is challenged by a range of xenobiotics [34]. The metabolism of glycine is absolutely dependent on an adequate micronutrient status, especially for the B vitamins riboflavin, pyridoxine, folate and cobalamin. The metabolic set for the pathways through which glycine is synthesized, and the regulated rate of its formation and turnover make this amino acid eligible to be identified as the 'pacemaker of metabolism', and thereby central to all considerations of metabolic programming.

Summary and conclusion

The complexity and order required for development demand that the resource is readily available as and when needed. This requires metabolic anticipation, and the placenta plays a critical role in anticipating the demands of the fetus for later growth and development against the opportunity offered by the mother's metabolism. Later attempts to respond to changed circumstances through adaptation may result in placental overgrowth. In these situations the addition placental acquisition may not be as efficient in fulfilling the necessary role. The nutrient demands for meeting any stress and the impact of stress on the handling of nutrients present competitive demands for the support the placenta provides to fetal growth and development. Structural and functional constraints in the placenta are likely to translate to constraints in fetal development, limiting metabolic capacity and creating a vulnerability to wider environmental exposure later in life, potentially leading to adverse health outcomes. The processes through which memory is captured and communicated between generations of cells is probably dependent on the sensing of overall nutritional status, and the factors that regulate and control the metabolism and exchange of one-carbon moieties are likely to be critical in this regard. The metabolic interactions of conditionally essential amino acids may represent the key metabolic sensors and regulators setting the opportunity for metabolic capacity and exchange.

References

1. **McCance RA**, **Widdowson EM**. The determinants of growth and form. *Proc Roy Soc Lond B Biol Sci* 1974; **185**: 1–17.
2. **Selye H**. *The Story of the Adaptation Syndrome*. Montreal: Acta inc, 1952.

3. **Mastorakos G**, **Ilias I**. Maternal and fetal hypothalamic-pituitary-adrenal axes during pregnancy and postpartum. *Ann N Y Acad Sci* 2003; **997**: 136–49.

4. **Murphy VE**, **Smith R**, **Giles WB**, **Clifton VL**. Endocrine regulation of human fetal growth: the role of the mother, placenta, and fetus. *Endocr Rev* 2006; **27**: 141–69.

5. **Jackson AA**, **Bhutta ZA**, **Lumbiganon P**. Nutrition as a preventative strategy against adverse pregnancy outcomes. Introduction. *J Nutr* 2003; **133**: 1589S–91S.

6. **WHO Technical Consultation**. *Towards the Development of a Strategy for Promoting Optimal Fetal Development*. Geneva: World Health Organization, 2003.

7. **Jackson AA**. Maternal and fetal demands for nutrients: significance of protein restriction. In: O'Brien PMS, Wheeler T, Barker DJP eds. *Fetal Programming: Influence on Development and Disease in Later Life*. London: Royal College of Obstetricians and Gynaecologists, 1999; 246–70.

8. **Jackson AA**. Perinatal nutrition: the impact on postnatal growth and development. In: Gluckman PD, Heyman MA eds. *Pediatrics and Perinatology*. 2nd ed. London: Arnold, 1996; 298–303.

9. **Hamilton WJ**, **Boyd JD**, **Mossman HW**. *Human Embryology: Prenatal Development of Form and Function*. Cambridge: W Heffer & Sons Ltd, 1946.

10. **Leese HJ**, **Isherwood-Peel SG**. Early embryo nutrition and disorders in later life. In: O'Brien PMS, Wheeler T, Barker DJP eds. *Fetal Programming: Influence on Development and Disease in Later Life*. London: Royal College of Obstetricians and Gynaecologists, 1999; 104–16.

11. **Karlberg J**, **Jalil F**, **Lam B**, **Low L**, **Yeung CY**. Linear growth retardation in relation to the three phases of growth. *Eur J Clin Nutr* 1994; **48**: S25–44.

12. **Kwong WY**, **Wild AE**, **Roberts P**, **Willis AC**, **Fleming TP**. Maternal undernutrition during the preimplantation period of rat development causes blastocyst abnormalities and programming of postnatal hypertension. *Development* 2000; **127**: 4195–202.

13. **Watkins AJ**, **Wilkins A**, **Cunningham C** et al. Low protein diet fed exclusively during mouse oocyte maturation leads to behavioural and cardiovascular abnormalities in offspring. *J Physiol* 2008; **586**: 2231–44.

14. **Jackson AA**. Nutrients, growth, and the development of programmed metabolic function. *Adv Exp Med Biol* 2000; **478**, 41–55.

15. **Duggleby SL**, **Jackson AA**. Protein, amino acid and nitrogen metabolism during pregnancy: how might the mother meet the needs of her fetus? *Curr Opin Clin Nutr Metab Care* 2002; **5**: 503–9.

16. **Catalano PM**, **Huston LP**, **Thomas AJ**, **Fung CM**. Effect of maternal metabolism on fetal growth and body composition. *Diabetes Care* 1998; **21**: B85–B90.

17. **Chapman AB**, **Abraham WT**, **Zamudio S** et al. Temporal relationships between hormonal and hemodynamic changes in early human pregnancy. *Kidney Int* 1998; **54**: 2056–63.

18. **Whittaker PG**, **Lind T**. The intravascular mass of albumin during human pregnancy: a serial study in normal and diabetic women. *Br J Obstet Gynaecol* 1993; **100**: 587–92.

19. **Lindheimer MD**, **Davison JM**. Osmoregulation, the secretion of arginine vasopressin and its metabolism during pregnancy. *Eur J Endocrinol* 1995; **132**: 133–43.

20. **Duvekot JJ**, **Cheriex EC**, **Pieters FA**, **Peeters LL**. Severely impaired fetal growth is preceded by maternal hemodynamic maladaptation in very early pregnancy. *Acta Obstet Gynecol Scand* 1995; **74**: 693–7.

21. **Duvekot JJ**, **Cheriex EC**, **Pieters FA** et al. Maternal volume homeostasis in early pregnancy in relation to fetal growth restriction. *Obstet Gynecol* 1995; **85**: 361–7.

22. **Scientific Advisory Committee on Nutrition**. *Folate and Disease Prevention*. London: The Stationery Office, 2006.

23. **Stein Z**, **Susser M**, **Sqaenger G**, **Marolla F**. *Famine and Human Development: the Dutch Hunger Winter of 1944–1945*. London: Oxford University Press, 1975.

24. **Salafia CM**, **Zhang J**, **Miller RK** et al. Placental growth patterns affect birth weight for given placental weight. *Birth Defects Res A Clin Mol Teratol* 2007; **79**: 281–8.

25. **Salafia CM**, **Zhang J**, **Charles AK** et al. Placental characteristics and birth weight. *Paediatr Perinat Epidemiol* 2008; **22**: 229–39.

26. **Baptiste-Roberts K**, **Salafia CM**, **Nicholson WK** et al. Maternal risk factors for abnormal placental growth: the national collaborative perinatal project. *BMC Pregnancy Childbirth* 2008; **8**: 44.

27. **Misra DP**, **Salafia CM**, **Miller RK**, **Charles AK**. Non-linear and gender-specific relationships among placental growth measures and the fetoplacental weight ratio. *Placenta* 2009; **30**: 1052–7.

28. **Yampolsky M**, **Salafia CM**, **Shlakhter O**, **Haas D**, **Eucker B**, **Thorp J**. Modeling the variability of shapes of a human placenta. *Placenta* 2008; **29**: 790–7.

29. **Salafia CM**, **Misra DP**, **Yampolsky M**, **Charles AK**, **Miller RK**. Allometric metabolic scaling and fetal and placental weight. *Placenta* 2009; **30**: 355–60.

30. Thame M, Osmond C, Bennett F, Wilks R, Forrester T. Fetal growth is directly related to maternal anthropometry and placental volume. *Eur J Clin Nutr* 2004; **58**: 894–900.

31. Thame M, Osmond C, Wilks R, Bennett FI, Forrester TE. Second-trimester placental volume and infant size at birth. *Obstet Gynecol* 2001; **98**: 279–83.

32. Thame M, Osmond C, Wilks RJ et al. Blood pressure is related to placental volume and birth weight. *Hypertension* 2000; **35**: 662–7.

33. Jackson AA. All that glitters. *British Nutrition Foundation, Nutrition Bulletin* 2001: **25**; 11–24.

34. Jackson AA. The glycine story. *Eur J Clin Nutr* 1991; **45**: 59–65.

35. McCormick CM, Mathews IZ, Thomas C, Waters P. Investigations of HPA function and the enduring consequences of stressors in adolescence in animal models. *Brain Cogn* 2010; **72**: 73–85.

36. Cottrell EC, Seckl JR. Prenatal stress, glucocorticoids and the programming of adult disease. *Front Behav Neurosci* 2009; **3**: 19.

37. Kyrou I, Tsigos C. Stress hormones: physiological stress and regulation of metabolism. *Curr Opin Pharmacol* 2009; **9**: 787–93.

38. Seckl JR, Meaney MJ. Glucocorticoid programming. *Ann N Y Acad Sci* 2004; **1032**: 63–84.

39. Meaney MJ, Szyf M, Seckl JR. Epigenetic mechanisms of perinatal programming of hypothalamic-pituitary-adrenal function and health. *Trends Mol Med* 2007; **13**: 269–77.

40. Burdge GC, Hanson MA, Slater-Jefferies JL, Lillycrop KA. Epigenetic regulation of transcription: a mechanism for inducing variations in phenotype (fetal programming) by differences in nutrition during early life? *Br J Nutr* 2007; **97**: 1036–46.

41. Burdge GC, Lillycrop KA, Jackson AA. Nutrition in early life, and risk of cancer and metabolic disease: alternative endings in an epigenetic tale? *Br J Nutr* 2009; **101**, 619–30.

42. Barker DJ, Osmond C, Golding J, Kuh D, Wadsworth ME. Growth in utero, blood pressure in childhood and adult life, and mortality from cardiovascular disease. *Br Med J* 1989; **298**: 564–7.

43. Levy L, Jackson AA. Modest restriction of dietary protein during pregnancy in the rat: fetal and placental growth. *J Dev Physiol* 1993; **19**: 113–18.

44. Langley SC, Jackson AA. Increased systolic blood pressure in adult rats induced by fetal exposure to maternal low protein diets. *Clin Sci (Lond)* 1994; **86**: 217–22.

45. Jackson AA, Langley-Evans SC, McCarthy HD. Nutritional influences in early life upon obesity and body proportions. In: Chadwick D J, Cardew G, eds. *The Origins and Consequence of Obesity*. Chichester: John Wiley & Sons. CIBA Foundation Symposium 1966; **201**: 118–37.

46. Langley-Evans SC, Phillips GJ, Benediktsson R et al. Protein intake in pregnancy, placental glucocorticoid metabolism and the programming of hypertension in the rat. *Placenta* 1996; **17**: 169–72.

47. Langley-Evans S, Jackson A. Intrauterine programming of hypertension: nutrient-hormone interactions. *Nutr Rev* 1996; **54**: 163–9.

48. Langley-Evans SC, Gardner DS, Jackson AA. Maternal protein restriction influences the programming of the rat hypothalamic-pituitary-adrenal axis. *J Nutr* 1996; **126**: 1578–85.

49. Bertram C, Trowern AR, Copin N, Jackson AA, Whorwood CB. The maternal diet during pregnancy programs altered expression of the glucocorticoid receptor and type 2 11beta-hydroxysteroid dehydrogenase: potential molecular mechanisms underlying the programming of hypertension in utero. *Endocrinology* 2001; **142**: 2841–53.

50. Szyf M, Weaver I, Meaney M. Maternal care, the epigenome and phenotypic differences in behavior. *Reprod Toxicol* 2007; **24**: 9–19.

51. Armitage JA, Khan IY, Taylor PD, Nathanielsz PW, Poston L. Developmental programming of the metabolic syndrome by maternal nutritional imbalance: how strong is the evidence from experimental models in mammals? *J Physiol* 2004; **561**: 355–77.

52. Jackson AA, Dunn RL, Marchand MC, Langley-Evans SC. Increased systolic blood pressure in rats induced by a maternal low-protein diet is reversed by dietary supplementation with glycine. *Clin Sci (Lond)* 2002; **103**: 633–9.

53. Brawley L, Torrens C, Anthony FW et al. Glycine rectifies vascular dysfunction induced by dietary protein imbalance during pregnancy. *J Physiol* 2004; **554**: 497–504.

54. Torrens C, Brawley L, Anthony FW et al. Folate supplementation during pregnancy improves offspring cardiovascular dysfunction induced by protein restriction. *Hypertension* 2006; **47**: 982–7.

55. Kwong WY, Adamiak SJ, Gwynn A, Singh R, Sinclair KD. Endogenous folates and single-carbon metabolism in the ovarian follicle, oocyte and pre-implantation embryo. *Reproduction* 2010; **139**: 705–15.

56. **Murgatroyd C**, **Patchev AV**, **Wu Y** et al. Dynamic DNA methylation programs persistent adverse effects of early-life stress. *Nature Neurosci* 2009; **12**: 1559–66.

57. **Wintour EM**. Water channels and urea transporters. *Clin Exp Pharmacol Physiol* 1997; **24**: 1–9.

58. **Sands JM**. Mammalian urea transporters. *Annu Rev Physiol* 2003; **65**: 543–66.

59. **Jackson AA**. Salvage of urea-nitrogen and protein requirements. *Proc Nutr Soc* 1995; **54**: 535–47.

60. **Waterlow JC**. The mysteries of nitrogen balance. *Nutr Res Rev* 1999; **12**: 25–54.

61. **Collins D**, **Winter DC**, **Hogan AM** et al. Differential protein abundance and function of UT-B urea transporters in human colon. *Am J Physiol Gastrointest Liver Physiol* 2010; **298**: G345–51.

62. **Schrier RW**. Vasopressin and aquaporin 2 (AQP2) in clinical disorders of water homeostasis. *Semin Nephrol* 2008; **28**; 289–96.

63. **Lang F**, **Busch GL**, **Ritter M** et al. Functional significance of cell volume regulatory mechanisms. *Physiol Rev* 1998; **78**: 247–306.

64. **Lillycrop KA**, **Slater-Jefferies JL**, **Hanson MA** et al. Induction of altered epigenetic regulation of the hepatic glucocorticoid receptor in the offspring of rats fed a protein-restricted diet during pregnancy suggests that reduced DNA methyltransferase-1 expression is involved in impaired DNA methylation and changes in histone modifications. *Br J Nutr* 2007; **97**: 1064–73.

65. **Lillycrop KA**, **Phillips ES**, **Jackson AA**, **Hanson MA**, **Burdge GC**. Dietary protein restriction of pregnant rats induces and folic acid supplementation prevents epigenetic modification of hepatic gene expression in the offspring. *J Nutr* 2005; **135**: 1382–6.

66. **Bird A**. DNA methylation patterns and epigenetic memory. *Genes Dev* 2002; **16**: 6–21.

67. **Gidekel S**, **Bergman Y**. A unique developmental pattern of Oct-3/4 DNA methylation is controlled by a cis-demodification element. *J Biol Chem* 2002; **277**: 34521–30.

68. **Fuks F**, **Hurd PJ**, **Wolf D**, **Nan X**, **Bird AP**, **Kouzarides T**. The methyl-CpG-binding protein MeCP2 links DNA methylation to histone methylation. *J Biol Chem* 2003; **278**: 4035–40.

69. **Turner BM**. Histone acetylation and an epigenetic code. *Bioessays* 2000; **22**: 836–45.

70. **Pham TD**, **MacLennan NK**, **Chiu CT** et al. Uteroplacental insufficiency increases apoptosis and alters p53 gene methylation in the full-term IUGR rat kidney. *Am J Physiol Regul Integr Comp Physiol* 2003; **285**: R962–70.

71. **Waterland RA**, **Jirtle RL**. Early nutrition, epigenetic changes at transposons and imprinted genes, and enhanced susceptibility to adult chronic diseases. *Nutrition* 2004; **20**: 63–8.

72. **Lillycrop KA**, **Phillips ES**, **Torrens C**, **Hanson MA**, **Jackson AA**, **Burdge GC**. Feeding pregnant rats a protein-restricted diet persistently alters the methylation of specific cytosines in the hepatic PPAR alpha promoter of the offspring. *Br J Nutr* 2008; **100**: 278–82.

73. **Torrens C**, **Poston L**, **Hanson MA**. Transmission of raised blood pressure and endothelial dysfunction to the F2 generation induced by maternal protein restriction in the F0, in the absence of dietary challenge in the F1 generation. *Br J Nutr* 2008; **100**: 760–6.

74. **Burdge GC**, **Slater-Jefferies J**, **Torrens C** et al. Dietary protein restriction of pregnant rats in the F0 generation induces altered methylation of hepatic gene promoters in the adult male offspring in the F1 and F2 generations. *Br J Nutr* 2007; **97**: 435–9.

75. **Benyshek DC**, **Johnston CS**, **Martin JF**. Glucose metabolism is altered in the adequately-nourished grand-offspring (F3 generation) of rats malnourished during gestation and perinatal life. *Diabetologia* 2006; **49**: 1117–19.

76. **Zambrano E**, **Martínez-Samayoa PM**, **Bautista CJ** et al. Sex differences in transgenerational alterations of growth and metabolism in progeny (F2) of female offspring (F1) of rats fed a low protein diet during pregnancy and lactation. *J Physiol* 2005; **566**: 225–36.

77. **Sturmey RG**, **Bermejo-Alvarez P**, **Gutierrez-Adan A** et al. Amino acid metabolism of bovine blastocysts: a biomarker of sex and viability. *Mol Reprod Dev* 2010; **77**: 285–96.

78. **Ozkaya MO**, **Nazıroğlu M**, **Barak C**, **Berkkanoglu M**. Effects of multivitamin/mineral supplementation on trace element levels in serum and follicular fluid of women undergoing in vitro fertilization (IVF). *Biol Trace Elem Res* 2010; Feb 24 [Epub ahead of print].

79. **Ozjya MO**, **Naziroglu M**. Multivitamin and mineral supplementation modulates oxidative stress and antioxidant vitamin levels in serum and follicular fluid of women undergoing in vitro fertilization. *Fertil Steril* 2010; Mar 11 [Epub ahead of print].

80. **Marei WF**, **Wathes DC**, **Fouladi-Nashta AA**. Impact of linoleic acid on bovine oocyte maturation and embryo development. *Reproduction* 2010; **139**: 979–88.

81. **Marei WF, Wathes DC, Fouladi-Nashta AA.** The effect of linolenic Acid on bovine oocyte maturation and development. *Biol Reprod* 2009; **81**: 1064–72.

82. **Hugentobler SA, Morris DG, Sreenan JM, Diskin MG.** Ion concentrations in oviduct and uterine fluid and blood serum during the estrous cycle in the bovine. *Theriogenology* 2007; **68**: 538–48.

83. **Rosenfeld CS, Roberts RM.** Maternal diet and other factors affecting offspring sex ratio: a review. *Biol Reprod* 2004; **71**: 1063–70.

84. **Bermejo-Alvarez P, Rizos D, Rath D** et al. Can bovine in vitro-matured oocytes selectively process X- or Y-sorted sperm differentially? *Biol Reprod* 2008; **79**: 594–7.

85. **Singh R, Sinclair KD.** Metabolomics: approaches to assessing oocyte and embryo quality. *Theriogenology* 2007; **68**: S56–62.

86. **Revelli A, Delle Piane L, Casano S** et al. Follicular fluid content and oocyte quality: from single biochemical markers to metabolomics. *Reprod Biol Endocrinol* 2009; **7**: 40.

87. **Piñero-Sagredo E, Nunes S, de Los Santos MJ, Celda B, Esteve V.** NMR metabolic profile of human follicular fluid. *NMR Biomed* 2010.

88. **Houghton FD, Hawkhead JA, Humpherson PG, Hogg JE, Balen AH, Rutherford AJ, Leese HJ.** Non-invasive amino acid turnover predicts human embryo developmental capacity. *Hum Reprod* 2002; **17**: 999–1005. Erratum in: *Hum Reprod* 2003; **18**: 1756–7.

89. **Baumann CG, Morris DG, Sreenan JM, Leese HJ.** The quiet embryo hypothesis: molecular characteristics favoring viability. *Mol Reprod Dev* 2007; **74**: 1345–53.

90. **Leese HJ, Baumann CG, Brison DR, McEvoy TG, Sturmey RG.** Metabolism of the viable mammalian embryo: quietness revisited. *Mol Hum Reprod* 2008; **14**: 667–72.

91. **Booth PJ, Watson TJ, Leese HJ.** Prediction of porcine blastocyst formation using morphological, kinetic, and amino acid depletion and appearance criteria determined during the early cleavage of in vitro-produced embryos. *Biol Reprod* 2007; **77**: 765–79.

92. **Eckert JJ, Houghton FD, Hawkhead JA** et al. Human embryos developing in vitro are susceptible to impaired epithelial junction biogenesis correlating with abnormal metabolic activity. *Hum Reprod* 2007; **22**: 2214–24.

93. **Ceelen M, van Weissenbruch MM, Vermeiden JP, van Leeuwen FE, Delemarre-van de Waal HA.** Pubertal development in children and adolescents born after IVF and spontaneous conception. *Hum Reprod* 2008; **23**: 2791–8.

94. **Ceelen M, van Weissenbruch MM, Roos JC** et al. Body composition in children and adolescents born after in vitro fertilization or spontaneous conception. *J Clin Endocrinol Metab* 2007; **92**: 3417–23.

95. **Ceelen M, van Weissenbruch MM, Vermeiden JP, van Leeuwen FE, Delemarre-van de Waal HA.** Growth and development of children born after in vitro fertilization. *Fertil Steril* 2008; **90**: 1662–73.

96. **Ceelen M, van Weissenbruch MM, Vermeiden JP, van Leeuwen FE, Delemarre-van de Waal HA.** Cardiometabolic differences in children born after in vitro fertilization: follow-up study. *J Clin Endocrinol Metab* 2008; **93**: 1682–8.

97. **Ceelen M, van Weissenbruch MM, Prein J** et al. Growth during infancy and early childhood in relation to blood pressure and body fat measures at age 8–18 years of IVF children and spontaneously conceived controls born to subfertile parents. *Hum Reprod* 2009; **24**: 2788–95.

98. **Scientific Advisory Committee on Nutrition.** *The Nutritional Wellbeing of the British Population.* London: The Stationery Office, 2008.

99. **Robinson SM, Crozier SR, Borland SE** et al. Impact of educational attainment on the quality of young women's diets. *Eur J Clin Nutr* 2004; **58**: 1174–80.

100. **Huda SS, Brodie LE, Sattar N.** Obesity in pregnancy: prevalence and metabolic consequences. *Semin Fetal Neonatal Med* 2010; **15**: 70–6.

101. **Denison FC, Roberts KA, Barr SM, Norman JE.** Obesity, pregnancy, inflammation and vascular function. *Reproduction* 2010; Mar 9 [Epub ahead of print].

102. **Nelson SM, Matthews P, Poston L.** Maternal metabolism and obesity: modifiable determinants of pregnancy outcome. *Hum Reprod Update* 2010; **16**: 255–75.

103. **Stewart FM, Freeman DJ, Ramsay JE** et al. Longitudinal assessment of maternal endothelial function and markers of inflammation and placental function throughout pregnancy in lean and obese mothers. *J Clin Endocrinol Metab* 2007; **92**: 969–75.

104. **Roberts VH, Smith J, McLea SA** et al. Effect of increasing maternal body mass index on oxidative and nitrative stress in the human placenta. *Placenta* 2009; **30**: 169–75.

105. **Farley D, Tejero ME, Comuzzie AG** et al. Fetoplacental adaptations to maternal obesity in the baboon. *Placenta* 2009; **30**: 752–60.

106. **Jones HN, Woollett LA, Barbour N** et al. High-fat diet before and during pregnancy causes marked upregulation of placental nutrient transport and fetal overgrowth in C57/BL6 mice. *FASEB J* 2009; **23**: 271–8.

107. **Fraser AM, Brockert JE, Ward RH**. Association of young maternal age with adverse reproductive outcomes. *N Engl J Med* 1995; **332**: 1113–17.

108. **Baker PN, Wheeler SJ, Sanders TA** et al. A prospective study of micronutrient status in adolescent pregnancy. *Am J Clin Nutr* 2009; **89**: 1114–24.

109. **Kirchengast S, Hartmann B**. Impact of maternal age and maternal somatic characteristics on newborn size. *Am J Hum Biol* 2003; **15**: 220–8.

110. **Wu T, Buck G, Mendola P**. Maternal cigarette smoking, regular use of multivitamin/mineral supplements, and risk of fetal death: the 1988 National Maternal and Infant Health Survey. *Am J Epidemiol* 1998; **148**: 215–21.

111. **Scholl TO, Hediger ML, Schall JI, Khoo CS, Fischer RL**. Maternal growth during pregnancy and the competition for nutrients. *Am J Clin Nutr* 1994; **60**: 183–8.

112. **Hediger ML, Scholl TO, Schall JI**. Implications of the Camden Study of adolescent pregnancy: interactions among maternal growth, nutritional status, and body composition. *Ann N Y Acad Sci* 1997; **817**: 281–91.

113. **Thame MM, Fletcher HM, Baker TM, Jahoor F**. Comparing the glucose kinetics of adolescent girls and adult women during pregnancy. *Am J Clin Nutr* 2010; **91**: 604–9.

114. **Thame M, Fletcher H, Baker T, Jahoor F**. Comparing the in vivo glycine fluxes of adolescent girls and adult women during early and late pregnancy. *Br J Nutr* 2010: 1–5. [Epub ahead of print]

115. **Wallace JM, Luther JS, Milne JS** et al. Nutritional modulation of adolescent pregnancy outcome – a review. *Placenta* 2006; **27**: S61–8.

116. **Wallace JM, Milne JS, Aitken RP**. Effect of weight and adiposity at conception and wide variations in gestational dietary intake on pregnancy outcome and early postnatal performance in young adolescent sheep. *Biol Reprod* 2010; **82**: 320–30.

117. **Wu G, Bazer FW, Davis TA** et al. Arginine metabolism and nutrition in growth, health and disease. *Amino Acids* 2009; **37**: 153–68.

118. **Jackson AA, Gibson NR, Lu Y, Jahoor F**. Synthesis of erythrocyte glutathione in healthy adults consuming the safe amount of dietary protein. *Am J Clin Nutr* 2004; **80**: 101–7.

Discussion

ROBERT BOYD: I want to challenge you on the use of the terms 'high' and 'low'. You made the assumption that mice living on 'normal' diets are normal and those on low-protein diets are abnormal. Surely, the mouse in the wild is normally on a low-protein diet, and there is normally not a good barrier function for maternal steroids. The survival of the tribe depends on the regulatory response to living on low protein. Conversely, with the sheep you are assuming that the adolescent goes to high protein: why is it the high protein and not the normal one?

JACKSON: I am trying not to make any assumptions; I am simply going with convention. This is how they are conventionally characterized. It may be that you are right that our point of reference is inappropriate. It is an argument that can only be resolved when we start from what the preferred outcome is and work back to what it is that enables that preferred outcome. I suppose if one looks at the preferred outcome, then in general terms the value that is accepted as the reference is the one that, by and large, has been associated with a better outcome. If we look at the ability to cope in the low-protein diet, then you have to switch on adaptive mechanisms. If you are unable to switch them on you increase vulnerability. The ability to cope with low-protein diets in a human requires an ability to engage in enhanced salvage of urea nitrogen through the colonic microflora. If the microflora find themselves in an inhospitable environment, then you lose this capability.

POSTON: Thank you for referring to our teenage pregnancy study. Since that study we have begun to concentrate on preconceptual nutrition because I am convinced that this is incredibly important in terms of placentation. In the most recent literature in relation to pre-eclampsia recent large studies from the Scandinavian cohorts and the USA on vitamin D status, folate status and multivitamin use, are pointing to poor preconceptual nutrition relating to placentation. If we look back in the literature a decade or so, there was less evidence for this. I think it is partly because of a shift in demography and a shift in nutritional behaviour in developed countries. There has been a gradual change in diet. If we start to concentrate on preconceptual care we may have real potential for reducing growth restriction in developed countries.

JACKSON: I would add two riders to that. I think in those studies the analysis should be stratified by adolescence and smoking. Those are two important considerations. Undoubtedly, this area of work has helped to cause that shift, but I think that another factor has contributed. Over the last decade in the international arena there have been substantial studies where multiple micronutrient supplements starting during pregnancy have been looked at, not only in terms of maternal well-being but also in terms of the well-being of the offspring. They have had some benefit in terms of maternal well-being, but staggeringly unimpressive effects on the offspring. There have even been some suggestions of potentially adverse effects in the offspring. The idea of starting multiple micronutrients in pregnancy against a background that is already set doesn't obviously produce any benefit in populations in which you would expect it to have a striking effect. Therefore, the shift back to preconceptual is in terms of perceptions, but also stems from practical experience.

POSTON: In relation to the teenage pregnancy study, we have just published a second paper (Jones *et al.*, BJOG 2010; 117: 200–11). The results seem to differ from those of Jackie Wallace (Wallace J *et al.*, J Reprod Fertil Suppl 1999; **54**: 385) in the adolescent sheep in which continued growth (and abundant energy supply) in adolescent pregnancy in associated with fetal growth restriction. Our results suggest that the pregnant adolescent who continues to grow has a larger fetus and better nutritional status than an adolescent who is no longer growing. We have no evidence to support the idea that the mother grows at the expense of the baby.

TYCKO: In your view, what is the most striking adverse effect of maternal obesity on pregnancy?

JACKSON: Maternal obesity is a statement about the mother's inability not only to regulate her energy, but also her nutrient environment. In terms of what she is making available to the placenta, the placenta is not only structured poorly but also has poor choices to make.

CETIN: I like your conclusion about glycine and serine. This is something that we need to investigate more. Concerning glycine, when you presented the data on rats fed low-protein diet, you showed the reverse effects of glycine on PPARs. We know that in rats very little of the glycine fed to the mother is transported into the placenta. How do you explain this?

JACKSON: I don't have a ready explanation. The levels of glycine that were given were substantial. It was 3% equivalent of total energy. The transport was probably a mass effect.

CETIN: I know PPAR is also involved in the transport of lipids. That dose would probably also modify placental lipid exchange.

JACKSON: Once we get down to the integrated responses they are so complex that interpreting them is difficult. We have data indicating that if, instead of adding folic acid to the low-protein diet it is added to the high-protein diet, we see a totally different response in terms of blood pressure, the epigenetic effects and in terms of the phenotypes.

MOFFETT: What is the effect of these prepregnancy factors you are talking about on the endometrium? This is an unexplored area. We don't know whether or not the menstrual cycle is affected. I think this is a particularly human problem, because the preparation for implantation begins before implantation in humans, and it doesn't in other species. Do you get normal decidualization, which normally starts in the secretory phase? Do you get normal influx of NK cells? Are the glands making the same nutrition that they should be?

CRITCHLEY: These are big unanswered questions. The data we have heard, about how glucocorticoid receptor can be modulated with folate, is of great interest. We forget that these receptors are temporally and cell site specific. For example, they are key receptors on NK and endothelial cells. If you can modulate the expression of these receptors, these are two cellular components of the endometrium that might be critical, and they haven't been studied. Has anyone localised these receptors?

JACKSON: There are tissue-specific effects. This is part of the complexity. There is a whole series of questions that needs to be addressed, and part of the challenge is to do this in a structured way.

CRITCHLEY: You could do this in pregnant women. There is a target organ that can be sampled. You could also do this in women who have had pregnancy failure.

JANSSON: I'd like to develop the concept of overnutrition. There are many epidemiological studies showing that the problem with obesity is fetal overgrowth. However, in most if not all of the animal models we are using, the fetal growth outcome is either no change, or, in many cases, growth restriction. Do you agree that even though we can programme any animal model, it is difficult to mimic the fetal growth pattern occurring with overnutriton in humans?

JACKSON: The animal model provides us with proof of concept and suggestions for mechanisms. When we come to human studies, the question is how we account for the critical variables, and how we measure those with assurance. For many studies, obesity is characterized on the basis of BMI, which tells us nothing about relative adiposity or leanness. There is every reason to think these factors are important. In addition to this, the level of activity and other considerations also play into how this body composition expresses itself. It is likely that if you go from one situation to another you have the possibility of quite variable responses, depending on how these factors are put together. The challenge is to recognize these important factors and allow for them in experimental design. In terms of characterizing nutritional status, I don't think that in most studies the characterization is sufficiently refined to enable us to ask questions in a general state. David Barker referred briefly to the studies where we measured nitrogen metabolism at 18 and 28 weeks in pregnant women, and the signals from this are very strong and powerful. We have a study in which we have gone from prepregnancy to early pregnancy to later pregnancy to see whether women carry or adjust their metabolic phenotype. These sorts of dynamic studies provide much more powerful statements of what is happening metabolically than simply the measurement of the structure. But the measurement of structure, if done well, can provide much more useful insights that we presently have.

SMITH: Is there a true effect of teenage pregnancy, and how is this defined? You used quite a wide range of ages, from 13 to 19; 13 and 14 could be regarded as adolescent, and 15 and beyond is slightly less extreme. Also, how can we separate this from the obvious socio-economic issues? In 2001 we did an analysis in the *BMJ* (Smith GC, Pell JP. Teenage pregnancy and risk of adverse perinatal outcomes associated with first and second births: population based retrospective cohort study. *BMJ* 2001; **323**: 476) looking at all Scottish births with non-smoking girls between 15 and 19, and outcome of first pregnancy. The lowest risk was at age 16. In an evolutionary context, postponing pregnancy until your twenties wouldn't be something regarded as physiological.

JACKSON: The only response I would make is that we have some studies in Jamaica, looking at adolescents and more mature women. We have dynamic measurements of amino acid metabolism, which are striking different between the two groups. I accept your point, but it doesn't say that there aren't important differences. They may be maturational, or socio-economic, or related to other factors. It is important to know what those factors are.

BARKER: In the Helsinki Birth Cohort we know when the mothers menstruated. The average age was 16.5 years, and many mothers didn't menstruate until they were 20. This is part of an evolutionary strategy to allow more time for growth.

FERGUSON SMITH: I wanted to refer to the earlier mechanistic part of your presentation. There is currently some excitement about epigenetic mechanisms and how they might control metabolic processes. Methylation can be quite a late event in the control of gene regulation. Most gene promoters are not methylated at all. Often, acquisition of methylation can occur long after the gene repression has occurred. We need to be careful whether we are attributing a change in methylation as a causal event, as opposed to a consequence of the change in gene expression.

JACKSON: There are factors later in life experience that also operate, and it gets increasingly complicated as you add layers in. Suffice to say, that for those studies I showed, you can look at the different promoters and get variable methylation of the individual promoters at each level. Whether or not this is causal is an open question. In principle, modifications of dietary intake within the usual range lead to substantial differences at one stage. Therefore, by implication, they have the potential to account for substantial differences at other stages. Do they? I don't know. Certainly, methylation has to be a player in the considerations. We don't get these changes for all nutrient exposures; we do for some and not for others, showing that there is some specificity to the actions.

Chapter 4

Nutrition and preimplantation development

Tom P. Fleming

Introduction

The concept of developmental origins of health and disease (DOHaD) has radically altered the fields of reproductive and developmental biology. The pursuit of developmental mechanisms used to be a rather *intrinsic* exercise, probing how the embryo or later stage progressed along the road to adult form, with attention paid to morphogenesis, body patterning, cell differentiation, regulated proliferation, intercellular communication and the like. Since David Barker and colleagues first demonstrated that human health across the world in different populations and circumstances can be determined to a large extent by prenatal environment and the quality of maternal nutrition [1,2], the scope of developmental biology has changed forever. There is now an appreciation that *extrinsic* factors interact with the core developmental process, providing 'plasticity' in phenotype which recognizes the biological continuum that exists from one generation to the next.

One enduring idea is that plasticity during development allows an organism to 'select' the phenotype best suited to its future environment from a given genotype, mediated through factors associated with nutrient availability. The matching of phenotype to environment makes good evolutionary sense, but if nutrient availability changes from that 'predicted' during the period of developmental plasticity, the mismatch can lead to an increased risk of adult disease [3]. In eutherian mammals, the extended care provided by the mother throughout pregnancy provides ample opportunity for plasticity, or developmental programming, to play its part in shaping adult phenotype. The associations identified by Barker and colleagues between the risk of adult chronic disease, including coronary heart disease, stroke, hypertension, type II diabetes, obesity, osteoporosis and certain cognitive disorders, and factors related to weight and body shape at birth mediated through maternal nutrition and physiology [1,2], now provide the developmental biologist with a road map to unravel mechanisms and causes of adverse long-term effects.

Whereas many of the chapters in this book focus on the critical role that the placenta may play in developmental programming during later pregnancy in humans and animal species, this chapter will consider the susceptibility of the very early stages of mammalian development before embryo implantation, and even before fertilization has occurred. The egg and preimplantation embryo do indeed show an environmental sensitivity, with both short-term and long-lasting consequences that have implications for maternal nutrition and DOHaD, and also for assisted reproductive technologies (ART) used in the clinical treatment of infertility and in domestic animal productivity.

The pre-conception period and developmental programming

The relationship identified between a mother's body composition and the quality of her oocytes is now well established. Maternal obesity and high body mass index (BMI) are associated with reduced fertility and an increased risk of miscarriage in both natural and assisted pregnancies [4]. Stringent modulation of caloric intake close to the time of ART treatment to alleviate this condition is not necessarily successful [5]. Obesity further associates with polycystic ovary syndrome and type II diabetes, which also show disturbed fertility. For example, in a mouse study, the eggs and very early embryos from diabetic mothers displayed compromised development even when transferred to normal foster mothers [6].

The Placenta and Human Developmental Programming, ed. Graham J. Burton, David J. P. Barker, Ashley Moffett and Kent Thornburg. Published by Cambridge University Press. © Cambridge University Press 2011.

In livestock species, protein and carbohydrate food intake has a positive effect on follicle development and ovulation rate as well as overall oocyte quality, gene expression and potential [7–10]. Conversely, maternal undernourishment during oocyte maturation in sheep associates with impaired insulin-related metabolism in later pregnancy, affecting growth [11]. This supportive influence of nutrition has also been demonstrated in a sheep somatic cell cloning model where oocytes, subsequently enucleated and fused with adult somatic cells, gave rise to increased pregnancies and improved fetal development if they derived from ewes fed high rather than low nutrition, implying a non-genomic contribution [12].

In a detailed analysis of the long-term outcomes of maternal protein undernutrition in mice, mediated through a low-protein diet (LPD) fed exclusively during the period of oocyte maturation (3.5 days) prior to conception, we found that although no effect on gestation length, litter size or sex ratio was apparent, both male and female offspring exhibited adult disease. This was evident in cardiovascular dysfunction and behavioural abnormalities associated with anxiety-related locomotory traits [13]. Significantly elevated systolic blood pressure (4.3% =3% and 2.9% higher for males and females, respectively, across the whole population and sustained during adult life) was accompanied by reduced capacity of isolated mesenteric arterial slices to relax in response to the vasodilatory reagents acetylcholine and isoprenaline, and reduced female kidney size but with elevated nephron numbers [13].

Why should the oocyte during maturation exhibit environmental sensitivity with such adverse health outcomes? At present, we simply do not know the answer. This is a period of intense activity at the cellular level, coinciding with resumption of meiosis, polar body extrusion, reorganization of the cytoplasmic constituents awaiting fertilization, and all facilitated through complex intercellular communication with surrounding granulosa cells [14,15]. These orchestrated events critical for reproductive success are all subject to environmental compromise. The longer-term effects of *in vitro* maturation (IVM) of oocytes, used across species for medical and biotechnological purposes, has been recently evaluated and compared with *in vivo* matured oocytes in mice. As we saw for the LPD model above, some adverse health outcomes from IVM did emerge, and interestingly these included a reduction in cardiac output and pulse rate [16]. Although cellular responses by oocytes to maternal nutrition are less well studied than those made by embryos (see below), changes in oocyte metabolism and energy availability appear to be a primary upstream mediator of altered developmental competence. Thus, the reduction in mouse ovulation rate and oocyte maturation mediated through dietary undernutrition is reversed specifically by glucose supplementation [17]. In addition, metabolic effects within the oocyte itself have been identified in altered mitochondrial localization and metabolism in response to a high long-chain n-3 polyunsaturated fatty acid diet in mice [18].

Environmental sensitivity of the preimplantation embryo in culture

The preimplantation embryo has received significant attention in recent years with respect to environmental effects on short- and long-term gestational development, and health into adulthood. This stage of development represents post-fertilization cleavage of the zygote, the activation of the embryonic genome and the morphogenesis of the blastocyst, with outer and inner cell populations comprising the trophectoderm (TE) and inner cell mass (ICM), respectively (Figure 4.1) [19,20]. Preimplantation embryos during either early cleavage or until blastocyst formation (some 3–4 days in rodents, 5–6 days in human) are cultured *in vitro* for IVF and ART treatments. Nutritional support during *in vitro* culture (IVC), despite improvements in embryo culture media and supplements [21], is not comparable with the *in vivo* environment and can compromise developmental potential. A clear demonstration of this phenomenon was the reported condition of 'large offspring syndrome' (LOS) evident in sheep and cattle following IVC in the presence of serum resulting in excess perinatal growth and myogenesis, organ defects and increased mortality around term [22,23]. Although LOS has been associated with undefined serum contamination and the presence of nutrient levels that are non-physiological, alterations in embryo potential in response to IVC conditions are complex and involve several cellular systems [24].

IVC of embryos usually results in a slower rate of development than in *in vivo* counterparts, and with fewer cells in TE or ICM lineages [25]; IVC embryos often exhibit poorer fetal development and growth after transfer [26]. IVC can alter the development and function of different bodily systems

Chapter 4. Nutrition and preimplantation development

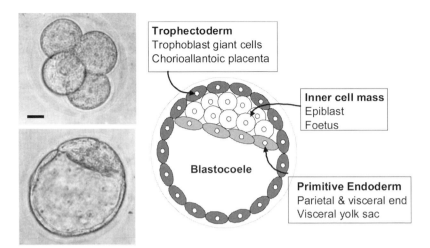

Figure 4.1 Preimplantation development. Cleavage of the mouse egg results in five or more rounds of reduction cell divisions leading to blastocyst formation with a blastocoele cavity generated by outer extraembryonic trophectoderm transepithelial transport. The inner cell mass subsequently segregates into internal epiblast, which gives rise to the fetus, and primitive endoderm at the interface with the blastocoele, a second extraembryonic lineage which gives rise to the visceral yolk sac endoderm as well as other lineages. Left: four-cell and blastocyst stages shown as photomicrographs. Right: blastocyst lineage diagram. Bar = 10 μm.

during late gestation and adult life, with the molecular identity of specific constituents of medium contributing to altered developmental programming in several cases. For example, IVC in the absence of granulocyte–macrophage colony-stimulating factor (GM-CSF) leads to restricted fetal growth and accelerated postnatal growth and adult body mass compared to *in vivo* controls; these effects are reversed by the inclusion of GM-CSF in IVC medium [27]. The beneficial effects of the presence of other factors in preimplantation embryo IVC, including amino acids, insulin and protein, on fetal growth have been reported [28,29]. In our own studies, mouse embryo IVC between two-cell and blastocyst stages in a standardized medium, followed by embryo transfer, induced relative hypertension in male and female offspring, independent of litter size, maternal origin or body weight compared with *in vivo* blastocyst formation followed by transfer [25]. This was accompanied by stimulated activity of enzymatic regulators of cardiovascular and metabolic physiology, namely serum angiotensin-converting enzyme (ACE) and the gluconeogenesis controller hepatic phosphoenolpyruvate carboxykinase (PEPCK) [25]. In other studies, mouse embryo IVC has been shown to alter postnatal behaviour, with deficits in memory and anxiety-related activities [30,31]. A relationship between either stimulated or retarded fetal growth and postnatal cardiovascular function has also been demonstrated in

the horse using between-breed transfer of thoroughbred and pony embryos to manipulate maternal uterine capacity [32].

An epigenetic contribution to developmental programming following IVC?

DNA methylation at specific CpG sites in gene regulatory domains functions in the suppression and control of gene expression in a tissue-specific manner and the maintenance of allele-specific expression from imprinted genes that commonly regulate growth during gestation [33,34]. DNA methylation patterns are extensively remodelled during preimplantation development, thereby providing a potential window for epigenetic sensitivity to environmental conditions [33]. Because DNA methylation patterns are preserved across cell divisions, such epigenetic changes may confer a long-lasting effect on gene expression and health into fetal and adult life.

As discussed above, LOS following IVC of sheep embryos in the presence of serum associates with abnormal expression and DNA methylation of the imprinted *IGF2R* gene during fetal development [35]. IVC of mouse embryos in the presence of serum has also been shown to disturb the monoallelic expression of imprinted genes in both fetal and placental lineages

[31,36,37]. Indeed, the sensitivity of the epigenetic status in embryos is such that embryo transfer alone, with minimal preceding culture, is sufficient to induce aberrant imprinted gene expression later in development [38]. Morgan et al. [39] have reported that mouse embryo IVC alters the level of expression of the epigenetically labile allele agouti viable yellow, causing a shift in the distribution of coat colour in offspring and implicating altered DNA methylation induced by culture conditions within this allele [39]. We have recently reviewed in more detail the potential role of DNA methylation in early embryo developmental programming [34].

Maternal nutrition and early embryo developmental programming

We have studied the importance of maternal diet specifically during preimplantation development, first in the rat [40] and more recently in the mouse [41]. Thus, mothers fed isocaloric LPD (9% casein) exclusively during preimplantation development (termed Emb-LPD; 0–4 days from mating) before returning to a normal protein diet (NPD; 18% casein) for the remainder of gestation, produced offspring (fed normal diet) that exhibited excess perinatal and postnatal growth and an increased risk of adult disease, notably sustained hypertension, anxiety-related behaviour and abnormal organ allometry, especially in females. The cardiovascular phenotype in adult offspring shows some similarities to that identified following LPD treatment targeted to oocyte maturation [13] and IVC of embryos [25] (see above). Accordingly, for male and female mouse offspring following Emb-LPD, the upward shift in systolic blood pressure (mm Hg) is 2.6% and 3.4%, respectively [41]. It should be noted that such an increase at a population level is large in terms of impact on disease. For example, in humas, a 3% increase in population SBP would lead to at least 9% more cases of cardiovascular disease, with some two million people in the UK entering the clinically hypertensive population [extrapolated from 42]. The increase in systolic blood pressure was accompanied by attenuated mesenteric artery vasodilation in response to isoprenaline and elevated lung ACE activity [43].

Data from the Emb-LPD model indicate that embryos respond to the nutritional challenge in both the embryonic and the extraembryonic lineages (see also below), affecting cellular processes that involve epigenetic and metabolic pathways. Thus, in the rat model, male conceptuses respond to Emb-LPD with reduced expression of the imprinted *H19* gene, a regulator of Igf2 expression, in both blastocysts and late gestation fetal liver, coincidentally with reduced expression of the *Igf2* gene [44]. Basal gene expression of hepatic metabolic enzymes is also altered in late gestation in a gender-specific manner in response to Emb-LPD with increased expression of PEPCK in males and increased expression of 11β-hydroxysteroid dehydrogenase type 1, which acts primarily as a reductase to produce active glucocorticoid in females [45]. The metabolic activity of embryos in response to maternal protein restriction is altered with LPD, causing a reduction in embryo mitochondrial membrane potential and reduced mitochondrial clustering around the nucleus suggestive of reduced metabolic activity [46].

Maternal nutritional restriction during the periconceptional period also leads to adverse developmental programming during fetal and postnatal life in domestic animal models. These studies usually involve treatment spanning both pre- and post-conception stages, and so the relative effects on oogenesis and embryogenesis are difficult to separate. However, they confirm a consistency across species of altered metabolic, endocrinological and physiological outcomes. Different models of periconceptional undernutrition (PCU) in sheep have shown that restriction is associated with accelerated maturation of the fetal HPA axis in late gestation, increased fetal arterial blood pressure, elevation in insulin secretion and early parturition [47–50]. These effects appear to be mediated at least in part through altered maternal endocrine status, and suggest a trade-off between a compensatory stimulation of the fetal stress axis to enhance survival in response to PCU and an association with later cardiovascular dysfunction [51,52]. Indeed, young adult sheep display long-term cardiovascular dysfunction, an increased risk of hypertension and impaired glucose tolerance following PCU [53,54]. The role of the placenta in coordinating the effects of PCU has also been proposed, in that placental 11β-hydroxysteroid dehydrogenase type 2 activity in late gestation is reduced following PCU, and may allow for increased placental transport of maternal glucocorticoids [55]. An important study demonstrating unambiguously an epigenetic pathway in developmental programming following maternal nutritional deficiency in sheep during the periconceptional period has been published by Sinclair and colleagues [56]. Here, a methyl-deficient diet with reduced vitamin B_{12}, folate and methionine resulted in altered DNA methylation of 4% of the fetal liver genome, and heavier, fatter and hypertensive offspring [56].

Early embryo developmental programming: the search for mechanisms

As indicated above, the association between embryo environment and adult health is complex and likely to encompass epigenetic and metabolic pathways. However, it is important to consider a *physiological* basis for these changes in the context of reproductive strategy. Further analysis of our mouse Emb-LPD model, made possible by generating a large sample size and tagging individual offspring from weaning, has revealed that the increased perinatal weight induced by Emb-LPD correlated positively and significantly with later adult obesity, hypertension and abnormal behaviour [41]. This implies that changes in growth rate are the primary mediator of later health outcomes and merit further investigation. We found that increased perinatal growth in Emb-LPD offspring was already induced in the blastocyst, as transfer of Emb-LPD and NPD blastocysts into opposite horns of NPD recipients led to heavier Emb-LPD conceptuses in late gestation. We conclude from these and other studies that early embryos must actively sense the quality of maternal nutrition and make compensatory responses by the blastocyst stage that modify how they develop and underlie the subsequent change in growth, postnatal phenotype and disease risk. Essentially, in a nutrient-deprived environment (Emb-LPD) embryo-activated responses will enhance nutrient retrieval from the mother (mechanisms discussed below) to protect fetal growth and maintain competitive fitness postnatally. These compensations do not need to influence gestational growth immediately, and may be targeted to late pregnancy when growth rate is maximal. In the Emb-LPD experimental design these responses are unnecessary, in that the diet is changed to NPD from implantation and overgrowth results. Indeed, when maternal LPD is maintained for the entire gestation period, embryo responses are appropriate in terms of growth control, with birth weight and postnatal growth being normal. Crucially, it is the *activation* of compensatory responses by preimplantation embryos, rather than whether they are appropriate in terms of growth control, that is critical for later health. Thus, maternal protein restriction, whether limited to the preimplantation period or maintained throughout gestation, predisposes offspring to adult disease, with the risk correlating positively with perinatal weight [41].

The role of the extraembryonic lineages in preimplantation responses to poor maternal diet

How might preimplantation embryos interact with the uterine environment to interpret the nutritional status of their mother and thereby regulate subsequent conceptus growth? Evidence is accumulating that early embryo responses to poor diet are mediated at least in part through the blastocyst extraembryonic lineages (trophectoderm, TE; primary endoderm, PE; Figure 4.1). This role appears to contribute to developmental plasticity and the matching of phenotype with nutrient availability, and can initiate from when the TE first surrounds the future fetal progenitor cells in the ICM before implantation.

The differentiation of the TE of the blastocyst results in an epithelial layer entirely surrounding the ICM from which the fetus will develop after implantation. During mouse blastocyst formation, the cells forming the TE acquire epithelial apicobasal polarity, form intercellular adhesive contacts mediated principally through E-cadherin, and differentiate circumferential tight junctions that seal outer cells together so that transepithelial transport between uterine and ICM environments can be closely regulated [19,20,57]. In this context, TE cells mature a polarized endocytotic pathway with preferential endocytosis occurring from their outer apical surfaces, with cytoplasmic endosomes and secondary lysosomes located along the apicobasal polarized cellular axis, and with distinct apical membrane recycling and transcellular basal surface delivery routes identified [58]. TE cells actively engage in endocytosis of uterine proteins mediated through an array of membrane receptors, which include the multiligand megalin and cubilin members of the LDL receptor gene family [59]. These receptors are known also to function significantly in the rodent visceral yolk sac, for which has been suggested a critical histiotrophic role in early gestational endocytosis and degradation of maternal proteins for delivery to the fetal circulation [60]. Thus, cubilin receptors in TE cells are responsible for the ingestion of uterine gland secretions comprising apolipoprotein and cholesterol [59]. This endocytotic pathway in TE is one way in which nutrient availability might be sensed by the blastocyst, providing information to guide developmental plasticity. Certainly, albumin endocytosis by mouse blastocyst TE cells has been shown to stimulate embryo development and also leads to increased fetal size in late gestation [29]. Indeed, in a broader context, the endocytotic character of TE and later trophoblast cells not only provides the early embryo with a means of histiotrophic nutrition but also may generate space within the maternal endometrium for subsequent embryo development [61].

Embryo interaction with the uterine maternal environment is also mediated through membrane transporters to regulate uptake of amino acids, including system B0,+ known to be critical in the regulation of cellular translation and growth [62]. In addition, the embryo is responsive to several maternal growth factor and cytokine pathways that promote proliferation and metabolic activity within early cell lineages and facilitate coordination of implantation [63].

These characteristics of TE cells are likely to contribute to preimplantation sensing of maternal nutrient environment. TE proliferation during cleavage has been shown to be sensitive to maternal diet during blastocyst formation in different species [40,46,64]. We are currently investigating the effects of Emb-LPD on the development and behaviour of the primary and secondary trophoblast, and the phenotype (microarray), structure and exchange topography (histomorphometry) and transport characteristics (amino acids; glucose) of the chorioallantoic placenta.

We have also explored the potential for the second extraembryonic lineage derived from the blastocyst, that of the primitive endoderm (PE) to contribute to maternal nutrient sensing and to promote compensatory responses to regulate growth in later gestation. The PE delaminates at the blastocoelic surface of the ICM and after implantation gives rise to the parietal and visceral endoderm (Figure 4.1). The visceral endoderm forms the absorptive and endocytotic surface of the yolk sac (visceral yolk sac endoderm, VYSE), which surrounds the rodent conceptus from early post-implantation development onwards throughout gestation, and functions in part in histiotrophic nutrition through endocytotic uptake and lysosomal

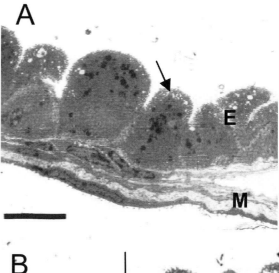

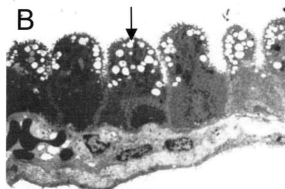

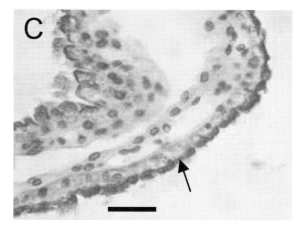

Figure 4.2 Visceral yolk sac endoderm in mouse. A. VYSE epithelial layer collected from day 17 late gestation conceptus following maternal NPD feeding and processed for transmission electron microscopy. Moderate apical endocytotic vesicles are evident (arrows) in VYSE cells (E); M, underlying mesothelial layer. B. YYSE layer from day 17 conceptus after maternal LPD feeding and with excess apical endocytotic vesicles (arrow). C. Immunocytochemical preparation at light microscope level of VYSE showing apical surface and cytoplasm staining (brown) for megalin endocytotic receptor (arrow). Bar = 10 μm (A,B); 50 μm (C).

digestion of maternal uterine lumen proteins and the delivery of released amino acids for fetal growth (Figure 4.2) [65]. The contribution of rodent VYSE cells to fetal nutrition extends from before the development of the chorioallantoic placenta until term, and in late gestation a significant proportion of fetal amino acids derive from VSCE histiotrophic activity [65]. We have found that VYSE derived from LPD-treated mice at day 17 of gestation exhibited an enhanced rate of fluid-phase endocytosis *in vitro*, around double that of control NPD VYSE [41]. This increased activity was accompanied by an increased density of endocytotic vesicles within their cells [Figs 4.2A, B] and, importantly, increased protein expression of the megalin receptor (*Lrp2* gene), the major multiligand apical surface and cytoplasm endocytotic receptor of VYSE (Figure 4.2C) [60]. We can therefore define one pathway by which compensatory responses activated by the blastocyst stage within the extraembryonic lineages regulating conceptus growth is regulated.

A human perspective

In recent years we have seen a consolidation of evidence from a number of different mammalian species that mature oocytes and preimplantation embryos interact with their environment, leading to changes in their metabolism and developmental potential. The human egg and embryo are, of course, not immune to these influences, and the data generated from animal models have been an important motivation for analysis of the periconceptional environment in human health. Certainly, exposure to nutritional deprivation, specifically during early gestation, as experienced by a number of mothers during the later period of the five-month Dutch Hunger Winter of 1944–5, has been shown to increase the risk of early-onset coronary heart disease and increased blood pressure in response to psychological stress, as well as increased BMI and glucose intolerance [66,67]. However, clear evidence as to whether nutritional effects on health are induced specifically during the human periconceptional period is understandably difficult to find, for ethical reasons.

Could periconceptional human nutrition also influence the development and functional activity of the extraembryonic lineages, as indicated above from the use of animal models? There is potential for this. Although the human yolk sac is proportionately smaller and organized structurally very differently from that of the rodent, a role in histiotrophic nutrition during the first trimester has been proposed [68]. Endometrial gland secretions during the first 10 weeks of pregnancy are able to permeate the placental intervillous space, are phagocytosed by the syncytiotrophoblast, and are subsequently found within the yolk sac epithelium in the exocoelomic cavity [69].

The importance of embryo environment on life-long health in humans has been most clearly demonstrated in ART. ART is associated with some adverse perinatal and postnatal outcomes, including low birth weight and an increased risk of imprinting disorders [70]. This has included evidence of an association between Beckwith–Wiedemann syndrome, Angelman syndrome and retinoblastoma with human ART [71]. However, most recently and relevant here, epidemiological studies have demonstrated increased systolic and diastolic blood pressure in IVF children compared to age-and gender-matched spontaneously conceived control children from subfertile parents [72]. Moreover, growth velocity in late infancy was higher in IVF children than controls, and childhood growth was predictive of later blood pressure [73]. These studies are consistent with our model of growth-related developmental plasticity induced by embryo environment correlating with later adverse cardiovascular health. They also underpin the need to identify the factors that induce such long-term effects, so that culture conditions used in ART can be made entirely safe for later health.

Conclusion

Mammalian eggs and embryos interact with their environment, causing changes in their metabolic and epigenetic status that affect their developmental potential. The concept of developmental plasticity in response to nutrient availability at the periconceptional stage indicates the capacity for extraembryonic lineages to initiate compensatory responses which may attune nutrient delivery to the needs of the developing fetus. The consistency of developmental programming effects at the time of oocyte maturation and preimplantation embryo formation across species, including the human, make it imperative to investigate mechanisms in more detail to protect against adverse disease risk.

Acknowledgements

I am grateful to the Medical Research Council, the Biotechnology and Biological Sciences Research

Council, the Gerald Kerkut Trust and the National Institutes of Health for financial support of research in my laboratory.

References

1. **Barker DJP**. *Mothers, Babies and Health in Later Life*, 2nd ed. Edinburgh: Churchill Livingstone, 1998.
2. **Barker DJ**. The origins of the developmental origins theory. *J Intern Med* 2007; **261**: 412–17.
3. **Gluckman PD**, **Hanson MA**. Living with the past: evolution, development, and patterns of disease. *Science* 2004; **305**: 1733–6.
4. **Metwally M**, **Ong KJ**, **Ledger WL**, **Li TC**. Does high body mass index increase the risk of miscarriage after spontaneous and assisted conception? A meta-analysis of the evidence. *Fertil Steril* 2008; **90**: 714–26.
5. **Tsagareli V**, **Noakes M**, **Norman RJ**. Effect of a very-low-calorie diet on in vitro fertilization outcomes. *Fertil Steril* 2006; **86**: 227–9.
6. **Wyman A**, **Pinto AB**, **Sheridan R**, **Moley KH**. One-cell zygote transfer from diabetic to nondiabetic mouse results in congenital malformations and growth retardation in offspring. *Endocrinology* 2008; **149**: 466–9.
7. **Ashworth CJ**, **Toma LM**, **Hunter MG**. Nutritional effects on oocyte and embryo development in mammals: implications for reproductive efficiency and environmental sustainability. *Philos Trans Roy Soc Lond B Biol Sci* 2009; **364**: 3351–61.
8. **Zak LJ**, **Xu X**, **Hardin RT**, **Foxcroft GR**. Impact of different patterns of feed intake during lactation in the primiparous sow on follicular development and oocyte maturation. *J Reprod Fertil* 1997; **110**: 99–106.
9. **Pisani LF**, **Antonini S**, **Pocar P** et al. Effects of pre-mating nutrition on mRNA levels of developmentally relevant genes in sheep oocytes and granulosa cells. *Reproduction* 2008; **136**: 303–12.
10. **Quesnel H**. Nutritional and lactational effects on follicular development in the pig. *Soc Reprod Fertil Suppl* 2009; **66**: 121–34.
11. **Jaquiery AL**, **Oliver MH**, **Rumball CW**, **Bloomfield FH**, **Harding JE**. Undernutrition before mating in ewes impairs the development of insulin resistance during pregnancy. *Obstet Gynecol* 2009; **114**: 869–76.
12. **Peura TT**, **Kleemann DO**, **Rudiger SR** et al. Effect of nutrition of oocyte donor on the outcomes of somatic cell nuclear transfer in the sheep. *Biol Reprod* 2003; **68**: 45–50.
13. **Watkins AJ**, **Wilkins A**, **Cunningham C** et al. Low protein diet fed exclusively during mouse oocyte maturation leads to behavioural and cardiovascular abnormalities in offspring. *J Physiol* 2008; **586**: 2231–44.
14. **Richards JS**, **Russell DL**, **Ochsner S**, **Espey LL**. Ovulation: new dimensions and new regulators of the inflammatory-like response. *Annu Rev Physiol* 2002; **64**: 69–92.
15. **Thomas FH**, **Vanderhyden BC**. Oocyte-granulosa cell interactions during mouse follicular development: regulation of kit ligand expression and its role in oocyte growth. *Reprod Biol Endocrinol* 2006; **4**: 19–26.
16. **Eppig JJ**, **O'Brien MJ**, **Wigglesworth K** et al. Effect of in vitro maturation of mouse oocytes on the health and lifespan of adult offspring. *Hum Reprod* 2009; **24**: 922–8.
17. **Yan J**, **Zhou B**, **Yang J** et al. Glucose can reverse the effects of acute fasting on mouse ovulation and oocyte maturation. *Reprod Fertil Dev* 2008; **20**: 703–12.
18. **Wakefield SL**, **Lane M**, **Schulz SJ** et al. Maternal supply of omega-3 polyunsaturated fatty acids alter mechanisms involved in oocyte and early embryo development in the mouse. *Am J Physiol Endocrinol Metab* 2008; **294**: E425–34.
19. **Fleming TP**, **Sheth B**, **Fesenko I**. Cell adhesion in the preimplantation mammalian embryo and its role in trophectoderm differentiation and blastocyst morphogenesis. *Front Biosci* 2001; **6**: D1000–7.
20. **Fleming TP**, **Wilkins A**, **Mears A** et al. The making of an embryo: short-term goals and long-term implications. *Reprod Fertil Dev* 2004; **16**: 325–37.
21. **Summers MC**, **Biggers JD**. Chemically defined media and the culture of mammalian preimplantation embryos: historical perspective and current issues. *Hum Reprod Update* 2003; **9**: 557–82.
22. **Young LE**, **Sinclair KD**, **Wilmut I**. Large offspring syndrome in cattle and sheep. *Rev Reprod* 1998; **3**: 155–63.
23. **Maxfield EK**, **Sinclair KD**, **Broadbent PJ** et al. Short-term culture of ovine embryos modifies fetal myogenesis. *Am J Physiol* 1998; **274**: E1121–3.
24. **Fleming TP**, **Kwong WY**, **Porter R** et al. The embryo and its future. *Biol Reprod* 2004; **71**: 1046–54.
25. **Watkins AJ**, **Platt D**, **Papenbrock T** et al. Mouse embryo culture induces changes in postnatal phenotype including raised systolic blood pressure. *Proc Natl Acad Sci USA* 2007; **104**: 5449–54.
26. **Mahsoudi B**, **Li A**, **O'Neill C**. Assessment of the long-term and transgenerational consequences of perturbing preimplantation embryo development in mice. *Biol Reprod* 2007; **77**: 889–96.
27. **Sjöblom C**, **Roberts CT**, **Wikland M**, **Robertson SA**. Granulocyte-macrophage colony-stimulating factor

28. **Lane M**, **Gardner DK**. Differential regulation of mouse embryo development and viability by amino acids. *J Reprod Fertil* 1997; **109**: 153–64.

29. **Kaye PL**, **Gardner HG**. Preimplantation access to maternal insulin and albumin increases fetal growth rate in mice. *Hum Reprod* 1999; **14**: 3052–9.

30. **Ecker DJ**, **Stein P**, **Xu Z**, **Williams CJ**, **Kopf GS**, **Bilker WB**, **Abel T**, **Schultz RM**. Long-term effects of culture of preimplantation mouse embryos on behavior. *Proc Natl Acad Sci USA* 2004; **101**: 1595–600.

31. **Fernández-Gonzalez R**, **Moreira P**, **Bilbao A** et al. Long-term effect of in vitro culture of mouse embryos with serum on mRNA expression of imprinting genes, development, and behavior. *Proc Natl Acad Sci USA* 2004; **101**: 5880–5.

32. **Giussani DA**, **Forhead AJ**, **Gardner DS** et al. Postnatal cardiovascular function after manipulation of fetal growth by embryo transfer in the horse. *J Physiol* 2003; **547**: 67–76.

33. **Reik W**. Stability and flexibility of epigenetic gene regulation in mammalian development. *Nature* 2007; **447**: 425–32.

34. **Lucas ES**, **Fleming TP**. Developmental programming and epigenetics: DNA methylation makes its mark. *Cell Tissue Biol Res* 2009; **1**: 15–23.

35. **Young LE**, **Fernandes K**, **McEvoy TG** et al. Epigenetic change in IGF2R is associated with fetal overgrowth after sheep embryo culture. *Nature Genet* 2001; **27**: 153–4.

36. **Doherty AS**, **Mann MR**, **Tremblay KD**, **Bartolomei MS**, **Schultz RM**. Differential effects of culture on imprinted H19 expression in the preimplantation mouse embryo. *Biol Reprod* 2000; **62**: 1526–35.

37. **Mann MR**, **Lee SS**, **Doherty AS** et al. Selective loss of imprinting in the placenta following preimplantation development in culture. *Development* 2004; **131**: 3727–35.

38. **Rivera RM**, **Stein P**, **Weaver JR** et al. Manipulations of mouse embryos prior to implantation result in aberrant expression of imprinted genes on day 9.5 of development. *Hum Mol Genet* 2008; **17**: 1–14.

39. **Morgan HD**, **Jin XL**, **Li A**, **Whitelaw E**, **O'Neill C**. The culture of zygotes to the blastocyst stage changes the postnatal expression of an epigenetically labile allele, agouti viable yellow, in mice. *Biol Reprod* 2008; **79**: 618–23.

40. **Kwong WY**, **Wild AE**, **Roberts P**, **Willis AC**, **Fleming TP**. Maternal undernutrition during the preimplantation period of rat development causes blastocyst abnormalities and programming of postnatal hypertension. *Development* 2000; **127**: 4195–202.

41. **Watkins AJ**, **Ursell E**, **Panton R** et al. Adaptive responses by mouse early embryos to maternal diet protect fetal growth but predispose to adult onset disease. *Biol Reprod* 2008; **78**: 299–306.

42. **Chobanian AV**, **Bakris GL**, **Black HR** et al. Seventh report of the Joint National Committee on Prevention, Detection, Evaluation, and Treatment of High Blood Pressure. *Hypertension* 2003; **42**: 1206–52.

43. **Watkins, AJ**, **Lucas ES**, **Torrens C** et al. Maternal low-protein diet during mouse pre-implantation development induces vascular dysfunction and altered renin-angiotensin-system homeostasis in the offspring. *Br J Nutrition* 2010; in press.

44. **Kwong WY**, **Miller DJ**, **Ursell E** et al. Imprinted gene expression in the rat embryo-fetal axis is altered in response to periconceptional maternal low protein diet. *Reproduction* 2006; **132**: 265–77.

45. **Kwong WY**, **Miller DJ**, **Wilkins AP** et al. Maternal low protein diet restricted to the preimplantation period induces a gender-specific change on hepatic gene expression in rat fetuses. *Mol Reprod Dev* 2007; **74**: 48–56.

46. **Mitchell M**, **Schulz SL**, **Armstrong DT**, **Lane M**. Metabolic and mitochondrial dysfunction in early mouse embryos following maternal dietary protein intervention. *Biol Reprod* 2009; **80**: 622–30.

47. **Bloomfield FH**, **Oliver MH**, **Hawkins P** et al. Periconceptional undernutrition in sheep accelerates maturation of the fetal hypothalamic-pituitary-adrenal axis in late gestation. *Endocrinology* 2004; **145**: 4278–85.

48. **Oliver MH**, **Hawkins P**, **Breier BH** et al. Maternal undernutrition during the periconceptual period increases plasma taurine levels and insulin response to glucose but not arginine in the late gestational fetal sheep. *Endocrinology* 2001; **142**: 4576–9.

49. **Kumarasamy V**, **Mitchell MD**, **Bloomfield FH** et al. Effects of periconceptional undernutrition on the initiation of parturition in sheep. *Am J Physiol Regul Integr Comp Physiol* 2005; **288**: R67–72.

50. **Edwards LJ**, **McMillen IC**. Periconceptional nutrition programs development of the cardiovascular system in the fetal sheep. *Am J Physiol Regul Integr Comp Physiol* 2002; **283**: R669–79.

51. **McMillen IC**, **MacLaughlin SM**, **Muhlhausler BS** et al. Developmental origins of adult health and disease: the role of periconceptional and foetal nutrition. *Basic Clin Pharmacol Toxicol* 2008; **102**: 82–9.

52. **Oliver MH**, **Hawkins P**, **Harding JE**. Periconceptional undernutrition alters growth trajectory and metabolic

and endocrine responses to fasting in late-gestation fetal sheep. *Pediatr Res* 2005; **57**: 591–8.

53. **Gardner DS**, **Pearce S**, **Dandrea J** et al. Peri-implantation undernutrition programs blunted angiotensin II evoked baroreflex responses in young adult sheep. *Hypertension* 2004; **43**: 1290–6.

54. **Todd SE**, **Oliver MH**, **Jaquiery AL**, **Bloomfield FH**, **Harding JE**. Periconceptional undernutrition of ewes impairs glucose tolerance in their adult offspring. *Pediatr Res* 2009; **65**: 409–13.

55. **Connor KL**, **Challis JR**, **van Zijl P** et al. Do alterations in placental 11beta-hydroxysteroid dehydrogenase (11betaHSD) activities explain differences in fetal hypothalamic-pituitary-adrenal (HPA) function following periconceptional undernutrition or twinning in sheep? *Reprod Sci* 2009; **16**: 1201–12.

56. **Sinclair KD**, **Allegrucci C**, **Singh R** et al. DNA methylation, insulin resistance, and blood pressure in offspring determined by maternal periconceptional B vitamin and methionine status. *Proc Natl Acad Sci USA* 2007; **104**: 19351–6.

57. **Eckert JJ**, **Fleming TP**. Tight junction biogenesis during early development. *Biochim Biophys Acta* 2008; **1778**: 717–28.

58. **Fleming TP**, **Pickering SJ**. Maturation and polarization of the endocytotic system in outside blastomeres during mouse preimplantation development. *J Embryol Exp Morphol* 1985; **89**: 175–208.

59. **Assémat E**, **Vinot S**, **Gofflot F** et al. Expression and role of cubilin in the internalization of nutrients during the peri-implantation development of the rodent embryo. *Biol Reprod* 2005; **72**: 1079–86.

60. **Christensen EI**, **Verroust PJ**. Megalin and cubilin, role in proximal tubule function and during development. *Pediatr Nephrol* 2002; **17**: 993–9.

61. **Bevilacqua E**, **Hoshida MS**, **Amarante-Paffaro A**, **Albieri-Borges A**, **Zago Gomes S**. Trophoblast phagocytic program: roles in different placental systems. *Int J Dev Biol* 2010; **54**: 495–505.

62. **Van Winkle LJ**, **Tesch JK**, **Shah A**, **Campione AL**. System B0,+ amino acid transport regulates the penetration stage of blastocyst implantation with possible long-term developmental consequences through adulthood. *Hum Reprod Update* 2006; **12**: 145–57.

63. **Hardy K**, **Spanos S**. Growth factor expression and function in the human and mouse preimplantation embryo. *J Endocrinol* 2002; **172**: 221–36.

64. **Kakar MA**, **Maddocks S**, **Lorimer MF** et al. The effect of peri-conception nutrition on embryo quality in the superovulated ewe. *Theriogenology* 2005; **64**: 1090–103.

65. **Beckman DA**. Mechanisms of amino acid supply to the rat conceptus in normal and abnormal development. *Reprod Toxicol* 1997; **11**: 595–9.

66. **Painter RC**, **de Rooij SR**, **Bossuyt PM** et al. Early onset of coronary artery disease after prenatal exposure to the Dutch famine. *Am J Clin Nutr* 2006; **84**: 322–7.

67. **Painter RC**, **de Rooij SR**, **Bossuyt PM** et al. Blood pressure response to psychological stressors in adults after prenatal exposure to the Dutch famine. *J Hypertens* 2006; **24**: 1771–8.

68. **Burton GJ**, **Hempstock J**, **Jauniaux E**. Nutrition of the human fetus during the first trimester – a review. *Placenta* 2001; **22** Suppl A: S70–7.

69. **Burton GJ**, **Watson AL**, **Hempstock J**, **Skepper JN**, **Jauniaux E**. Uterine glands provide histiotrophic nutrition for the human fetus during the first trimester of pregnancy. *J Clin Endocrinol Metab* 2002; **87**: 2954–9.

70. **Hansen M**, **Bower C**, **Milne E**, **de Klerk N**, **Kurinczuk JJ**. Assisted reproductive technologies and the risk of birth defects – a systematic review. *Hum Reprod* 2005; **20**: 328–38.

71. **Manipalviratn S**, **DeCherney A**, **Segars J**. Imprinting disorders and assisted reproductive technology. *Fertil Steril* 2009; **91**: 305–15.

72. **Ceelen M**, **van Weissenbruch MM**, **Vermeiden JP**, **van Leeuwen FE**, **Delemarre-van de Waal HA**. Cardiometabolic differences in children born after in vitro fertilization: follow-up study. *J Clin Endocrinol Metab* 2008; **93**: 1682–8.

73. **Ceelen M**, **van Weissenbruch MM**, **Prein J** et al. Growth during infancy and early childhood in relation to blood pressure and body fat measures at age 8–18 years of IVF children and spontaneously conceived controls born to subfertile parents. *Hum Reprod* 2009; **24**: 2788–95.

Discussion

SFERRUZZI-PERRI: You talked about developmental programmes, and that you found differences in the trophectoderm between your embryos on different protein diets. Do you know much about trophectoderm cell fate? What determines which trophoblasts will form the exchange region of the placenta or the junctional zone? Are there differences between the architectures of the placentas between these different dietary protocols?

FLEMING: We don't know at this stage the detailed relationship between early trophectoderm changes mediated through diet and later organization and function of the placenta. When our current datasets on the different placental phenotypes derived from periconceptional diets have come through, we hope to be able to say more.

GIUSSANI: These treatments might impact on the shape of the placenta, as well as having fetal consequences.

FLEMING: Perhaps. We are coming at this from a slightly different angle to other groups. Our responsibility is to try to understand what is really deriving from the early embryonic period. Then comparison can then be made to other models. The whole point of these animal models is ultimately to advise for the human.

HUNT: I really enjoyed your studies on the effects on the trophectoderm, and whether there will be differentiation occurring during this time. I have a question about the other kinds of cells in the inner cell mass: the mesenchyme cells. In the human placenta we know that the fibroblasts and macrophages develop from precursor cells. These grow and develop markers and evidence of differentiation, just as the trophectoderm does. Are you thinking about the signalling between the mesenchyme and the endothelium, or is that just too difficult to study in the mouse preimplantation blastocyst?

FLEMING: We haven't looked at this. I agree with what you are saying about trying to characterize the process of differentiation of cells, both in the trophoblast and also from within the epiblast. That is ahead of us. Our strategy has been that if there is a preimplantation window, we have to demonstrate that it has real health issues. We are the main group that has been able to do this. We are collaborating with groups using myographical techniques, particularly looking at our cardiovascular phenotypes. We are also trying to catalogue those more overt changes that we can see in the very early embryonic lineages, to develop *in vitro* models. It may be from these models that we will be able to look more closely at markers for differentiation. We are also a lab that has spent a lot of time looking at the differentiation itself of the trophectoderm, and the formation of a functional epithelium to generate a blastocoele cavity, for example. We are studying this in relation to the maternal environment. We know, for example, that PKC-mediated pathways are important in this differentiation process and we have some circumstantial evidence that they may be susceptible to environmental conditions.

HUNT: Would you be thinking of looking at the pre-blastocyst stages to see whether there is dietary-induced differences occurring at the two-cell and four-cell stages?

FLEMING: It may well be that embryonic genome activation is affected by diet if we go back to the two-cell stage in the mouse. We are also looking at how the oviduct, as well as the uterine environment may be a mediator of these processes. We know that these changes in developmental potential have been induced by the blastocyst stage, but we wouldn't rule out the possibility that induction of these processes may occur prior to this.

BURTON: I have a comment on Akt. Akt tracks very closely with the mTOR pathway. The *Akt1* knockout mouse has IUGR. We found a positive linear correlation between Akt levels and placental weight, but they particularly have an absence in the junctional zone of the placenta, and not so much effect in the labyrinth. This is the endocrine part of the mouse placenta, and we don't fully understand what impact these endocrine actions have. This is a grey area that needs research.

FALL: Do you see differences in the responses between male and female embryos?

FLEMING: Yes. In our rat model, the male and female blastocysts tend to have different sensitivities with respect to *H19* expression in relation to diet. We find in our mouse model that for the postnatal outcomes, it is the females who are more sensitive, particularly to some of the behavioural effects and also the growth effects. If you induce increased growth as a compensatory process in both the males and females, then postnatally in a normal diet the females are unable to compensate back down again, but males can. So, there are differences in the dietary consequences of those induced processes between males and females and, generally speaking, we see a greater sensitivity and a bigger effect in females.

BARKER: I would like to pick up on Caroline Fall's point. What we have just heard is quite amazing, because the epidemiological data suggest that the male fetus is more sensitive. Are you saying that the female embryo is more sensitive to the mother's diet than the male?

FLEMING: No. The male embryos tend to develop slightly faster and reach the blastocyst stage earlier and this may contribute to a greater environmental sensitivity during this period. However, the cardiovascular phenotype we see is clearly there in both the male and female progeny. It is just that in some datasets we see a more striking response in the female. The kind of growth consequences we see induced during gestation, and their association with later disease, is evident in both the males and the females. It is really probably a matter of detail.

BARKER: You used the word 'compensatory'. This word comes up in thinking about the sheep model of placental enlargement. It is also comes up in relation to postnatal growth. The road to premature death from coronary heart disease is a road in which a relatively small baby remains small in infancy, but then starts to gain weight rapidly. There is a lot known in domestic husbandry about the adverse effects on fish, turkeys and pigs of rapid postnatal growth. The word compensatory is coming up for the embryo, for the placenta in mid gestation and for the child. So what is so fundamental about 'compensatory' being bad? Is it that rapid growth has huge costs, or is it that the insults that trigger such growth cause damage?

FLEMING: My first feeling about any of these potentially compensatory processes is that they are helping the animal survive. Is the low protein diet the normal one for the mouse in the wild? Probably. Compensation may be a word that we use loosely that really reflects the norm. It may be there is suppression of compensatory processes in a well fed, well looked after rodent and we see this as normal. We may be looking at them the wrong way round. The last thing a mouse has to worry about is whether or not late in life it has high blood pressure. So, it is better to interpret compensation as an evolutionary strategy for survival and competitiveness.

BARKER: If you make salmon hold back their growth by putting them in cold water, and then restore them to normal temperatures, they die within a few weeks. Fish farmers know that there is something about rapid growth that has huge costs.

FLEMING: There are so many different ways we can analyse these animals. We don't know enough yet.

RICHARD BOYD: We heard at the beginning of this meeting about buffering. Does the oviduct buffer the early conceptus from the effects of maternal environment, or in fact amplify them?

FLEMING: There may be an amplifying effect here. If you want to construct a building you need the blueprints set before you start. The whole process of placentation needs to be structured and planned from this very early stage, I would argue. This relates to the earlier points made: we need to go to that early postimplantation period and understand the differentiation of the cells, their expression capacity, and how they are behaving. In the context of the placenta, the work with the secondary trophoblast that we are starting may give us some leads in this respect.

JACKSON: I want to make a point about the term 'compensatory' growth. The suggestion is that it is mimicking normal growth. But growth is coordinated and structured in space and time, and if you miss the time then the nature of growth is different. It isn't growth; it is an increase in size that it out of time and against a different background. It is dangerous to presume that 'compensatory' growth is capturing something that is missed. You gain shape and size, but this is likely to be of a different quality to what would have been done.

SIBLEY: That fits with placental compensation, as Abby Fowden showed in her work with Phil Coan (Coan PM, Angiolini E, Sandovici I, *et al*. Adaptations in placental nutrient transfer capacity to meet fetal growth demands depend on placental size in mice. *J Physiol* 2008 Sep 15; **586**: 4567–76), where there are changes in morphology out of tune with changes in transporter activity to compensate for the small placenta.

PIJNENBORG: I am interested in your embryo transfer experiments. I would like to encourage you to look also to the maternal side. Rats and mice don't decidualize spontaneously. Decidualization is normally induced by implanting blastocysts, but can also be artificially induced in pseudopregnant animals by applying a local traumatic stimulus to the uterus. We did such experiments several years ago on diabetic rats and found that the decidualization is much less than in normal animals. How would decidualization occur in animals with low-protein diets?

FLEMING: That is interesting. We could talk about this.

Chapter 5

Maternofetal transport pathways during embryogenesis and organogenesis

Eric Jauniaux and Graham J. Burton

Introduction

Abnormalities of human placentation are associated with diseases that are either unique to the human species, such as pre-eclampsia or hydatidiform mole, or very rare in other species, such as miscarriage. There is increasing evidence showing that failures of placentation are associated with an imbalance of free radicals, which will further affect placental development and function and may subsequently have an influence on both the fetus and its mother [1]. Maternal metabolic disorders, for example diabetes, which are associated with an increased production of free radical species, are associated with a higher incidence of miscarriage and fetal structural defects [2]. Furthermore, the teratogenicity of drugs such as thalidomide has also been shown to involve free radical-mediated oxidative damage [3], confirming that the human fetus can be irreversibly damaged by oxidative stress [1].

In the past it has been assumed that the principal function of the placenta is to supply the fetus with as much oxygen as possible, and to a large extent that is certainly the case in the second half of pregnancy when fetal body mass must increase rapidly. The appearance of maternal erythrocytes within the lacunar spaces of the syncytiotrophoblastic layer of the primitive placenta shortly after implantation has been widely interpreted by many anatomists as evidence of early onset of the maternal circulation. In 1986, Hustin and Schaaps proposed a new concept that has revolutionized our understanding of human placentation [4,5]. Their anatomical study of hysterectomy specimens with pregnancy *in situ* showed that the maternal circulation to the human placenta is not fully established until 10–12 weeks of pregnancy. Using hysteroscopy, they also showed that during the first trimester the intervillous space is filled by a clear fluid, and hypothesized that this could be a mixture of uterine gland secretions and filtered maternal blood. Our study of the Boyd collection of hysterectomy specimens with pregnancy *in situ* has shown that from the time of implantation, the extravillous trophoblast not only invades the uterine tissues but also forms a continuous shell at the level of the deciduas [6]. The cells of this shell anchor the placenta to the maternal tissue, but also form plugs in the tips of the uteroplacental arteries. The shell and the plugs act like a labyrinthine interface that filters maternal blood, permitting a slow seepage of plasma but no true blood flow into the intervillous space. Overall, these data indicate that human placentation is, in fact, not truly haemochorial in early pregnancy.

The placental syncytiotrophoblast is extremely sensitive to oxidative stress, partly because it is the outermost tissue of the conceptus and so exposed to the highest concentrations of oxygen coming from the mother, and partly because it contains surprisingly low concentrations of the principal antioxidant enzymes, particularly in early pregnancy [7]. Using high-resolution ultrasound/Doppler techniques, we found that the intervillous circulation starts in the periphery of the placenta at around 9 weeks of gestation, and that it becomes continuous and diffuse only after 12 weeks [8]. We also demonstrated that the pO_2 measured within the human placenta *in vivo* is <20 mmHg at 7–10 weeks' gestation [9,10]. It subsequently rises to >50 mmHg at 11–14 weeks as the maternal intraplacental circulation becomes fully established, and the placental concentration of antioxidant enzymes increases over the same period [9].

On the fetal side, before 10 weeks of gestation, the placental villi display only a few capillaries and most fetal erythrocytes are nucleated [11]. This well-established anatomical finding suggests that the fetal

---

The Placenta and Human Developmental Programming, ed. Graham J. Burton, David J. P. Barker, Ashley Moffett and Kent Thornburg. Published by Cambridge University Press. © Cambridge University Press 2011.

blood is extremely viscous, which is supported by umbilical artery Doppler features showing high resistance to blood flow [12]. Consequently, before 14 weeks the fetoplacental blood flow must be limited. Furthermore, during the first trimester the villous membrane has twice the thickness it will have in the second trimester, thereby reducing its diffusing capacity [13]. Finally, the early placenta and fetus are separated by the exocoelomic cavity, which occupies most of the space inside the gestational sac and does not contain an oxygen carrier [14]. These findings have led us to propose that the first-trimester placenta limits, rather than facilitates, oxygen supply to the fetus during the period of organogenesis [15]. The earliest stages of development therefore take place in a low-oxygen environment, reflecting to some extent the evolutionary path [16].

In vivo and anatomical data indicate that at the end of the first trimester the trophoblastic plugs are progressively dislocated, allowing maternal blood to flow progressively freely and continuously within the intervillous space [4–6,8,11,12,16]. During the transitional phase of 10–14 weeks' gestation, two-thirds of the primitive placenta disappears, the exocoelomic cavity is obliterated by the growth of the amniotic sac, and maternal blood flows progressively throughout the entire placenta [14,16]. These events bring the maternal blood closer to the fetal tissues, facilitating nutrient and gaseous exchange between the maternal and fetal circulations. However, the cytotrophoblast shell and plugs, which limit the entry of maternal blood into the placenta during the first trimester, must also limit direct exchange of nutrients between the maternal and the fetal circulations, suggesting that alternative nutritional pathways must operate. This chapter reviews the key roles of the different layers of the maternofetal interface in supplying essential nutrients to the developing fetus before the placental circulations are fully established.

The decidua

In many 'eutherian' or placental mammals the uptake of secretions derived from the endometrial glands by the placental membranes provides an important pathway for maternofetal exchange in early pregnancy [17,18]. These secretions contain a complex array of carbohydrates, proteins and lipids, and have been referred to variously as 'uterine milk' or histiotroph. They are particularly important in equids and ruminants, where there is a relatively long interval between the arrival of the conceptus within the uterine cavity and implantation [19,20]. The endometrial glands are known to play a pivotal role in regulating placental development in these species. Ablation of the glands in neonatal ewes by the administration of high doses of 19-norprogestin results in the conceptus failing to thrive and proliferate [21].

In the human, the contribution of the uterine glands has largely been ignored once decidualization and implantation are complete, mainly because it was assumed that the maternal circulation is fully established within the placenta soon after implantation. Furthermore, the invasive form of implantation of the human blastocyst, removing it from the uterine lumen and thus supposedly from uterine secretions by day 7–10 post-fertilization, supported this old concept. These well-established anatomical features have led to the conclusion that histiotrophic nutrition is of little importance in human maternofetal nutrition following implantation.

Our study of placenta-*in situ* specimens from the Boyd collection has shown that the uterine glands are still well developed and highly active at 6 weeks of pregnancy [6]. Although there is considerable individual variation, they gradually regress, in terms of both their length and the height of their epithelium, as the first trimester advances [17,18,22]. At 6 weeks of gestation, the glandular epithelial cells closely resemble those during the luteal phase of the cycle, with large accumulations of glycogen within the apical portions of the cell [18,22,23]. In the normal menstrual cycle, these accumulations disperse around days 23–24, but their persistence indicates that the corpus luteum of pregnancy maintains the glands in a highly active state during early gestation. In contrast, by the start of the second trimester the endometrium beneath the placenta is very thin, the glandular epithelium is cuboidal and secretory organelles are no longer predominant.

Our data have also demonstrated that the uterine glands deliver secretions directly into the intervillous space (Figure 5.1) until at least 10 weeks of gestation [6], suggesting that they may at least contribute to the formation of the intervillous fluid described by Hustin and Schaaps [4]. The secretions of the decidual glands are rich in carbohydrates, glycoproteins and lipids, suggesting that they provide an important source of nutrients for energy and elements for anabolic pathways within the fetoplacental unit

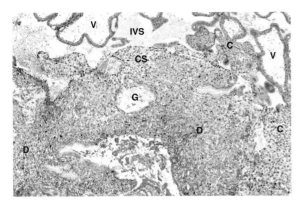

Figure 5.1 Photomicrograph of the maternofetal interface at 6 weeks of pregnancy, showing early villi (V), cytotrophoblast cell columns (C) and the cytotrophoblastic shell (CS) and the underlying decidua (D). Endometrial glands (G) can be seen discharging secretions into the intervillous space (IVS). (See colour plate section.)

[18,24,25]. The observation that glycodelin, a protein that is not expressed within placental tissues and so must be of maternal origin [26], enters the lysosomal digestive pathway within the syncytiotrophoblast supports this hypothesis. Tocopherol transfer protein is a transport protein first identified in the liver, but which has recently also been reported in the syncytiotrophoblast of the human placenta [27–29]. The high level of expression in the glandular epithelium suggests that histiotroph may be an important route for the transfer of antioxidants during early pregnancy, increasing the defences of the fetoplacental tissues against oxidative stress associated with onset of the maternal intraplacental circulation [29].

The glands should not be viewed only as a source of nutrients, however, for the epithelial cells and their secretions are strongly immunoreactive for a wide variety of growth factors and cytokines [18]. These could have important effects on the developing placental tissues. Thus, epidermal growth factor is present in the secretions, and when applied to villous explants from early pregnancy stimulates proliferation of the cytotrophoblast cells [30]. Equally, in the horse, the expression of mRNA encoding epidermal growth factor increases greatly in the gland epithelial cells during early pregnancy, and there is a close temporal and spatial relationship between this rise and proliferation in the overlying trophoblast tissues [31]. An endocrine-mediated servomechanism by which the conceptus signals to the endometrial glands and stimulates the production of 'uterine milk' proteins has been described in the sheep, rabbit and pig [32]. We have speculated that a similar mechanism may operate in the human, involving lactogenic hormones derived from the placenta and/or the decidua [33]. Although there is no experimental or clinical evidence to support this hypothesis at present, it is clear that normal development and functioning of the glands is essential for a successful pregnancy. Reduced levels of secretion around the time of implantation have been linked to miscarriage [34,35]. It is possible, therefore, that less severe deficits may enable a pregnancy to be established, but with some compromise of placental development, resulting in abnormal organogenesis and developmental programming. Further research into the morphogenesis of the glands, and the regulation of their function in early pregnancy, is clearly required.

The early human placenta

The formation of the placenta begins between 13 and 15 days after ovulation, corresponding to stage 6 of embryonic development and to the end of the 4th week after the last menstrual period [36]. The primary villi are composed of a central mass of cytotrophoblast surrounded by a thick layer of syncytiotrophoblast. During the following week of gestation they acquire a central mesenchymal core from the extraembryonic mesoderm and become branched, forming the secondary villi. The appearance of embryonic blood vessels within the mesenchymal core transforms the secondary villi into tertiary villi. At the end of the 5th gestational week all three primitive types of placental villi can be found but tertiary villi progressively predominate. Towards the end of pregnancy the villi present a surface area of 12–14 m^2, providing an extensive and intimate interface for maternofetal exchange [36].

Up to the 9–10th week post menstruation, which corresponds to the last week of the embryonic period (stages 19–23), villi cover the entire surface of the chorionic sac. As the gestational sac grows during fetal life, the villi associated with the decidua capsularis surrounding the amniotic sac degenerate, forming the chorion laeve, whereas the villi associated with the decidua basalis proliferate, forming the definitive placenta [36]. We have shown that the underlying uteroplacental circulation in the centre of the primitive placenta is plugged (Figure 5.2), whereas in the periphery the mouth of the spiral arteries are never plugged by the trophoblastic shell, allowing limited

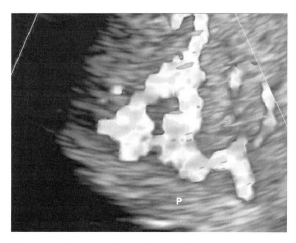

Figure 5.2 Ultrasound Doppler images of a gestational sac at 6 weeks of gestation showing the uterine circulation under the placental bed. Note that the maternal circulation stops before entering the centre of the primitive placenta (P). (See colour plate section.)

maternal blood flow to enter the marginal zone of placenta from 8–9 weeks of gestation [8]. This leads to higher local oxygen concentrations at a stage of pregnancy when the trophoblast possesses low concentrations and activities of the main antioxidant enzymes superoxide dismutase, catalase and glutathione peroxidase [9]. Focal trophoblastic oxidative damage and progressive villous degeneration trigger the formation of the fetal membranes that remodel the uteroplacental interface.

The placental adnexae

In contrast to the arrangement in most mammalian species, in primates, and in the human in particular, the secondary yolk sac floats within the extraembryonic coelom – also called the chorionic cavity or exocoelomic cavity (ECC) – lying between the placenta and the amniotic cavity [14]. The ECC develops during the 4th week after the last menstrual period [36]. It surrounds the blastocyst, which is composed of two cavities separated by the bilaminar embryonic disc, i.e. the amniotic cavity and the primary yolk sac. At the end of the 4th week of gestation the developing exocoelomic cavity splits the extraembryonic mesoderm into two layers, the somatic mesoderm, lining the trophoblast, and the splanchnic mesoderm covering the secondary yolk sac (SYS) and the embryo [36].

The ECC is the first space to appear and largest anatomical space inside the gestational sac between 5 and 9 weeks of gestation and there is never an anatomical barrier between the mesenchyme of the placental fetal plate and the ECC. In 1958 MacKay et al. were the first to report on the protein composition of the chorionic fluid or coelomic fluid (CF) obtained by aspiration of ECC in pregnancy in situ after hysterectomy [38]. These original anatomical and biological findings clearly indicating that the ECC is a fluid cavity were completely ignored by most authors of the 1970s and 1980s. Some even believed that it was a thin virtual space containing a gelatinous substance that could not be aspirated [39,40]. Our investigation into the composition of the CF has shown that it is the result of an ultrafiltrate of maternal serum [14] with the addition of specific placental and secondary yolk sac bioproducts. Higher concentrations of human chorionic gonadotrophin (hCG), oestradiol, oestriol and progesterone in CF samples than in maternal serum strongly indicate the presence of a direct pathway between the villous trophoblast and the ECC [29,41–45]. Our experiments have also demonstrated that inulin [46], diazepam [47] and most anaesthetics drugs [48,49] are transferred to and accumulate in the ECC, suggesting that the turnover of the CF is extremely slow [46,47]. Morphologically, this may be via the villous stromal channels and the loose mesenchymal tissue of the chorionic plate. The presence of an electrical potential difference between the maternal blood and the ECC indicates that from the earliest stage of pregnancy there must be a transport mechanism operating in the placenta that drives net transfer in the maternal to fetal direction against the electrochemical gradient [50].

Molecules such as thyroid hormones, immunoglobulins (Igs), complement factors, relaxin and iron are not synthesized by the fetoplacental unit during the first trimester but play an essential role in fetal development. These molecules are detectable in CF samples from 5 weeks of gestation, indicating an efficient materno-embryonic transfer activity, probably from as soon as the tertiary placental villi and the ECC are formed [14]. Our investigation of the distribution and transfer pathways of antioxidant molecules inside the first-trimester gestational sac has shown similar concentrations of ascorbic and uric acid in CF and maternal plasma (Table 5.1). In addition, detectable levels of α- and γ-tocopherol in CF indicate that they may play also an essential

Table 5.1 Comparison of the mean (SEM) values of the different antioxidant molecules in maternal plasma, coelomic fluid and amniotic fluid [from ref 29].

Variables (μmol/L)	Maternal plasma	Coelomic fluid	Amniotic fluid
GSH	17.1(2.3)[a]	0.10(0.02)	ND
α-Tocopherol	22.3(1.1)[a]	0.60(0.07)[b]	0.09 (0.03)
γ-Tocopherol	1.69(0.13)[a]	0.02(0.01)	ND
Ascorbic acid	37.6(4.7)	31.9(1.6)[b]	10.0(3.5)
DHA	5.0(0.9)[a]	3.0(0.5)	1.7(0.6)
Uric acid	171.8(10.4)	162.8(10.5)[b]	66.3(7.4)

Notes: DHA= dehydroascorbic acid.
[a] significant difference between maternal plasma and coelomic fluid.
[b] significant difference between coelomic and amniotic fluid.

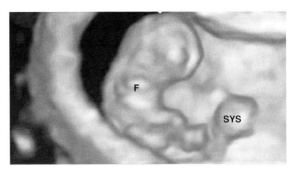

Figure 5.3 Three-dimensional ultrasound view of a gestational sac at 8 weeks of gestation showing the fetus (F) and the secondary yolk sac (SYS). (See colour plate section.)

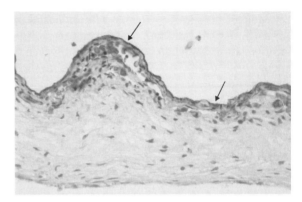

Figure 5.4 Immunohistochemical localization of the GLUT1 transporter protein to the outer mesothelial layer of the human yolk sac at 11 weeks of gestational age. (See colour plate section.)

role in the antioxidant capacity of fetal tissues at a time when the fetus is most vulnerable to oxidative stress [29]. In addition, higher concentrations of myoinositol, sorbitol, erythritol, fructose and ribitol in CF than in maternal serum support the concept that metabolism in the human conceptus during early pregnancy occurs under heavily reducing conditions [44]. They also suggest that polyols play an important physiological role in the development of the human conceptus, possibly drawing water and solutes across the placenta and so expanding the gestational sac.

The secondary yolk sac (SYS) forms at the beginning of the 5th week post menstruation and develops rapidly, so that by 37th menstrual day it is larger than the amniotic cavity [36,37]. From the 6th week of gestation it appears as a spherical and cystic structure covered by numerous superficial small vessels that mergeat the base of the vitelline duct (Figure 5.3). This connects the yolk sac to the ventral part of the embryo, the gut and main blood circulation. The wall of the SYS is formed by an external mesothelial layer lining the ECC, a vascular mesenchyme and an endodermal layer facing the yolk sac cavity. The extraembryonic human circulation is first established within the vitelline duct artery via the dorsal aorta [37]. The endodermal layer of the secondary yolk sac is known to synthesize several serum proteins in common with the fetal liver, such as α-fetoprotein (AFP), albumin, prealbumin and transferrin. With rare exceptions, the synthesis of most of these proteins is confined to the embryonic compartments, and the contribution of the yolk sac to the maternal protein pool is limited.

The distribution of the placental-specific protein hCG in yolk sac and coelomic fluid samples, and the absence of hCG mRNA expression in yolk sac tissue, suggests the SYS has an absorptive function [41]. The detection of glycodelin of glandular origin within mesothelial layer of the SYS [22] indicates that maternal proteins are able to reach the yolk sac, where they could be absorbed, further supports this hypothesis. Recently, we have shown by immunohistochemistry the presence of specific transporter proteins in the external mesothelial layer of the SYS, in particular the folate receptor-α [45], tocopherol transfer protein [29] and GLUT1 (Figure 5.4). Although there are no physiological data indicating the activity of these receptors *in vivo*, their presence suggests that the SYS plays an important role in both macro- and micronutrient exchange prior to vascularization of the chorionic villi. The functional capacity of this pathway may therefore be critical for determining events when the major organ systems are differentiating, laying the foundations for future health or disease. For example, deficiency in the transport of folate during the early weeks of pregnancy could have a profound effect on genome methylation patterns, and hence on developmental programming.

During the 10th week of gestation the yolk sac starts to degenerate and rapidly ceases to function [37].

Conclusion

The anatomy of the materno-fetal interface in the first trimester is the result of the need for a delicate balance between the metabolic requirements of the developing fetus and the potential harmful effects of oxygen during embryogenesis and organogenesis. The involvement of the endometrial glands in the supply of nutrients and growth factors during the first trimester indicates that the endometrium plays a much more active role in support of the conceptus during pregnancy than previously anticipated. The pre- and periconceptional development of the endometrium, and the effects of nutrient and endocrine stressors on this development, therefore deserve further research.

The results of our investigations indicate that most maternal and placental molecules are transferred into the ECC, and subsequently into the fetal gut and circulation via the SYS. These findings suggest that the yolk sac membrane is an important zone of transfer between the extraembryonic and embryonic compartments, and that the main flux of molecules occurs from outside the yolk sac, i.e. from the exocoelomic cavity in the direction of the lumen, and subsequently to the embryonic gut and circulation. They also indicate that ECC is a physiological liquid extension of the early placenta, and an important interface in fetal nutritional pathways during the most of the first trimester. When, after 10 weeks of gestation, the cellular components of the wall of the secondary yolk start to degenerate, this route of transfer becomes no longer functional, and most exchanges between the mother and the fetal circulation must then take place at the level of the chorionic plate.

Overall, our findings are consistent with a gradual shift from essentially histiotrophic nutrition of the human conceptus during the early first trimester to haemotrophic nutrition towards the start of the second trimester. We have speculated that reliance on histiotroph during the period of organogenesis may protect the fetus from teratogenic damage by reactive oxygen species, for all mammalian embryos studied so far appear to rely heavily on anaerobic pathways during this period of development. Once organogenesis is complete, the oxygen concentration within the fetoplacental unit rises as placental attachment and development occurs or, as in the case of the human, the maternal circulation to the placenta is fully established.

References

1. **Burton GJ, Hempstock J, Jauniaux E**. Oxygen, early embryonic metabolism and radical mediated embryopathies. *Reprod Biomed Online* 2003; **6**: 84–96.
2. **Wiznitzer A, Furman B, Mazor M, Reece EA**. The role of prostanoids in the development of diabetic embryopathy. *Semin Reprod Endocrinol* 1999; **17**: 175–81.
3. **Parman T, Wiley MJ, Wells PG**. Free radical-mediated oxidative DNA damage in the mechanism of thalidomide teratogenicity. *Nature Med* 1999; **5**: 582–5.
4. **Hustin J, Schaaps JP**. Echographic and anatomic studies of the maternotrophoblastic border during the first trimester of pregnancy. *Am J Obstet Gynecol* 1987; **157**: 162–8.
5. **Schaaps JP, Hustin J**. In vivo aspect of the maternal-trophoblastic border during the first trimester of gestation. *Troph Res* 1988; **3**: 39–48.
6. **Burton GJ, Jauniaux E, Watson AL**. Maternal arterial connections to the placental intervillous space during the first trimester of human pregnancy: the Boyd collection revisited. *Am J Obstet Gynecol* 1999; **181**: 718–24.
7. **Watson AL, Palmer ME, Skepper JN, Jauniaux E, Burton GJ**. Susceptibility of human placental syncytiotrophoblast mitochondria to oxygen-mediated damage in relation to gestational age. *J Clin Endocrinol Metab* 1998; **83**: 1697–705.
8. **Jauniaux E, Hempstock J, Greenwold N, Burton GJ**. Trophoblastic oxidative stress in relation to temporal and regional differences in maternal placental blood flow in normal and abnormal early pregnancy. *Am J Pathol* 2003; **162**: 115–25.
9. **Jauniaux E, Watson AL, Hempstock J et al.** Onset of placental blood flow and trophoblastic oxidative stress: a possible factor in human early pregnancy failure. *Am J Pathol* 2000; **157**: 2111–22.
10. **Rodesch F, Simon P, Donner C, Jauniaux E**. Oxygen measurements in endometrial and trophoblastic tissues during early pregnancy. *Obstet Gynecol* 1992; **80**: 283–5.

11. **Jauniaux E**, **Jurkovic D**, **Campbell S**. In vivo investigations of anatomy and physiology of early human placental circulations. *Ultrasound Obstet Gynecol* 1991; **1**: 435–45.

12. **Jauniaux E**, **Jurkovic D**, **Campbell S**, **Hustin J**. Doppler ultrasound study of the developing placental circulations: correlation with anatomic findings. *Am J Obstet Gynecol* 1992; **166**: 585–7.

13. **Jauniaux E**, **Burton GJ**, **Moscoso GJ**, **Hustin J**. Development of the early human placenta: a Morphometric study. *Placenta* 1991; **12**: 269–76.

14. **Jauniaux E**, **Gulbis B** 2000 Fluid compartments of the embryonic environment. *Hum Reprod Update* 2000; **6**: 268–78.

15. **Jauniaux E**, **Gulbis B**, **Burton GJ**. The human first trimester gestational sac limits rather than facilities oxygen transfer to the foetus: a review. *Placenta* 2003; **24**: S86–S93.

16. **Jauniaux E**, **Poston L**, **Burton GJ**. Placental-related diseases of pregnancy: involvement of oxidative stress and implications in human evolution. *Hum Reprod Update* 2006; **12**: 747–55.

17. **Burton GJ**, **Hempstock J**, **Jauniaux E**. Nutrition of the human fetus during the first trimester: a review. *Placenta-Trophoblast Res* 2001; **22**: S70–S76.

18. **Hempstock J**, **Cindrova-Davies T**, **Jauniaux E**, **Burton GJ**. Endometrial glands as a source of nutrients, growth factors and cytokines during the first trimester of human pregnancy: a morphological and immunohistochemical study. *Reprod Biol Endocrinol* 2004; **20**: 58.

19. **Wooding FBP**, **Flint APF**. *Placentation*. In *Marshall's Physiology of Reproduction*. Lamming GE ed. London: Chapman & Hall, 1994; 233–460.

20. **Mossman HW**. *Vertebrate Fetal Membranes: Comparative Ontogeny and Morphology, Evolution, Phylogenetic Significance, Basic Functions, Research Opportunities*. London: Macmillan, 1987.

21. **Gray CA**, **Taylor KM**, **Ramsey WS** et al. Endometrial glands are required for preimplantation conceptus elongation and survival. *Biol Reprod* 2001; **64**: 1608–13.

22. **Burton GJ**, **Watson AL**, **Hempstock J**, **Skepper JN**, **Jauniaux E**. Uterine glands provide histiotrophic nutrition for the human fetus during the first trimester of pregnancy. *J Clin Endocrinol Metab* 2002; **87**: 2954–9.

23. **Dockery P**, **Li TC**, **Rogers AW**, **Cooke ID**, **Lenton EA**. The ultrastructure of the glandular epithelium in the timed endometrial biopsy. *Hum Reprod* 1988; **3**: 826–34.

24. **Bell SC**. Secretory endometrial/decidual proteins and their function in early pregnancy. *J Reprod Fertil* 1988; **36** Suppl.: 109–25.

25. **Beir-Hellwig K**, **Sterzik K**, **Bonn B**, **Beir HM**. Contribution to the physiology and pathology of endometrial receptivity: the determination of protein patterns in human uterine secretions. *Hum Reprod* 1989; **4** Suppl. : 115–20.

26. **Seppälä M**, **Jukunen M**, **Riitinen L**, **Koistinen R**. Endometrial proteins: a reappraisal. *Hum Reprod* 1992; **1** Suppl.: 31–8.

27. **Gordon MJ**, **Campbell FM**, **Dutta-Roy AK**. alpha-Tocopherol-binding protein in the cytosol of the human placenta. *Biochem Soc Trans* 1996; **24**: 202S.

28. **Kaempf-Rotzoll DE**, **Horiguchi M**, **Hashiguchi K** et al. Human placental trophoblast cells express alpha-tocopherol transfer protein. *Placenta* 2003; **24**: 439–44.

29. **Jauniaux E**, **Cindrova-Davies T**, **Johns T** et al. Distribution and transfer pathways of antioxidant molecules inside the first trimester human gestational sac. *J Clin Endocrinol Metab* 2004; **89**: 1452–8.

30. **Maruo T**, **Matsuo H**, **Murata K**, **Mochizuki M**. Gestational age-dependent dual action of epidermal growth factor on human placenta early in gestation. *J Clin Endocrinol Metab* 1992; **75**: 1362–7.

31. **Lennard SN**, **Gerstenberg C**, **Allen WR**, **Stewart F**. Expression of epidermal growth factor and its receptor in equine placental tissues. *J Reprod Fertil* 1998; **112**: 49–57.

32. **Spencer TE**, **Johnson GA**, **Burghardt RC**, **Bazer FW**. Progesterone and placental hormone actions on the uterus: insights from domestic animals. *Biol Reprod* 2004; **71**: 2–10.

33. **Burton GJ**, **Jauniaux E**, **Charnock-Jones DS**. Human early placental development: potential roles of the endometrial glands. *Placenta* 2007; **28** Suppl A: S64–9.

34. **Hey NA**, **Li TC**, **Devine PL** et al. MUC1 in secretory phase endometrium: expression in precisely dated biopsies and flushings from normal and recurrent miscarriage patients. *Hum Reprod* 1995; **10**: 2655–62.

35. **Dalton CF**, **Laird SM**, **Estdale SE**, **Saravelos HG**, **Li TC**: Endometrial protein PP14 and CA-125 in recurrent miscarriage patients; correlation with pregnancy outcome. *Hum Reprod* 1998; **13**: 3197–202.

36. **Boyd JD**, **Hamilton WJ**. *The Human Placenta*. Cambridge: Heffer and Sons, 1970.

37. **Jones CPJ**, **Jauniaux E**. Ultrastructure of the materno-embryonic interface in the first trimester of pregnancy. *Micron* 1995; **2**: 145–73.

38. **MacKay DG**, **Richardson MV**, **Hertig AT**. Studies of the function of the early human trophoblast. III: a

study of the protein structure of mole fluid, chorionic and amniotic fluid by paper electrophoresis. *Am J Obstet Gynecol* 1958; **75**: 699–707.

39. **McCarthy T**, **Suanders P**. The origin and circulation of the amniotic fluid. In: Fairweather DVI, Eskes TKAB eds. *Amniotic Fluid: Research and Clinical Application*. Amsterdam: Elsevier, 1978; 1–18.

40. **Larsen WJ**. *Human Embryology*. Edinburgh: Churchill Livingstone, 1993.

41. **Gulbis B**, **Jauniaux E**, **Cotton F**, **Stordeur P**. Protein and enzyme pattern in fluid cavities of the first trimester gestational sac: relevance to the absorptive role of the secondary yolk sac. *Mo Hum Reprod* 1998; **4**: 857–62.

42. **Jauniaux E**, **Gulbis B**, **Jurkovic D** *et al*. Relationship between protein levels in embryological fluids and maternal serum and yolk sac size during early human pregnancy. *Hum Reprod* 1994; **9**: 161–6.

43. **Calvo RM**, **Jauniaux E**, **Gulbis B** *et al*. Fetal tissues are exposed to biologically relevant free tryroxine concentrations during early phases of development. *J Clin Endocrinol Metab* 2002; **87**: 1768–77.

44. **Jauniaux E**, **Hempstock J**, **Teng C**, **Battaglia FC**, **Burton GJ**. Polyol concentrations in the fluid compartments of the human conceptus during the first trimester of pregnancy: maintenance of redox potential in a low oxygen environment. *J Clin Endocrinol Metab* 2005; **90**: 1171–5.

45. **Jauniaux E**, **Johns J**, **Gulbis B**, **Spasic-Boskovic O**, **Burton GJ**. Transfer of folic acid inside the first trimester gestational sac and the effect of maternal Smoking. *Am J Obstet Gynecol* 2007; **197**: 58e1–58e6.

46. **Jauniaux E**, **Lees C**, **Jurkovic D**, **Campbell S**, **Gulbis B**. Transfer of inulin across the first trimester human placenta. *Am J Obstet Gynecol* 1997; **176**: 33–6.

47. **Jauniaux E**, **Jurkovic D**, **Lees C**, **Campbell S**, **Gulbis B**. In vivo study of diazepam transfer across the first trimester human placenta. *Hum Reprod* 1996; **11**: 889–92.

48. **Jauniaux E**, **Gulbis B**, **Shannon C** *et al*. Placental propofol transfer and fetal sedation during maternal general anaesthesia in early pregnancy. *Lancet* 1998; **352**: 290–1.

49. **Shannon C**, **Jauniaux E**, **Gulbis B**, **Sitham M**, **Bromley L**. Placental transfer of fentanyl in early human pregnancy. *Hum Reprod* 1998; **13**: 2317–20.

50. **Ward S**, **Jauniaux E**, **Shannon C**, **Rodeck C**, **Boyd R**, **Sibley C**. Electrical potential difference between exocoelomic fluid and maternal blood in early pregnancy. *Am J Physiology* 1998; **274**: R1492–R5.

Discussion

SIBLEY: You showed a picture of the coelomic cavity, and you said that macromolecules (proteins) are important. I would argue that the stroma is quite a significant barrier for protein movement, because proteins are large molecules and the stroma could be intermeshed.

JAUNIAUX: It is quite loose at the time of pregnancy. The chorionic or exocoelomic cavity becomes a liquid extension of the villous stroma, if you will, and thus all molecules produced by the villous trophoblast are likely to be found there independently of their molecular weight.

SIBLEY: In terms of the secretions, amino acids will be secreted from the uterine glands, as will glucose, fluids and ions. We know that the trophoblast has the normal array of transporters at this stage. I don't think it is that big a difference to what happens during the second and third trimesters when bloodflow starts. The basic mechanisms are similar. There are slightly different ways of getting things into the intervillous space, but beyond that I am not sure that there is a huge difference.

JAUNIAUX: There is at the level of the umbilico-placental circulation which is very limited within the placenta itself until 9–10 weeks. Most of the circulation between the fetus and its adnexae goes through the vitelline circulation of the secondary yolk sac.

SIBLEY: I take your point on that. You can still have all the other mechanisms in the chorioallantoic membrane that happen in the second and third trimesters.

JAUNIAUX: The main difference is that there is no true maternal circulation in the placenta and probably very little fetal circulation.

ROBERT BOYD: The big difference is actually the diffusion distance from one site to another. If you look at this under ultrasound, is there movement? Is the coelomic cavity stirred?

JAUNIAUX: Yes. The secondary yolk sac moves but the turnover of the coelomic fluid is probably very slow.

CETIN: Another difference is proteins. I think this is the major difference. Later on, it is free amino acids that go across.

LOKE: Since the whole purpose of placentation is to deliver oxygen to the fetus, it always seems to me paradoxical that we must control the amount of oxygen actually delivered, at least in the early stages of development. Is this a legacy of our evolutionary past; that we all used to live in an anaerobic environment?

JAUNIAUX: Yes. The early fetus needs oxygen, but only what is strictly necessary for its metabolism and at higher doses it is quite toxic. It takes several months before the fetus is capable and needs assimilating more oxygen.

HUNT: The primitive yolk sac is hugely understudied. It is difficult to get to. There is some old literature that suggests that mesenchymal cells of the developing placenta arise from the yolk sac. Would these cells be bearing any proteins that you would see as important to placental development?

JAUNIAUX: I would love to study this. As you say, it is impossible to get these specimens. Getting secondary yolk sac is extremely difficult. You need to get them intact, not only to be able to culture them but also to get the fluid that is inside.

MOORE: It is important to take account of calorimetry. The developing embryo is, of course, very small, hence the nutrient supply through these morphological sources, in relation to the size of the developing embryo, is sufficient. But the fetus is going to grow exponentially, so the requirements in the later phase are much higher.

JAUNIAUX: We tried to correlate the concentration of some of these molecules between the mother compartment and the coelomic fluid. The only two molecules that correlated were thyroxine hormone (T_4) and pre-albumin. There was no correlation for any of the other molecules. It is understandable: thyroid hormones are not being produced by the first trimester fetus, and so it entirely depends on the mother's thyroxine. The link between nutrition and prealbumin needs to be further explored. This is more complex to understand.

JANSSON: We reported evidence to suggest that whereas the term syncytiotrophoblast only relies on one glucose transporter isoform, in the 6–8-week gestation syncytiotrophoblast there are four different glucose transporters expressed, including the high-affinity Glut3. This supports the idea that this is a tissue in a low-oxygen environment exquisitely dependent on glucose uptake. Coming back to Colin Sibley's question about the transport mechanism being the same, did I understand correctly that the amino acid concentration in the coelomic cavity was higher than maternal serum?

JAUNIAUX: Yes.

JANSSON: That proves that it has to be some sort of active transport across the trophoblast, rather than phagocytosis and protein breakdown. It suggests that the transporters are probably active and functioning the same in the trophoblast.

JAUNIAUX: I agree, but there is probably a very slow turnover. It is not like serum. There is no liver and kidney clearing the serum; it is just accumulating and being picked up by the yolk sac.

ROBERT BOYD: There is also another point. Eric demonstrated that the coelomic cavity is electrically negative with regard to the outside. With charged ions, this will set up an asymmetrical distribution.

RICHARD BOYD: It is striking that lysine, arginine, histidine and ornithine were your highest ratios. The one that interested me was at the other end, the tryptophan data.

CETIN: How did you do this study? For me it seems difficult to have such precision. Which week was it?

JAUNIAUX: Five to 12 weeks. We obtained the fluid by transvaginal puncture. It is a direct puncture, and we discard the first millilitre. We look at the level of contamination with maternal cells. There is only a trace in some of the samples. From 7 to 8 weeks we also started puncturing the amniotic cavity to find out the composition of the amniotic fluid. Very little actually crosses from the coelomic cavity into the amniotic cavity. The membrane seems to be very thin and doesn't accumulate many of these molecules before the end of term, which is probably when the kidney starts producing urine.

CETIN: You may have some contamination between tophoblast, coelomic cavity and yolk sac. I don't want to say it is maternal.

JAUNIAUX: The difficulty came when the other side of the team wanted to do prenatal diagnosis on these samples. There were very few live cells. You could do a DNA analysis and look for mutations such as sickle cell anaemia, but you couldn't really get fetal cells for culture.

MOFFETT: I want to return to the decidua. The NK cells are always thought to be involved in vascular transformation in the arterial blood supply. But they are also centred underneath the glands. In the gut it is now known that the NK cells are crucial for physiological regulation of mucosal integrity as well as defence. What are the NK cells doing to glands? This is a question waiting to be explored.

JAUNIAUX: I agree. My main interest is in understanding the mechanism of miscarriages. More than half of miscarriages have no explanation: they are not linked to a chromosomal abnormality. It could be because of failure of implantation because of poor development of the endometrial tissue. There is a lot to study.

THORNBURG: You have a wonderful model that not is not available to most people. You could help us in the future by focusing more clearly on how the mother influences the constituents in these fluids, and perhaps you will open a new world of programming mechanisms.

JAUNIAUX: One of the problems is getting the samples with information about the patient. These samples are from units where the patient arrived a couple of hours before the procedure, and we have no idea of the weight, height and only have limited patient history.

BURTON: These glands are producing an enormous range of growth factors as well as nutrients. Epidermal growth factor and a number of other growth factors are being discharged into the intervillous space. They could have an important role in regulating cell proliferation and differentiation in the placenta. From the work in the sheep by Tom Spencer (Gray CA, Taylor KM, Ramsey WS, et al. Endometrial glands are required for preimplantation conceptus elongation and survival. *Biol Reprod* 2001; **64**: 1608–13.) We know that these glands are crucially important: if they are ablated, the conceptus will not survive. We also know that in the sheep the conceptus itself signals to the glands' upregulating factors such as EGF. It would be exciting to see whether this is also true in humans. We don't know enough about the endometrial function in early pregnancy, or how this is set up prior to pregnancy.

CRITCHLEY: These are important sets of data. I am seeking clarification about the source of the patients. Were these patients undergoing investigation in early pregnancy, or undergoing termination? If the latter, at least in my unit, if it was a medical termination they would have been exposed 36–48 hours beforehand to a progesterone receptor antagonist, and then prostaglandin. If it was surgical they may have had prostaglandin. These compounds will have an effect on early response genes and chemotactic factors within 6 hours of administration. Studying women in early pregnancy is different, and if you are studying normal ongoing pregnancy you have to look at it within those limitations.

JAUNIAUX: It is straightforward: we never use samples from medical terminations, just surgical terminations, and we did not routinely use prostaglandins to prime the cervix before 11–12 weeks. If we need to do something at a later stage we used a hygroscopic dilator (dilapan).

BARKER: I would like some clarification on the increased placental: birth weight ratio in maternal anaemia. If the fetus can be damaged by oxygen, and so limits its invasion, how does this big placenta come about in maternal anaemia? It sounds paradoxical.

CETIN: It could be something happening later on, in the second trimester.

BURTON: The onset of the oxygenation correlates almost precisely with the reduction in the sensitivity to teratogenesis at 8 weeks post-fertilization.

BARKER: Maternal anaemia could be useful up to 10 weeks, but thereafter it could become disadvantageous.

JAUNIAUX: The vast majority of mothers are not anaemic before 12 weeks of gestation. It is something we diagnose clinically during the second half of pregnancy.

BARKER: What about the situation in India. Is every mother anaemic there?

FALL: About 60%.

JANSSON: One possibility is that the anaemic patient has a hyperdynamic circulation with vasodilation of the vascular bed. Nutrient flow might increase in these patients, and this could drive placental growth.

LOKE: I have sat through this morning's papers and heard everyone talking about how maternal nutrients affect fetal and placental development, but I have a basic question. Given that there is a finite amount of nutrients, how is it decided who gets priority, the mother, the baby or the placenta?

SIBLEY: You can't make that relationship on the basis of just placental size.

FOWDEN: It would change, depending on the stage of pregnancy. The fetal demands for nutrients are still very low in early pregnancy. The fetus starts to signal in late gestation, which is when you can get a different dimension.

Chapter 6

Imprinted genes and placental growth
Implications for the developmental origins of health and disease

Benjamin Tycko and Rosalind John

Introduction

The discovery in 1991 that a growth factor gene, *Igf2*, and the gene encoding its antagonistic clearance receptor, *Igf2r*, are expressed monoallelically from opposite parental alleles in mice owing to parental imprinting, initiated a burst of research activity which has by now largely defined the repertoire of imprinted genes. Since 2006, when we last reviewed this area, progress has continued, with established laboratories filling in key experimental gaps and new investigators entering the field. Here we provide an updated overview of this area and propose some lines of investigation that seem particularly compelling for future work. Given that an important group of imprinted genes have been shown to act as rheostats for placental growth, we consider whether mice engineered to over- and underexpress specific imprinted genes in the placenta might be developed as experimental models for exploring the developmental origins of health and disease (DOHAD) hypothesis. Such models could become useful for testing interventions to ameliorate the adverse cardiovascular, metabolic and psychological effects in adults that can result from a poor *in utero* environment. We illustrate this direction of work with our studies of the imprinted *Phlda2* gene, which regulates placental and, secondarily, fetal growth in mouse models and appears likely to have these same functions in human pregnancies.

As the main site of materno-fetal exchange, the placenta, and earlier in gestation the yolk sac, takes nutritional resources from the mother and delivers them to the growing fetus. Here, we argue that to understand the contribution to common adult diseases of adverse exposures of the fetus it is crucial to continue working towards a more complete understanding of how genetic and epigenetic influences converge to regulate the growth and function of these key nutritive organs. In an impressive confirmation of a hypothetical scenario, the 'conflict' or 'kinship' model proposed largely on theoretical grounds by evolutionary biologists two decades ago, experimental findings from organisms as diverse as plants and mammals have now made it clear that the organs responsible for transferring maternal nutrients to the growing conceptus are regulated in their growth and development in part by a special class of genes – those subject to parental imprinting (a.k.a. genomic imprinting). Imprinted genes are regulated by parent-of-origin-specific epigenetic marks, notably DNA methylation, leading to monoallelic expression of these genes in the offspring. Imprinted genes in mice and humans have been discovered by various means, including observations of non-mendelian parent-of-origin-dependent inheritance of phenotypes in knockout (KO) mice or human genetic diseases, chromosome walking and imprinting analysis of flanking genes near known imprinted loci, microarray-based expression profiling in androgenetic and gynogenetic versus normal conceptuses and, less productively so far, direct screens for allele-specific expression or allele-specific DNA methylation. Imprinted genes were rapidly uncovered in the 1990s and early 2000s, but the rate of discovery of new examples has slowed and the number of imprinted genes now appears to be reaching an asymptote at about 100 (http://igc.otago.ac.nz/home.html). Thus, although genomic imprinting is a potent gene regulatory mechanism, only a small minority (<1%) of mammalian genes are imprinted. This conclusion is consistent with long-standing observations from classic mouse genetics and clinical genetics in humans that uniparental disomies (UPDs) for some chromosomes

---

The Placenta and Human Developmental Programming, ed. Graham J. Burton, David J. P. Barker, Ashley Moffett and Kent Thornburg. Published by Cambridge University Press. © Cambridge University Press 2011.

Chapter 6. Imprinted genes and placental growth

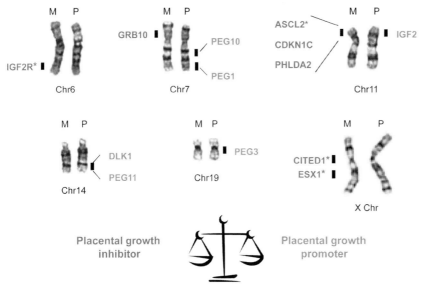

Figure 6.1 Chromosomal locations and affects on placental growth of selected imprinted genes. Maternally expressed genes are shown next to the maternal chromosome homologue (M), and paternally expressed genes next to the paternal homologue (P). Genes promoting placental growth (placental stunting in knockout mice) are in green; genes inhibiting placental growth (placentomegaly in knockout mice) are in red. *ASCL2* is the human homologue of *Mash2/Ascl2*, which is essential for placental development in mice. The severe disruption of placental development in *Ascl2* KO mice makes this gene difficult to assign with respect to its effect on growth of the intact organ, although future experiments with transgenic lines might be informative. Several of these genes, including *ASCL2*, *IGF2R*, *CITED1* and *ESX1* are imprinted in mice, but are never or only rarely functionally imprinted in humans (imprinted genes on the mouse X-chromosome may in part be explained by selective paternal X-inactivation in that species). This fact raises the interesting hypothesis that imprinting is slowly being lost in humans relative to rodents, possibly due to weaker intergenomic conflict in our species. (See colour plate section.)

can produce parent-of-origin-dependent phenotypic abnormalities indicating the presence of imprinted genes on these chromosomes, whereas UPDs for other entire chromosomes do not produce a detectably abnormal phenotype [1]. We last reviewed the role of imprinted genes in placental growth and pregnancy outcome in early 2006 [2]. New findings since then warrant an update on this important topic, which is given in this chapter and in the accompanying chapter by Radford and Ferguson-Smith.

Genomic imprinting and imprinted genes in the placenta

The imprinted genes discussed here are indicated with their positions on the human chromosomes in Figure 6.1. All of these genes are conserved in mice and humans, although there are a few differences in imprinting status in the two species, with a slightly greater number of genes imprinted in mice than in humans. To facilitate discussion of tissue-specific gene expression and the biological functions of these genes, we illustrate the key anatomical compartments of the mouse and the human placenta in Figure 6.2. The following descriptions of imprinted genes expressed in placental tissues are updated where necessary with new information since our 2006 review [2]. *Igf2* (insulin-like growth factor 2), is highly expressed in the placenta, with *Igf2* mRNA abundant in the vasculogenic mesenchyme, labyrinthine layer and junctional zone of the mouse placenta and in homologous structures of the human placenta [3–6]. Genetic studies in mice identified a placenta-specific promoter, called P0, which drives imprinted expression of *Igf2* mRNA specifically in the labyrinth [6]. The closely linked and oppositely imprinted gene *H19*, encoding a non-translated RNA, is also strongly expressed in the human and mouse placenta [4]. A nearby, but independently regulated, imprinted domain on mouse (distal) chromosome 7 and human chromosome band 11p15.5 contains several imprinted genes, including *Cdkn1c* (encoding the $p57^{Kip2}$ cyclin-cdk inhibitor), *Slc22a18* (a.k.a. *Impt1*, *Orctl2*, *Slc22a1l*, encoding a metabolite transporter) and *Phlda2* (a.k.a. *Ipl*, *Tssc3*, encoding a small PH-domain protein), which all share the placenta as one

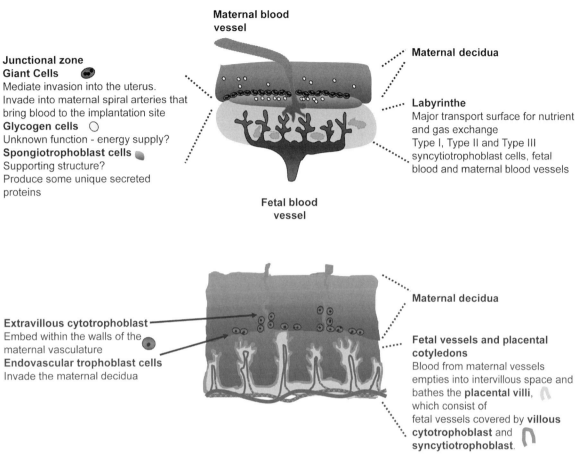

Figure 6.2 Structure of the placenta in human and mouse. The placenta is composed of trophoblast and fetal mesenchyme and blood vessels, which are the embryonic component, and deciduas and maternal vessels, which are the maternal component. In humans, the fetal blood vessels are covered by a layer of cytotrophoblast cells and an adjacent layer of multinucleated syncytiotrophoblast. These structures, called placental villi, are bathed in maternal blood. In mouse, the functionally equivalent region is termed the labyrinth layer. Here fetal blood vessels are encased in a bilayer of syncytiotrophoblast and a layer of mononuclear trophoblast cells which are in contact with the maternal blood. Both humans and mice possess invasive cell types. In humans, extravillous cytotrophoblast cells, which appear to arise from anchoring villi, invade the maternal decidua. Some of these cells embed within the walls of the maternal vasculature and are called endovascular trophoblast cells. Mice possess giant trophoblast cells, which are aligned between the junctional zone and the decidua and are behaviourally somewhat equivalent to the invasive extravillous cytotrophoblast, both being polyploid cell types. In the mouse there is a specialized region termed the junctional zone, which is composed primarily of spongiotrophoblast cells for which there is no clear human parallel. Glycogen cells, which arise within the junctional zone, also possess invasive properties migrating into the maternal decidua at E16.5 where they may provide an easily exploitable energy source to support late gestational growth. Adopted from Rossant J, Cross JC. Placenta development: lessons from mouse mutants. *Nat Rev Genet* 2001; **2** (7): 538–48 [92]. (See colour plate section.)

of their strongest sites of expression, in both mice and humans [7–12]. *Phlda2*, *Slc22a18* and *Cdkn1c* are expressed in the labyrinthine layer of the mouse placenta [13–16]. Human PHLDA2 protein marks the homologous human tissue – the cytotrophoblast of the free chorionic villi [12], whereas the product of the *CDKN1C* gene, p57^{KIP2}, is present in these same cells as well as in several other placental cell types [11,12].

The functional imprinting of human and mouse *Phlda2* and *Slc22a18*, that is, their allelic expression bias, is stronger in the placenta than in several other organs [7,8]. From a mechanistic point of view, it is interesting that the imprint at each of these three genes, as well as that of *Ascl2/Mash2* in the same chromosomal domain (distal mouse chromosome 7 and human chromosome band 11p15.5), is less sensitive to DNA-demethylating (*Dnmt1*-hypomorphic or null) genetic backgrounds than the imprint at *Igf2/H19* [17–19]. In addition, there is evidence for allele-specific histone modifications, without allele-specific DNA methylation, in promoter regions of several imprinted genes within this chromosomal region [20,21]. Since we last reviewed this topic, additional work has further suggested that functional imprinting is somewhat less dependent on promoter methylation and more dependent on histone modifications in the placenta than in other tissues [22]. Nonetheless, the Kvdmr1 imprinting centre, which does show allele-specific DNA methylation in the placenta, has been proven in mice to act at a distance to initiate the imprinting of these and several other genes in *cis* [23,24], and recent data from *Dnmt3l* KO mice, which fail to properly establish maternal methylation patterns, imply that differential DNA methylation is genetically necessary for the acquisition of differential histone marks at this and several other imprinted loci [25].

Turning to imprinted loci on other chromosomes (Figure 6.1), the *Peg10* gene, encoding a major placental transcript with distant homology to retroviral gag-pol proteins, was identified as paternally expressed/maternally silenced in mice [26], and this gene is also imprinted in humans [27]. By immunohistochemistry the PEG10 protein in the human placenta is specific for proliferating cytotrophoblast of the free chorionic villi [2,28]. *PEG10* maps to human chromosome 7, which is rich in imprinted genes, grouped in several separate imprinted domains. One of these, telomeric to the *PEG10* domain, contains the *MEST* gene (a.k.a. *PEG1*). This gene is imprinted in mice and humans, as well as other mammals that have been examined, with the paternal allele active/maternal allele repressed, sometimes with mRNA isoform-specific imprinting [29–32]. *MEST* encodes a protein with a 'hydrolase fold', and its localization to the endoplasmic reticulum of adipocytes has suggested a possible function as a lipase or acyltransferase mediating lipid accumulation under conditions of positive energy balance [33]. *Peg1/Mest* mRNA is expressed in mesenchymal haemangioblast precursor cells and endothelial cells of the mouse placenta, and in mesenchymal cells, as well as villous and invasive cytotrophoblast in the human placenta [34].

As described below, a number of additional imprinted genes, *Ascl2/Mash2, Igf2r, Grb10, Peg3* and *Peg11*, are associated with placental phenotypes, and the placental expression of some of these genes has also been studied in some detail. The *Ascl2* mRNA, albeit not placenta-restricted, is highly expressed in this organ [35]; it is expressed from the maternal allele in the spongiotrophoblast and labyrinth of mouse placentas early in gestation [36] and in extravillous cytotrophoblast in humans, where it usually escapes imprinting [37,38]. *PEG3*, encoding a zinc finger protein, is strongly expressed from the paternal allele in human and mouse placentas [39,40] and its mRNA was localized to human villous cytotrophoblast, whereas in the mouse *Peg3* mRNA was found in all types of trophoblast [39]. *Peg11/Rtl1*, another important retrotransposon-derived imprinted gene distantly related to *Peg10*, appears to be most highly expressed in placental endothelial cells [41].

Multiple imprinted genes act as rheostats for placental growth

There are a number of hypotheses to explain the evolution of imprinting, and one of these, intragenomic conflict, seems to be standing the test of time. This model, first proposed in the early 1990s based on the seminal reports of opposite growth phenotypes in mice mutant for the oppositely imprinted *Igf2* and *Igf2r* genes [42,43], posits opposite selective pressures for expression of maternal versus. paternal alleles, and predicts that paternally silenced/maternally expressed genes will limit the growth of each conceptus, whereas maternally silenced/paternally expressed genes will promote growth of the conceptus and increase demands on maternal resources [44,45]. As a nutritive organ, the placenta is the anatomical site *par excellence* for conflict between maternal and paternal alleles. In terms of the mother, excessive growth of the placenta of any single conceptus would deplete nutritional resources and might therefore limit her current and future reproductive success. In fact, the placenta is known to consume up to 30% of the calories delivered by the mother, simply to maintain its own function [46]. In contrast, as promiscuous mating is the rule among mammals, and as paternal investment in nourishment and care of the offspring tends to be much less than maternal, the reproductive success of paternal alleles does not depend strongly on the number of offspring of any given female, and might be increased by imprinting paternal alleles in such a way as to promote the growth of every placenta. As indicated by the colour-coding in Figure 6.1, *in vivo* data for the functions of imprinted genes strongly support this theory. Manipulations of imprinted genes to produce loss of expression (knockouts) or, conversely, over-expression (transgenic mice and/or deletions of regulatory elements causing loss of imprinting and a 2× gene dosage), have indicated a growth rheostat function for the placenta, the fetus or both for many of these genes. Whether the few examples of loci lacking conservation of imprinting in humans compared to mice might indicate less intense intragenomic conflict in our own species is an interesting question. In any case, data from both species show that among imprinted genes with measureable effects on fetal and/or placental growth, the direction of the effects has invariably met the predictions of the conflict hypothesis.

Igf2

Loss of *Igf2* expression via deletion of the active paternal allele causes placental stunting [47], and over-expression of *Igf2* via loss of imprinting consequent upon the deletion of *H19* and its upstream sequences causes placentomegaly as well as fetal overgrowth [48]. Similarly, in one type of Beckwith–Wiedemann syndrome (BWS) the human *IGF2* gene undergoes loss of imprinting, with conversion from monoallelic to biallelic *IGF2* mRNA expression, secondary to gain of DNA methylation at critical *H19* upstream sequences. These epimutations, and more rarely DNA microdeletions affecting the same imprinting centre, underlie somatic and placental overgrowth *in utero* and susceptibility to Wilms' tumors after birth in this subgroup of BWS cases ([49] and references therein).

Ascl2

Mouse conceptuses in which the active maternal allele of the *Ascl2/Mash2* helix–loop–helix transcription factor gene has been deleted are non-viable, and the mid-gestation lethality is due to a failure of placental development, with a complete lack of spongiotrophoblast [50].

Phlda2

Deletion of the active maternal allele of *Phlda2* produces placentomegaly, with disproportionate overgrowth of the junctional zone [14]. Conversely, over-expression of *Phlda2* in transgenic and loss-of-imprinting mouse models causes placental stunting with a reduced thickness of this same anatomical layer [15,16]. Fetal growth and viability are not affected in the *Phlda2* knockout conceptuses with enlarged placentas, but there is fetal growth retardation late in gestation, presumably secondary to placental insufficiency, in transgenic conceptuses that over-express this gene [15,16]. These opposite phenotypes from over- and underexpression of *Phlda2* are shown in Figures 6.3 and 6.4, which highlight their opposite effects on fetal demand for maternal resources. The Phlda2 protein binds to phosphoinositide lipids via its PH domain [51]. Proteins containing PH domains function in cell signalling, cytoskeletal regulation and intracellular vesicular transport. Because the overgrowth phenotype after *Phlda2* deletion seems to be non-cell-autonomous, affecting the layer of the placenta (junctional zone) adjacent to the *Phlda2*-expressing layer (labyrinth), one model would have the Phlda2 protein regulating growth factor secretion. However, data from genetic crosses argue against an interaction of *Phlda2* with *Igf2* [14], and *Igf2* levels are not altered in placentas over-expressing *Phlda2* [16]. As *Phlda2* expression is very strong as early as the egg cylinder stage of development [52] an alternative model is that its deletion promotes a greater allocation of spongiotrophoblast precursor cells at this early stage, leading to the observed disproportionate expansion of this layer later in development. In future work it may be possible to test this model using trophoblast stem cells differentiated *in vitro*.

Cdkn1c

As predicted from the inhibitory effect of the p57^{Kip2} cyclin-cdk inhibitor on the cell cycle, the maternally expressed *Cdkn1c* gene mediates growth inhibition. Moderate placental overgrowth with trophoblast hyperplasia is found in *Cdkn1c* knockout conceptuses [13,53] and double-mutant mice carrying a combination of maternal *Cdkn1c* deletion and *Igf2* loss of imprinting (maternal *H19* deletion), which could be generated from the single knockout lines by meiotic recombination, produce conceptuses with striking placentomegaly [54]. In studying the pregnancies in the single KO mice with *Cdkn1c* deficiency Kanayama *et al.* found some features of pre-eclampsia in the mothers [53], but another laboratory, despite confirming the placental overgrowth, did not reproduce the same pre-eclamptic features [55]. Reduced expression of *Cdkn1c* accounts for fetal overgrowth and omphalocoele in a second, somewhat larger subgroup of BWS. Interestingly, these individuals are not predisposed to

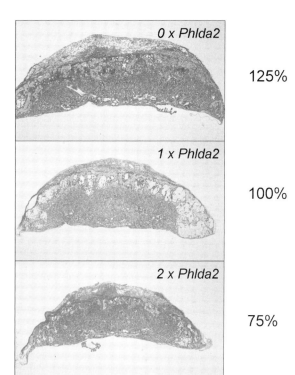

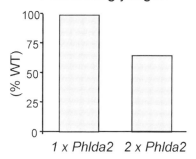

Figure 6.3 *Phlda2*: an example of an imprinted gene acting as a rheostat for placental growth. Shown are haemotoxylin/eosin midline sections of E14.5 placenta from *Phlda*−/+ (0 X *Phlda2*), wild type and 5D3 transgenic (2 X *Phlda2*) embryos and biochemically determined placental glycogen shown as a percentage of wild type. Deletion of the maternal allele of *Phlda2* (0 X *Phlda2*) produces placentomegaly, with disproportionate overgrowth of the junctional zone whereas overexpression of *Phlda2* in the transgenic model with a double dose of *Phlda2* (2 X *Phlda2*) causes placental stunting, with a reduced volume of the junctional zone. In addition, there is a 35% reduction in stored glycogen at E14.5. Although no definitive data are available concerning the function of placental glycogen stores, one possibility is that they provide an easily exploitable source of energy to support very late embryonic growth or parturition. (See colour plate section.)

Wilms' tumors [56], a finding that stands in contrast to children with BWS due to epigenetic dysregulation of *IGF2/H19*.

Igf2r

This gene is strongly imprinted and only expressed from the maternal allele in mice and in marsupials [43,57], but *IGF2R* expression in humans is usually found to be biallelic, indicating loss of the functional imprint in this gene, even though its regulatory imprinting centre is conserved [58,59]. Complete elimination of *Igf2r* expression by deletion of the active maternal allele in mice causes fetal overgrowth (with variable late fetal lethality) and placentomegaly [60]. Over-expression of this gene via loss of imprinting consequent upon deletion of *Air*/region-2 imprinting control sequences produces fetal growth retardation, albeit without a detectable effect on placental growth [61].

Grb10/Meg1

Grb10 encodes a cytoplasmic adaptor protein in the insulin and IGF-1 signalling pathways, and probably in other related pathways, which acts by binding to the cytoplasmic domains of tyrosine kinase receptors via its C-terminal SH2 domain. Mouse conceptuses with deletion of the active maternal allele of *Grb10* have overgrowth of both the placenta and the fetus [62]. More recently, this analysis has been extended to show that following disruption of the maternal *Grb10* allele the volume of the labyrinthine layer of the placenta is increased in a manner consistent with a cell-autonomous function of the Grb10 signalling protein [63]. Conversely, Ishino and colleagues found that paternal deletion of the *Meg1/Grb10* DMR (differentially methylated imprinting control region) causes loss of imprinting and over-expression of *Grb10*, resulting in fetal growth restriction with reduced placental size, as well as postnatal growth retardation [64].

Peg1/Mest

Mouse conceptuses lacking expression of *Mest* consequent upon the deletion of the active paternal allele show fetal and placental growth retardation [65].

Peg3

Deletion of the active paternal allele of *Peg3* (encoding a Kruppel-type zinc finger protein that can function as

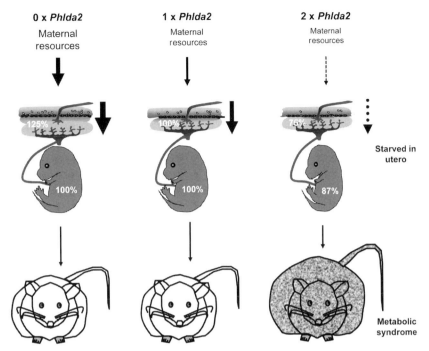

Figure 6.4 Phlda2 KO and Tg mice as a possible example of a genetic system for modeling DOHAD. Embryonic growth and placental support are carefully balanced so that the genetic growth potential of the fetus is fully supported by the nutrient capacity of the placenta. Phlda2-deficiency (0 X Phlda2) increases placental weight and glycogen stores but does not benefit embryonic growth – an example whereby excessive growth of the placenta depletes maternal nutritional resources and potentially limits her current and future reproductive success, i.e. reduced maternal fitness. In contrast, excess Phlda2 (2 X Phlda2) restricts placental growth, which in turn restricts late embryonic growth. The consequences later in life can include a phenotype similar to human metabolic syndrome. (See colour plate section.)

a transcriptional repressor) causes fetal and placental stunting when deleted in the conceptus, and abnormal nurturing behavior when deleted in the dam [66].

Peg10 and Peg11

Peg10, a retrotransposon-derived gene, produces an imprinted placental transcript in both mice and humans which encodes a gag-pol-related protein. Deletion of the paternal allele in mice produces severe placental and fetal stunting, with loss of the junctional zone of the placenta [67]. There is evidence that *PEG10* has a general role in cell proliferation, not only in trophoblast but also in hepatocellular carcinomas, where it is often reactivated to high levels [68,69]. Consistent with this growth-positive role, *PEG10* is a direct downstream target gene of the c-MYC transcription factor [28]. Interestingly, a second retrotransposon-derived gene, *Peg11/Rtl1*, which is only distantly related to *Peg10* and which maps to a different chromosome, was also shown by a knockout in mice to be essential for normal placental development [41]. Additional studies of imprinting in other eutherian mammals [70] and in marsupials [71] have shown that *PEG10* became established as an imprinted locus in the mammalian genome surprisingly early, before the divergence of eutherian and metatherian (marsupial) branches. Indeed, the term 'placental mammals', which is usually applied only to eutherians, is somewhat misleading, as marsupials do have a rudimentary non-invasive yolk sac placenta that acts early in the pregnancy, before its nutritive function is taken over by the mammary gland in the pouch. Imprinted genes have not yet been found in the prototherian (non-placental) platypus, supporting the idea that the evolution of mammalian imprinting was driven partly by evolution of the placenta.

Plagl1/Zac1/LOT1

KO of the imprinted *Plagl1* gene in mice leads to fetal growth restriction, the lack of placental abnormalities in these mice suggesting that the growth retardation is due mostly to embryo-autonomous deficiencies in growth, not placental defects [72]. The *Plagl1/Zac1* KO experiment nicely illustrates the predictive power of the conflict hypothesis. Prior analyses of cancer cells had suggested *Zac1* as a candidate tumour suppressor gene, encoding a transcription factor with proapoptotic and cell-cycle-blocking activities in cell lines, and hence predicted to be growth restraining. However, the intragenomic conflict theory makes the opposite prediction that paternally expressed imprinted genes, like *Zac1*, should be growth promoting. The observed growth-retarded phenotype of Zac1-deficient embryos shows that the *Zac1* gene acts as a growth promoter in normal development, precisely as predicted by intragenomic conflict.

Esx1

This X-linked gene, encoding a homeobox transcription factor, is interesting as an example of a locus that is not conserved in terms of its imprinting, being imprinted in mice but not in humans [73,74], probably at least in part by virtue of its position on the X chromosome, which is non-randomly inactivated on the paternal copy in the mouse. *Esx1*-deficient mouse placentas are heavier than normal, with fluid-filled cysts and an excessive number of glycogen cells [74].

Cited1

Similar to *Esx1*, *Cited1* is imprinted in mice by virtue of its location on the X chromosome [75]. *Cited1* is expressed in all the trophoblast-derived placental cells. *Cited1*-deficient placentas are normal in weight, but the border between the junctional zone and the labyrinth is irregular and the junctional zone is increased in volume, with a concomitant decrease in the labyrinthine zone and a disruption of fetal vasculature.

Altered expression of imprinted and non-imprinted genes in human intrauterine growth restriction

Intrauterine growth restriction (IUGR) is a common medical condition that often leads to expensive neonatal hospitalization and predisposes to serious postnatal complications. Growth restriction *in utero* can have various causes, including fetal genetic or epigenetic aberrations and confined placental mosaicism, as well as maternally transmitted infections, maternal autoimmune disease, and maternal cigarette smoking. However, many cases of IUGR are idiopathic, and maternal vascular underperfusion (uteroplacental insufficiency) is an important factor. Such vascular insufficiency manifests clinically by abnormalities in Doppler ultrasound measurements of umbilical and uterine artery blood flow, and pathologically by increased syncytiotrophoblast knots, distal villous hypoplasia, increased intervillous fibrin, chronic infarcts and maternal decidual vasculopathy in the placenta. Given their frequent function as growth rheostats, imprinted genes are interesting candidates for a role in IUGR. In fact, specific forms of severe IUGR in humans can result from UPDs for chromosomes containing imprinted genes, notably maternal UPD7 in the Silver–Russell syndrome. This syndrome can also be caused, even more commonly, by epimutations on chromosome band 11p15, specifically a loss of DNA methylation upstream of *H19* on the paternal allele, resulting in loss of *IGF2* expression [76].

In contrast, in non-syndromic IUGR evidence so far indicates that the imprinting process *per se* is not disrupted. Nonetheless, there are now some consistent data indicating unbalanced mRNA expression of a subset of imprinted genes in IUGR placentas. In our survey for abnormalities of gene expression in growth-retarded placentas associated with fetal IUGR matching 14 IUGR placentas with 15 non-IUGR placentas spanning a similar range of gestational age, taking tissue samples of chorionic villi and measuring mRNA expression using Affymetrix U133A microarrays, we found that six imprinted genes were differentially expressed, with increased expression of *PHLDA2* and decreased expression of *MEST*, *MEG3*, *GATM*, *GNAS* and *PLAGL1* in the IUGR placentas [77]. Less significantly (greater case-to-case variation), we found *IGF2* mRNA decreased and *CDKN1C* mRNA increased in the IUGR placentas. Importantly, for most of the imprinted genes, the magnitude of the transcriptional dysregulation was modest (less than twofold increased or decreased), whereas in contrast a number of non-imprinted genes (e.g. *IGF1*, *LEP* and others) were more strongly dysregulated [2,77]. These data suggest unbalanced expression of imprinted genes as one component of the physiological response of placental tissues to maternal vascular underperfusion in IUGR. The upregulation of *PHLDA2* mRNA in both IUGR and small-for-gestational-age (SGA) placentas has been confirmed by independent studies, one of which also studied the *PLAGL1* gene and confirmed its downmodulation in the IUGR placentas [78,79]. As noted above, there is no evidence that this dysregulated expression reflects aberrant imprinting *per se*, as none of these three studies found evidence for altered DNA methylation at imprinting centres in cases of IUGR. Instead, the observed changes in gene expression probably reflect conventional transcriptional modulation. Understanding this transcriptional response may help in identifying molecular mechanisms of growth restriction in IUGR. Interestingly, the identities of the 'IUGR signature' genes and the directions of their altered expression suggest that some components of this response (e.g. increased expression of *PHLDA2*, decreased expression of *IGF1*, etc.) are adverse with respect to placental growth, raising the possibility of a 'vicious cycle' of altered gene expression leading to progressive placental insufficiency [77]. This cycle will need to be interrupted for successful treatment of IUGR.

Manipulating imprinted and non-imprinted genes in mice to create models for DOHAD

Epidemiological studies have identified the phenomenon of fetal programming in humans, but the mechanisms leading to diseases later in life are unknown. Given this lack of knowledge it is not yet possible to design preventive therapies. The stakes are high, as adverse fetal programming seems to underlie some of the most common medical conditions that plague modern societies. Key to the phenomenon is the presence of asymmetrical growth with relative sparing of the brain in combination with IUGR followed by accelerated postnatal weight gain. It is this category of growth-restricted infant that appears to show an increased susceptibility to abnormal glucose and insulin metabolism, dyslipidaemia and hypertension, with the subsequent development of obesity, type 2 diabetes and cardiovascular disease, with variations in the combination, time of onset and severity of disease between individuals (see also Chapter 2). A well-developed and highly sophisticated body of experiments has demonstrated that fetal programming can be reproduced in a number of animal models. As discussed in other elsewhere (see Chapters 4 and 14) these models mostly involve either manipulating the maternal state (diet, stress) with a secondary effect on fetal growth, or direct surgical manipulation such as uterine artery ligation to reduce the delivery of oxygen and nutrients to the fetus [80–82]. Animal models are particularly useful in studying specific interventions under controlled conditions that allow cross-laboratory comparisons. This is important as even subtle changes in protocols, such as the timing of the intervention, can lead to differences in adult phenotypes.

Despite the success of these interventional models, studies on fetal programming would benefit from a genetic mouse model of placental insufficiency. In addition to providing control and experimental animals within a single pregnancy, no preterm intervention is required, which reduces variation between animals. In addition, embryos can be studied from the start of growth restriction so that early changes in tissue structure and function can be determined. Once precise measurements have been obtained, a number of interventions can be applied to either increase or reduce the severity of the IUGR and its postnatal consequences. These include changes in diet or the use of different strain backgrounds. Another compelling reason for adopting a genetic model of IUGR is that crosses can be made with other knockout or transgenic models to 'rescue' or aggravate the IUGR phenotype. Owing to their action in the placenta, there are a number of genetic models involving imprinted genes that already seem promising for investigating DOHAD. When studying targeted deficiency phenotypes, an additional advantage of an imprinted model is that 50% of the fetuses carry the mutation, rather than the 25% generated in a classic genetic cross, thereby reducing the number of experimental animals required.

The Igf2P0$^{+/-}$ model [6]

This model is based on paternal inheritance of a targeted deletion of a placental-specific promoter that drives *Igf2* expression exclusively in the labyrinthine trophoblast. Despite the restricted expression of this transcript, placental growth is proportionately restrained, with a reduction of both the labyrinthine and the junctional zones. Placental growth is impaired from E12.5 with a 35% reduction in weight by E18 [83]. *Igf2P0$^{+/-}$* embryos are 4% lighter by E16.5, 12–14% lighter at E18.5 and 21% lighter at P0, with postnatal catch-up growth by 3 months. Viability is unaffected. A key advantage to this model for DOHAD is that the primary genetic defect is localized to the placenta.

The Phlda2 overexpression model [15,16]

This model is based on temporally and spatially accurate overexpression of *Phlda2* and *Slc22a18* from a genomic fragment spanning the locus. *Phlda2* is expressed predominantly in extraembryonic tissues until mid-gestation and at low levels in the embryonic kidney, liver and lung and adult kidney. *Slc22a18* is highly expressed in tissues with metabolite transport function, including extraembryonic membranes, the placenta, liver, kidney and intestine. In this model, placental weight is reduced, with a maximum reduction of 26% at E13.5, and defects in the ectoplacental cone are visible as early as E10.5. There is a specific reduction in the volume of the junctional zone and loss of stored glycogen with additional mislocalization of the glycogen cells. Alterations in embryonic weight are not apparent until E16.5, when transgenic embryos are 10% lighter than controls. The loss of growth potential is progressive, with a 13% reduction in weight at E18.5. As with classic placental insufficiency, there is evidence of catch-up growth by 3 weeks. However, in these mice both *Phlda2* and *Slc22a18* are over expressed in the placenta, the embryo and the adult. Although we are able to exclude a role for *Slc22a18* in the placental phenotype [16], we cannot confidently assign a placental origin for any adult phenotype. However, it may be possible to overcome this problem using transgenes with more restricted expression. A critical advantage of the *Phlda2* model not shared by any other genetic model is that over expression of the human gene has been reported in a large proportion of placentae from human IUGR and small-for-gestational-age infants [77,78]. In our preliminary investigations we have found that the IUGR in our mice is asymmetrical, and there is some indication that adult mice are more inclined towards glucose intolerance and weight gain as they age (S. J. Tunster and R. John, unpublished data). This suggests that the *Phlda2*-altered mice may represent one of the first genetic models for adverse fetal programming that reproduces a known human condition.

The Cited1 model [75]

Cited1-deficient embryos inheriting the targeted allele from their mother, albeit similar in weight to wild-type embryos at E16.5, are 20% lighter by E18.5, indicating a progressive loss of growth potential as gestation proceeds. Adult mice also exhibit renal medullary dysplasia suggestive of fetal programming. Although *Cited1* is expressed in both embryonic and adult tissues, including the progenitors of the heart, limb and axial skeleton, and in adult melanocytes, heart and mammary gland, the kidney defect occurs as a consequence of decreased tissue oxygenation due to placental insufficiency [84]. This discrimination is possible because *Cited1* is not a directly imprinted gene but shows 'imprinted' expression in the placenta owing to its location on the X chromosome. In the mouse, the X is inactivated randomly in the embryo, but only the paternal X chromosome is inactivated in extraembryonic tissues. Therefore, heterozygous females that inherit the targeted allele from their mother are mosaic for *Cited1* expression in the embryo and null for *Cited1* in extraembryonic ectoderm, whereas animals inheriting the targeted allele paternally are similarly mosaic for *Cited1* in the embryo but have wild-type levels of *Cited1* expression in their placenta, expression being from the active X. By comparing the two scenarios, adult phenotypes can to some extent be assigned to either the placental defect or to haploinsufficiency.

The Esx1 model [74]

Despite this lack of conservation of imprinting, the phenotype in *Esx1* KO mice is interesting as a model for IUGR. Heterozygous females that inherit the *Esx1* mutation from their mother are born 20% smaller than normal. Defects in the morphogenesis of the labyrinthine layer are observed as early as 11.5 days of gestation, and subsequently, vascularization abnormalities develop at the maternofetal interface, causing the fetal growth retardation [74]. *Esx1* expression is detected only in extraembryonic tissues and in male germ cells. As noted above, *Esx1*-deficient placenta are heavier than normal, with fluid-filled cysts and excessive numbers of glycogen cells. *Esx1*-deficient embryos inheriting the targeted allele from their mother are similar in weight to wild-type embryos at E13.5 but show a noticeable reduction in weight by E16.5, and are born 20% lighter. By maturity the weight of the mutants is similar to that of controls, indicating catch-up growth. With a similar rational to *Cited1*, an intrinsic influence on any identified adult phenotypes can be excluded.

Mouse models of IUGR involving non-imprinted genes

There are examples of interesting models created by knocking out non-imprinted placenta-specific or placenta-predominant genes. *Syncytin-A* is an endogenous retroviral gene specifically expressed in the labyrinth of the mouse placenta [85]. Homozygous null embryos fail to form the syncytial trophoblast layer, resulting in poor placental vascularization and a reduction in placental transport function. Colony-stimulating factor 2 (Csf2) deficiency results in a severe placental defect, with mutant placentas exhibiting a 22% increase in volume of the junctional zone and 13% reduction of the labyrinthine layer at E14.5, followed by a 7% reduction in fetal weight by E16.5 [86]. The severity of the placental phenotypes in both these models precludes further analysis, but it might be possible to identify a scenario that is compatible with survival. The *SynA* targeted allele is flanked by *loxP* sites, suggesting an approach based on partial loss of expression. Studies using heterozygous animals may also be successful. Homozygous loss of expression of the *glucose transporter isoform-3* (*Slc2a3/Glut3*) gene, which encodes the trophoblastic facilitative glucose transporter, is embryo lethal [87]. However, heterozygous animals display late-gestation fetal growth (22% lighter at E20.5) with catch-up growth by 10 weeks of age, permitting studies on these mice as they age. As with other murine models where expression is not restricted to the placenta, caution must be applied when interpreting adult phenotypes.

Conclusion and future directions

The strong role of imprinted genes in placental and fetal growth that we have reviewed here suggests obvious connections to DOHAD, including practical implications for making mouse models. Arguably, the ideal genetic model should meet four criteria. First, we must be able to confidently assign any adult phenotype to a defect in the placenta, to exclude any possibility of a metabolic phenotype being caused directly by the engineered change. Very few genes are expressed only in the placenta, and even some legendary placenta-specific genes are, in fact, expressed in the adult animal, an example being *Ascl2*. Second, the growth profile of slow-down/catch-up must be present in combination with typical asymmetrical (brain-sparing) IUGR. Third, the defect must be compatible with survival to adulthood. Finally, the engineered change should ideally also be one reported in human IUGR.

In the case of the *Igf2* gene, it was possible to develop a placental-specific model by engineering a very specific alteration. It should therefore be possible to build on these, and other models in which there are both placental and embryonic defects, to meet the most important criterion, a placental origin, using more sophisticated approaches for placenta-specific gain and loss of gene function, for example combining conditional alleles with a placental-specific Cre recombinase such as the *CYP19-Cre* transgene [88] or a lentivirus-mediated approach [89,90]. Inducible gene promoters may also be useful, but caution must be employed as in at least one of these systems, the Tet-on/off system, the stimulant doxycycline can by itself cause placental defects limiting the usefulness of this approach [91].

Owing to differences in their site of action within the placenta, creating genetic models of placental insufficiency with modifications to the different compartment of the placenta is also now a realistic goal. This compartment-specific approach may provide important information relating a specific impairment at the maternofetal interface, such as a poor connection between fetal and maternal vasculature, abnormal hormonal secretion or glycogen deficiency, to the severity and timing of fetal growth restriction, the potential for catch-up growth and the presence of metabolic abnormalities in the adult. Ultimately, large-scale screens for alterations in gene expression, and also in epigenetic markers such as DNA methylation or histone modification, in developmentally programmed tissue will enable the identification of specific alterations unique to each model and also those alterations that are overlapping. The next challenge will be to demonstrate a direct cause and effect relationship between these early alterations and suboptimal health later in life. Lastly, and most importantly, having convenient, fast and highly reproducible mouse models will allow testing of therapies both to ameliorate IUGR and to protect against its late adverse consequences.

Acknowledgments

This work was supported by grants from the March of Dimes to BT and from the Biotechnology and Biological Sciences Research Council to RMJ. The authors would like to thank Dr Simon Tunster for helpful discussions and the provision of unpublished images.

References

1. **Cattanach BM, Beechey, Peters**. Interactions between imprinting effects: summary and review. *Cytogenet Genome Res* 2006; **113**(1–4): 17–23.
2. **Tycko B**. Imprinted genes in placental growth and obstetric disorders. *Cytogenet Genome Res* 2006; **113**(1–4): 271–8.
3. **Redline RW** et al. Differential expression of insulin-like growth factor-II in specific regions of the late (post day 9.5) murine placenta. *Mol Reprod Dev* 1993; **36**(2): 121–9.
4. **Lustig O** et al. Expression of the imprinted gene H19 in the human fetus. *Mol Reprod Dev* 1994; **38**(3): 239–46.
5. **Han VK, Carter AM**. Spatial and temporal patterns of expression of messenger RNA for insulin-like growth factors and their binding proteins in the placenta of man and laboratory animals. *Placenta* 2000; **21**(4): 289–305.
6. **Constancia M** et al. Placental-specific IGF-II is a major modulator of placental and fetal growth. *Nature* 2002; **417**(6892): 945–8.
7. **Dao D** et al. IMPT1, an imprinted gene similar to polyspecific transporter and multi-drug resistance genes. *Hum Mol Genet* 1998; **7**(4): 597–608.
8. **Qian N** et al. The IPL gene on chromosome 11p15.5 is imprinted in humans and mice and is similar to TDAG51, implicated in Fas expression and apoptosis. *Hum Mol Genet* 1997; **6**(12): 2021–9.
9. **Lee MP, Feinberg AP**. Genomic imprinting of a human apoptosis gene homologue, TSSC3. *Cancer Res* 1998; **58**(5): 1052–6.
10. **Chilosi M** et al. Differential expression of p57kip2, a maternally imprinted cdk inhibitor, in normal human placenta and gestational trophoblastic disease. *Lab Invest* 1998; **78**(3): 269–76.
11. **Castrillon DH** et al. Discrimination of complete hydatidiform mole from its mimics by immunohistochemistry of the paternally imprinted gene product p57KIP2. *Am J Surg Pathol* 2001; **25**(10): 1225–30.
12. **Saxena A** et al. The product of the imprinted gene IPL marks human villous cytotrophoblast and is lost in complete hydatidiform mole. *Placenta* 2003; **24**(8–9): 835–42.
13. **Takahashi K, Kobayashi T, Kanayama N**. p57(Kip2) regulates the proper development of labyrinthine and spongiotrophoblasts. *Mol Hum Reprod* 2000; **6**(11): 1019–25.
14. **Frank D** et al. Placental overgrowth in mice lacking the imprinted gene Ipl. *Proc Natl Acad Sci USA* 2002; **99**(11): 7490–5.
15. **Salas M** et al. Placental growth retardation due to loss of imprinting of Phlda2. *Mech Dev* 2004; **121**(10): 1199–210.
16. **Tunster SJ, Tycko B, John RM**. The imprinted Phlda2 gene regulates extraembryonic energy stores. *Mol Cell Biol* 2009; **30**(1): 295–306.
17. **Dao D** et al. Multipoint analysis of human chromosome 11p15/mouse distal chromosome 7: inclusion of H19/IGF2 in the minimal WT2 region, gene specificity of H19 silencing in Wilms' tumorigenesis and methylation hyper-dependence of H19 imprinting. *Hum Mol Genet* 1999; **8**(7): 1337–52.
18. **Caspary T** et al. Multiple mechanisms regulate imprinting of the mouse distal chromosome 7 gene cluster. *Mol Cell Biol* 1998; **18**(6): 3466–74.
19. **Tanaka M** et al. Parental origin-specific expression of Mash2 is established at the time of implantation with its imprinting mechanism highly resistant to genome-wide demethylation. *Mech Dev* 1999; **87**(1–2): 129–42.
20. **Lewis A** et al. Imprinting on distal chromosome 7 in the placenta involves repressive histone methylation independent of DNA methylation. *Nature Genet* 2004; **36**(12): 1291–5.
21. **Umlauf D** et al. Imprinting along the Kcnq1 domain on mouse chromosome 7 involves repressive histone methylation and recruitment of Polycomb group complexes. *Nature Genet* 2004; **36**(12): 1296–300.
22. **Kacem S, Feil R**. Chromatin mechanisms in genomic imprinting. *Mamm Genome* 2009; **10**: 54–56.
23. **Shin JY, Fitzpatrick GV, Higgins MJ**. Two distinct mechanisms of silencing by the KvDMR1 imprinting control region. *EMBO J* 2008; **27**(1): 168–78.
24. **Fitzpatrick GV, Soloway PD, Higgins MJ**. Regional loss of imprinting and growth deficiency in mice with a targeted deletion of KvDMR1. *Nature Genet* 2002; **32**(3): 426–31.
25. **Henckel A** et al. Histone methylation is mechanistically linked to DNA methylation at imprinting control regions in mammals. *Hum Mol Genet* 2009; **18**(18): 3375–83.
26. **Ono R** et al. A retrotransposon-derived gene, PEG10, is a novel imprinted gene located on human chromosome 7q21. *Genomics* 2001; **73**(2): 232–7.
27. **Okita C** et al. A new imprinted cluster on the human chromosome 7q21-q31, identified by human-mouse monochromosomal hybrids. *Genomics* 2003; **81**(6): 556–9.

28. **Li CM** et al. PEG10 is a c-MYC target gene in cancer cells. *Cancer Res* 2006; **66**(2): 665–72.
29. **Kaneko-Ishino T** et al. Peg1/Mest imprinted gene on chromosome 6 identified by cDNA subtraction hybridization. *Nature Genet* 1995; **11**(1): 52–9.
30. **Kobayashi S** et al. Human PEG1/MEST, an imprinted gene on chromosome 7. *Hum Mol Genet* 1997; **6**(5): 781–6.
31. **McMinn J** et al. Imprinting of PEG1/MEST isoform 2 in human placenta. *Placenta* 2006; **27**(2–3): 119–26.
32. **Pedersen IS** et al. Promoter switch: a novel mechanism causing biallelic PEG1/MEST expression in invasive breast cancer. *Hum Mol Genet* 2002; **11**(12): 1449–53.
33. **Nikonova L** et al. Mesoderm-specific transcript is associated with fat mass expansion in response to a positive energy balance. *FASEB J* 2008; **22**(11): 3925–37.
34. **Mayer W** et al. Expression of the imprinted genes MEST/Mest in human and murine placenta suggests a role in angiogenesis. *Dev Dyn* 2000; **217**(1): 1–10.
35. **Stepan H** et al. Structure and regulation of the murine Mash2 gene. *Biol Reprod* 2003; **68**(1): 40–4.
36. **Nakayama H** et al. Developmental restriction of Mash-2 expression in trophoblast correlates with potential activation of the notch-2 pathway. *Dev Genet* 1997; **21**(1): 21–30.
37. **Alders M** et al. The human Achaete-Scute homologue 2 (ASCL2,HASH2) maps to chromosome 11p15.5, close to IGF2 and is expressed in extravillus trophoblasts. *Hum Mol Genet* 1997; **6**(6): 859–67.
38. **Miyamoto T** et al. The human ASCL2 gene escaping genomic imprinting and its expression pattern. *J Assist Reprod Genet* 2002; **19**(5): 240–4.
39. **Hiby SE** et al. Paternal monoallelic expression of PEG3 in the human placenta. *Hum Mol Genet* 2001; **10**(10): 1093–100.
40. **Kim J** et al. The human homolog of a mouse-imprinted gene, Peg3, maps to a zinc finger gene-rich region of human chromosome 19q13.4. *Genome Res* 1997; **7**(5): 532–40.
41. **Sekita Y** et al. Role of retrotransposon-derived imprinted gene, Rtl1, in the feto-maternal interface of mouse placenta. *Nature Genet* 2008; **40**(2): 243–8.
42. **DeChiara TM, Robertson EJ, Efstratiadis A**. Parental imprinting of the mouse insulin-like growth factor II gene. *Cell* 1991; **64**(4): 849–59.
43. **Barlow DP** et al. The mouse insulin-like growth factor type-2 receptor is imprinted and closely linked to the Tme locus. *Nature* 1991; **349**(6304): 84–7.
44. **Haig D, Graham C**. Genomic imprinting and the strange case of the insulin-like growth factor II receptor. *Cell* 1991; **64**(6): 1045–6.
45. **Moore T, Haig D**. Genomic imprinting in mammalian development: a parental tug-of-war. *Trends Genet* 1991; **7**(2): 45–9.
46. **Gluckman PD**. Endocrine and nutritional regulation of prenatal growth. *Acta Paediatr Suppl* 1997; **423**: 153–7; discussion 158.
47. **Baker J** et al. Role of insulin-like growth factors in embryonic and postnatal growth. *Cell* 1993; **75**(1): 73–82.
48. **Eggenschwiler J** et al. Mouse mutant embryos overexpressing IGF-II exhibit phenotypic features of the Beckwith-Wiedemann and Simpson-Golabi-Behmel syndromes. *Genes Dev* 1997; **11**(23): 3128–42.
49. **Riccio A** et al. Inherited and Sporadic Epimutations at the IGF2-H19 locus in Beckwith-Wiedemann syndrome and Wilms' tumor. *Endocr Dev* 2009; **14**: 1–9.
50. **Tanaka M** et al. Mash2 acts cell autonomously in mouse spongiotrophoblast development. *Dev Biol* 1997; **190**(1): 55–65.
51. **Saxena A** et al. Phosphoinositide binding by the pleckstrin homology domains of Ipl and Tih1. *J Biol Chem* 2002; **277**(51): 49935–44.
52. **Frank D** et al. A novel pleckstrin homology-related gene family defined by Ipl/Tssc3, TDAG51, and Tih1: tissue-specific expression, chromosomal location, and parental imprinting. *Mamm Genome* 1999; **10**(12): 1150–9.
53. **Kanayama N** et al. Deficiency in p57Kip2 expression induces preeclampsia-like symptoms in mice. *Mol Hum Reprod* 2002; **8**(12): 1129–35.
54. **Caspary T** et al. Oppositely imprinted genes p57(Kip2) and igf2 interact in a mouse model for Beckwith-Wiedemann syndrome. *Genes Dev* 1999; **13**(23): 3115–24.
55. **Knox KS, Baker JC**. Genome-wide expression profiling of placentas in the p57Kip2 model of pre-eclampsia. *Mol Hum Reprod* 2007; **13**(4): 251–63.
56. **Bliek J** et al. Epigenotyping as a tool for the prediction of tumor risk and tumor type in patients with Beckwith-Wiedemann syndrome (BWS). *J Pediatr* 2004; **145**(6): 796–9.
57. **Killian JK** et al. M6P/IGF2R imprinting evolution in mammals. *Mol Cell* 2000; **5**(4): 707–16.
58. **Xu Y** et al. Functional polymorphism in the parental imprinting of the human IGF2R gene. *Biochem Biophys Res Commun* 1993; **197**(2): 747–54.

59. **Yotova IY** *et al.* Identification of the human homolog of the imprinted mouse Air non-coding RNA. *Genomics* 2008; **92**(6): 464–73.

60. **Ludwig T** *et al.* Mouse mutants lacking the type 2 IGF receptor (IGF2R) are rescued from perinatal lethality in Igf2 and Igf1r null backgrounds. *Dev Biol* 1996; **177**(2): 517–35.

61. **Wutz A** *et al.* Non-imprinted Igf2r expression decreases growth and rescues the Tme mutation in mice. *Development* 2001; **128**(10): 1881–7.

62. **Charalambous M** *et al.* Disruption of the imprinted Grb10 gene leads to disproportionate overgrowth by an Igf2-independent mechanism. *Proc Natl Acad Sci USA* 2003; **100**(14): 8292–7.

63. **Charalambous M** *et al.* Maternally-inherited Grb10 reduces placental size and efficiency. *Dev Biol* 2009; **337**(1): 1–8.

64. **Shiura H** *et al.* Paternal deletion of Meg1/Grb10 DMR causes maternalization of the Meg1/Grb10 cluster in mouse proximal Chromosome 11 leading to severe pre- and postnatal growth retardation. *Hum Mol Genet* 2009; **18**(8): 1424–38.

65. **Lefebvre L** *et al.* Abnormal maternal behaviour and growth retardation associated with loss of the imprinted gene Mest. *Nature Genet* 1998; **20**(2): 163–9.

66. **Li L** *et al.* Regulation of maternal behavior and offspring growth by paternally expressed Peg3. *Science* 1999; **284**(5412): 330–3.

67. **Ono R** *et al.* Deletion of Peg10, an imprinted gene acquired from a retrotransposon, causes early embryonic lethality. *Nature Genet* 2006; **38**(1): 101–6.

68. **Tsou AP** *et al.* Overexpression of a novel imprinted gene, PEG10, in human hepatocellular carcinoma and in regenerating mouse livers. *J Biomed Sci* 2003; **10**(6 Pt 1): 625–35.

69. **Okabe H** *et al.* Involvement of PEG10 in human hepatocellular carcinogenesis through interaction with SIAH1. *Cancer Res* 2003; **63**(12): 3043–8.

70. **Bischoff SR** *et al.* Characterization of conserved and nonconserved imprinted genes in swine. *Biol Reprod* 2009; **81**(5): 906–20.

71. **Renfree MB** *et al.* Genomic imprinting in marsupial placentation. *Reproduction* 2008; **136**(5): 523–31.

72. **Varrault A** *et al.* Zac1 regulates an imprinted gene network critically involved in the control of embryonic growth. *Dev Cell* 2006; **11**(5): 711–22.

73. **Grati FR** *et al.* Biparental expression of ESX1L gene in placentas from normal and intrauterine growth-restricted pregnancies. *Eur J Hum Genet* 2004; **12**(4): 272–8.

74. **Li Y, Behringer RR**. Esx1 is an X-chromosome-imprinted regulator of placental development and fetal growth. *Nature Genet* 1998 **20**(3): 309–11.

75. **Rodriguez TA** *et al.* Cited1 is required in trophoblasts for placental development and for embryo growth and survival. *Mol Cell Biol* 2004; **24**(1): 228–44.

76. **Gicquel C** *et al.* Epimutation of the telomeric imprinting center region on chromosome 11p15 in Silver-Russell syndrome. *Nature Genet* 2005; **37**(9): 1003–7.

77. **McMinn J** *et al.* Unbalanced placental expression of imprinted genes in human intrauterine growth restriction. *Placenta* 2006; **27**(6–7): 540–9.

78. **Apostolidou S** *et al.* Elevated placental expression of the imprinted PHLDA2 gene is associated with low birth weight. *J Mol Med* 2007; **85**(4): 379–87.

79. **Diplas AI** *et al.* Differential expression of imprinted genes in normal and IUGR human placentas. *Epigenetics* 2009; **4**(4): 235–40.

80. **Vuguin PM**. Animal models for small for gestational age and fetal programming of adult disease. *Horm Res* 2007; **68**(3): 113–23.

81. **Fowden AL** *et al.* Programming placental nutrient transport capacity. *J Physiol* 2006; **572**(Pt 1): 5–15.

82. **Holemans K, Aerts L, Van Assche FA**. Fetal growth restriction and consequences for the offspring in animal models. *J Soc Gynecol Invest* 2003; **10**(7): 392–9.

83. **Coan PM** *et al.* Adaptations in placental nutrient transfer capacity to meet fetal growth demands depend on placental size in mice. *J Physiol* 2008; **586**(Pt 18): 4567–76.

84. **Sparrow DB** *et al.* Placental insufficiency associated with loss of Cited1 causes renal medullary dysplasia. *J Am Soc Nephrol* 2009; **20**(4): 777–86.

85. **Dupressoir A** *et al.* Syncytin-A knockout mice demonstrate the critical role in placentation of a fusogenic, endogenous retrovirus-derived, envelope gene. *Proc Natl Acad Sci USA* 2009. ; **106**(29): 12127–32.

86. **Robertson SA** *et al.* Fertility impairment in granulocyte-macrophage colony-stimulating factor-deficient mice. *Biol Reprod* 1999; **60**(2): 251–61.

87. **Ganguly A** *et al.* Glucose transporter isoform-3 mutations cause early pregnancy loss and fetal growth

restriction. *Am J Physiol Endocrinol Metab* 2007; **292**(5): E1241–55.

88. **Wenzel PL**, **Leone G**. Expression of Cre recombinase in early diploid trophoblast cells of the mouse placenta. *Genesis* 2007; **45**(3): 129–34.

89. **Georgiades P** *et al.* Trophoblast-specific gene manipulation using lentivirus-based vectors. *Biotechniques* 2007; **42**(3): 317–8, 320, 322–5.

90. **Okada Y** *et al.* Complementation of placental defects and embryonic lethality by trophoblast-specific lentiviral gene transfer. *Nature Biotechnol* 2007; **25**(2): 233–7.

91. **Moutier R** *et al.* Placental anomalies and fetal loss in mice, after administration of doxycycline in food for tet-system activation. *Transgenic Res* 2003; **12**(3): 369–73.

92. **Rossant J**, **Cross JC**. Placenta development: lessons from mouse mutants. *Nat Rev Genet* 2001; **2**(7): 538–49.

Discussion

ROBERT BOYD: If you nutritionally disadvantage your *Phlda2*-excluded ones which have an overlarge placenta, is it of any benefit in those circumstances?

TYCKO: This would be an interesting experiment, but we haven't looked.

MOORE: It seems to me that one of the values of imprinting is in relation to its role in screening genes regulating fetal growth. In humans, there may be many more imprinted genes than the 100 or so that are known. This gives us ways to try to identify what some of the genetic influences on fetal growth are and how they operate.

TYCKO: *Phlda2* has been reported, both by us and also Gudrun Moore's lab (Apostolidou S, Abu-Amero S, O'Donoghue K, Frost J, Olafsdottir O, Chavele KM, Whittaker JC, Loughna P, Stanier P, Moore GE. Elevated placental expression of the imprinted PHLDA2 gene is associated with low birth weight. *J Mol Med* 2007 Apr; 85(4): 379–87. Epub 2006 Dec 16. PubMed PMID: 17180344) as being a marker for low birth weight. In other words, it is expressed at a higher level in human low birth weight placenta. But that is not due to loss of imprinting. It seems to be due to some other mechanism that might indeed be genetic. There are new data showing that DNA methylation can be influenced by adjacent haplotypes on the chromosome (1: Tycko B. Mapping allele-specific DNA methylation: a new tool for maximizing information from GWAS. *Am J Hum Genet* 2010 Feb 12; 86(2): 109–12. PubMed PMID: 20159108), and so it is possible that imprinted genes are regulated in a purely epigenetic parent-of-origin-specific manner, but also might be modulated by haplotypes (i.e. a genetic influence on the expression of imprinted genes). There are some data on haplotypes in people modulating the expression of *IGF2* and affecting birth weight (e.g. Nagaya K, Makita Y, Taketazu G, Okamoto T, Nakamura E, Hayashi T, Fujieda K. Paternal allele of IGF2 gene haplotype CTG is associated with fetal and placental growth in Japanese. *Pediatr Res* 2009 Aug; 66(2): 135–9. PubMed PMID: 19390492).

THORNBURG: These placentas lose efficiency as they get bigger. This raises the question about how the fetus grows. Many people tacitly assume that a bigger placenta has more surface area, and there is more ability to grow on the part of the fetus. Why these fetuses don't grow in proportion to the placentas is an important question. There may be information in your model that could speak to this point.

TYCKO: I think it would be interesting to study the detailed structure of these placentas using methods such as microvascular injection.

FOWDEN: Part of the efficiency change in this model is a change in the junctional zone. It is an apparent efficiency change rather than a real one. The labyrinth, which is the exchange surface, is the same size, which is why the fetus doesn't overgrow.

TYCKO: In the knockout, the placenta is wasting energy on the expanded junctional zone.

FOWDEN: It also explains why in over-expression where there is a smaller junctional zone, you don't have the endocrine component and the adaptation of maternal metabolism to favour nutrient supply to the fetus, so in that situation the fetus does get smaller.

THORNBURG: Efficiency can't just be confined to the volume of the junctional zone: it is also how much fetus you get from it. Are you saying that in this case they are the same?

FOWDEN: The labyrinthine zone, which is the exchange surface, is the same. So efficiency shouldn't be calculated as gram of fetus per gram of placenta, but rather gram fetus per centimetre squared of exchange surface.

FERGUSON SMITH: From an imprinting perspective, there are a number of imprinting mutations that give rise to a larger placenta. I only know one that results in a larger fetus. When you start looking at diffusional characteristics and detailed morphometric analyses, these large placentae are structurally not sound and may be functionally compromised.

Chapter 6. Imprinted genes and placental growth

SFERRUZZI-PERRI: So the Phlda2 protein is involved in regulating the uptake and trafficking of proteins within the cell. The junctional zone does a lot of hormone synthesis. Do you think it is a larger non-functional junctional zone? Even though it is larger it might be accumulating misfolded proteins.

TYCKO: The problem here is that the effect appears non-cell autonomous. The very high Phlda2 expression is in the labyrinth; it occurs early in placental development. The job is being done early and by a layer of cells that is adjacent to the layer of cells that is affected most strongly in the phenotype. Presumably it is affecting signalling within the labyrinth and then has a secondary effect on the junctional cells. (Added in proof: alternatively, as this gene is expressed from early stages it could be that it affects net proliferation of a common precursor cell of both placental compartments and thereby influences junctional zone growth in a cell autonomous manner. At present there are no data to distinguish between these two possibilities.) In the human placenta Phlda2 is a strongly expressed protein, specifically in the cytotrophoblast of the free chorionic villi.

SIBLEY: I'm trying to get my head back to where we started, with David Barker's comments on the ovality of the placenta. I don't know if mouse placentas or other haemochorial placentas have ovality like a human placenta. If Phlda2 is having a role in the spongiotrophoblast and hormones, one question would be are those placentas more oval in David's terms than they are normal.

TYCKO: They do look a little different in their length versus width. I never thought it was important.

GIUSSANI: What about the resulting phenotype of the offspring?

TYCKO: Ros John has some evidence which she hasn't published yet that in these stunted pregnancies, where the stunting effect is disproportional on the placenta as opposed to the fetus, there is a metabolic syndrome-like phenotype in the offspring. This model potentially might be able to be developed into a useful IUGR mouse model. A general discussion of mouse models of IUGR might be interesting.

GIUSSANI: What about postnatal growth?

TYCKO: She doesn't have data on this.

LOKE: I am not clear in my mind whether a big placenta is *per se* a good thing, or whether it is a manifestation of some disruption of imprinting.

TYCKO: This comes back to the question of Ros John's data on the effect on the offspring, in the two models (overgrown and undergrown placentas). The question of to what degree a large placenta is a bad thing can be addressed directly by making these kinds of observations, testing the effect on the offspring of these two models where they are nicely reciprocal. There is the normal placenta, then the model of placental stunting, and the model of placental overgrowth, all due to the influence of a single imprinted gene. There may be other examples like this.

FERGUSON SMITH: There are quite a few examples of this. Big placentas don't normally associate with an effect on fetal growth while small placentas clearly do.

TYCKO: The question then is, is the big placenta not just innocuous, but actually dangerous?

JAUNIAUX: In Beckwith–Wiedemann syndrome, you have a situation in which there is a gigantic placenta. But you can find exactly the same phenomenon in perfectly normal pregnancies (i.e. the fetus looks perfectly normal). It is the stem villi mesenchyme that is abnormal in those cases and we have called this placental anomaly mesenchymal dysplasia.

TYCKO: Maybe there is a slightly different genetic or epigenetic mechanism there.

ROBERT BOYD: The big question is what that baby is like 5, 15, 25 years later, both in terms of growth and cardiovascular function.

JANSSON: Coming back to the significance of the regulation of placental imprinted genes, if they were important, wouldn't you expect that their expression wouldn't be changed or altered in cases of growth restriction? When people have looked they find little change of expression. When they do have changes they are not due to changes of imprinting or methylation. Doesn't this suggest that they might not be as important in the human as they are in the mice?

TYCKO: Silver–Russell syndrome is syndromic severe IUGR, clearly due to an imprinting defect. It may primarily be due to maternal vascular insufficiency. We looked at this type of IUGR, and a number of imprinted genes showed imbalanced expression. The signature genes that are the most strongly dysregulated in sporadic IUGR are not the imprinted ones. They are genes such as *LEPTIN*. But there are reproducible changes in expression imprinted genes in cases of sporadic human IUGR. *Phlda2* is probably the best example so far of this.

RICHARD BOYD: In terms of resource allocation, are there imprinted genes in mammary physiology that are relevant?

TYCKO: Yes. *Peg3* affects maternal nurturing behaviour. Looking for mammary gland specific-genomic imprinted is relevant, especially in marsupials. Even the chemical nature of the mother's milk in marsupials changes drastically through gestation. Changing the type of milk you nurture the conceptus on will lead to IUGR models. Just dissecting out mammary epithelium and doing imprinting analysis would be an interesting experiment.

HUNT: Some years ago we looked at TNF receptor knockout mice, and they too had aberrant placental morphology. But, to our dismay, there was no effect on the fetus. We did about 60 pregnancies, but when they grew up we were able to assess the animals from different combinations, and we discovered that the males were very sensitive to the loss of the TNF receptors but the females had no problems. Have you segregated according to gender?

TYCKO: No, but we should.

CRITCHLEY: Have you had the opportunity to stain a human growth-restricted placenta for the protein?

TYCKO: This was published in the *Placenta* Journal where we did the protein as well (McMinn J, Wei M, Schupf N, Cusmai J, Johnson EB, Smith AC, Weksberg R, Thaker HM, Tycko B. Unbalanced placental expression of imprinted genes in human intrauterine growth restriction. Placenta 2006 Jun-Jul; 27(6–7): 540–9. Epub 2005 Aug 24. PubMed PMID: 16125225). The cell-type specific expression remains accurate in the IUGR placentas. It is just overexpressed to a moderate degree in the IUGR-associated placentas. Glycogen becomes very interesting. What is the counterpart of the junctional zone and the glycogen cell in the human placenta? This is a key question.

CETIN: What did you find in human IUGR placenta?

TYCKO: Phlda2 protein is correctly localized. It is still seen primarily in the cytotrophoblast of the free chorionic villi, where it is supposed to be. It is just modestly overexpressed. This other question about glycogen seems to be annoyingly open.

CETIN: Glycogen in general is different in the human IUGR placenta. For example, in diabetic placenta there are definitely differences in glycogen.

ROBERT BOYD: David Barker, in any of your large datasets, is there a correlation between breast feeding duration/success and placental shape and polarity?

BARKER: That is a very good question, but we haven't examined it. We should do. I have a question for you: if it isn't always good to have a big placenta, which our data clearly suggest, is it ever good to have a big placenta?

ROBERT BOYD: I don't think we have an answer for this.

SIBLEY: Bigger babies do have bigger placentas, and a lot of big babies are quite normal. So it has to be something to do with placental growth and how this relates to the fetus.

FOWDEN: Having an efficient small placenta is a benefit.

Chapter 7

Genomic imprinting
Epigenetic control and potential roles in the developmental origins of postnatal health and disease

Elizabeth J. Radford and Anne C. Ferguson-Smith

Introduction

The 'thrifty phenotype' hypothesis of metabolic programming posits that fetal developmental plasticity moulds growth and metabolism appropriate to nutritional availability *in utero*, producing in times of nutritional scarcity a metabolic phenotype that minimizes energy expenditure and maximizes storage of scarce calories [1]. This improves the chances of survival to term and to reproduction, but at a cost of maladaptive consequences should the postnatal environment be discordant. Such developmental programming allows phenotypic adaptation to environmental changes over a single generation which may or may not persist over a few generations – a flexible, fast counterpart to natural selection. How mechanistically could an environmental change such as *in utero* nutritional availability alter the developmental programme of gene expression? Genetic mutation and selection is far too slow, rare and permanent. However, epigenetic alterations, which can result in stably inherited phenotypes without an associated change in coding sequence [2] are perfectly poised to form a cellular memory of an *in utero* insult.

Imprinted genes are a unique class of approximately 80 genes that which are expressed predominantly from one chromosome in a parental-origin-dependent manner and are attractive candidates for key players in developmental programming for the following reasons: first, imprinted genes have been shown to regulate the development of key metabolic organs, function postnatally in the control of metabolic axes, and to be exquisitely vulnerable to dosage modulation. Second, imprinted genes are fundamentally reliant on epigenetic mechanisms of expression control, and it has been suggested that this may render them particularly sensitive to environmental changes *in utero*.

Third, as germ cell reprogramming of the chromatin elements that control imprinting in the future generation occurs at least partially *in utero*, it has been postulated that deregulation of this process may be involved in transgenerational developmental programming. In this chapter we will explore what imprinted expression is, and the epigenetic mechanisms that control it. We will examine the reasons why imprinted genes have been proposed as key developmental programming genes, and the existing evidence for their involvement in the developmental origins of adult disease.

Imprinted genes are required for successful development of both the embryo and extraembryonic tissues

In 1984, three ground breaking papers were published showing that being diploid was not sufficient for normal development, and that chromosomes 'remember' their parental origin. This led the authors to propose that some form of 'imprinting' during gamete development differentiates the parental genomes, rendering the maternal and paternal chromosomes functionally non-equivalent [3–5]. By pronuclear transplantation the authors produced androgenetic and parthenogenetic/gynogenetic conceptuses that contained two paternal or two maternal pronuclei, respectively. Both combinations resulted in prenatal lethality, but the associated developmental phenotypes were strikingly different. Parthenogenetic embryos (and also gynogenetic, containing maternal genomes from two different conceptuses) were correctly patterned but growth restricted, and this was associated with very poor development of the extraembryonic tissues. In

The Placenta and Human Developmental Programming, ed. Graham J. Burton, David J. P. Barker, Ashley Moffett and Kent Thornburg. Published by Cambridge University Press. © Cambridge University Press 2011.

Chapter 7. Genomic imprinting

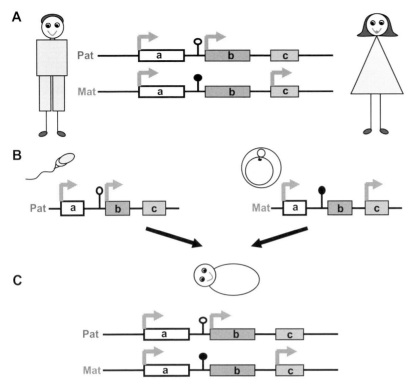

Figure 7.1 Genomic imprinting. (A) The majority of genes in the genome are like gene a, they are expressed or repressed biallelically from both parentally inherited chromosomes. A small group of genes are imprinted, they are expressed monoallelically according to their parental origin. In this case gene b is expressed only from the paternally inherited chromosome and gene c from the maternally inherited chromosome. This process is controlled by epigenetic marks which distinguish the two chromosomes, represented here by unfilled and filled lollipops. (B) In the mammalian gametes – sperm and egg – these epigenetic marks must be erased and re-established according to the sex of the individual. This ensures that all potential offspring (C) will inherit the appropriate complement of epigenetic marks. After fertilization, these marks are stable and heritable in all the somatic cells and maintain the imprinted state. (See colour plate section.)

contrast, androgenetic conceptuses were characterized by hyperplastic extraembryonic tissues but highly retarded, abnormal or even absent embryos. The dramatic phenotypes of these experiments clearly demonstrate the importance of having both maternally and paternally inherited chromosomes for normal embryonic and placental growth and development. This suggested the presence of genes that were expressed differently from maternally and paternally inherited chromosomes – imprinted genes.

Subsequent balanced translocation experiments produced uniparental disomies: euploid embryos where both homologues of a specific chromosomal region are inherited from one parent and absent from the other. Systematic production of uniparental disomies covering all chromosomal regions and analysis of the developmental phenotypes associated with each disomy, allowed those areas of the genome that harboured imprinted genes to be mapped [6]. This revealed that imprinted genes are not dispersed uniformly across the genome, but instead are largely grouped together in coordinately regulated clusters consisting of several protein-coding genes and at least one non-coding RNA gene (Figures 7.1 and 7.2). The parental allele-specific pattern of gene expression at each cluster is regulated by imprinting control regions (ICRs) through long-range *cis*-acting mechanisms. ICRs are characterized by differing epigenetic marks on the two parentally inherited chromosomes [7].

Mechanisms of imprinting control

Genomic imprinting is regulated by epigenetic mechanisms. We favour the definition of epigenetic as 'the structural adaptation of chromosomal regions so as to register, signal or perpetuate altered activity states' [2]. An epigenetic mark is a modification to the DNA

Chapter 7. Genomic imprinting

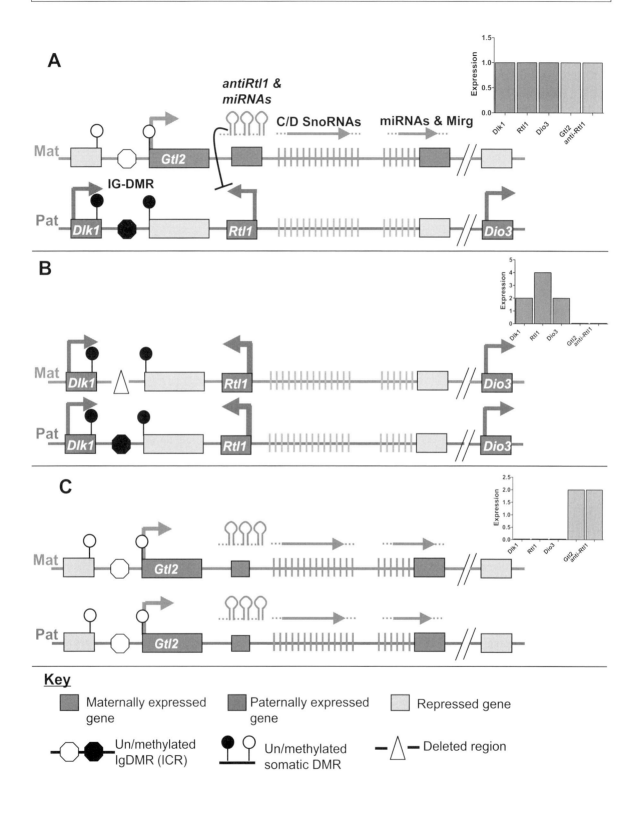

and/or chromatin that brings about a potentially heritable change in gene expression without altering the underlying sequence. DNA methylation and the post-translational modification of core histones are important epigenetic modifications. Non-coding RNAs are increasingly recognized as playing a key role in mediating altered genome activity states. These will be discussed below in the context of imprinting control.

DNA methylation

DNA methylation involves the addition of a methyl group to the fifth carbon in the cytosine pyrimidine ring. In vertebrate somatic tissues this largely occurs symmetrically on the two strands in CpG dinucleotides, although recently non-CpG methylation has been identified in embryonic stem cells [8]. All further discussion here will refer to CpG methylation.

To date, all ICRs identified have differentially methylated regions (DMRs) on the two parental chromosomes. These differential methylation marks originate from the gametes and, in normal circumstances, are maintained throughout life. ICRs can be divided into those which are methylated on the paternally inherited copy and those with maternally inherited methylation. The four paternally methylated ICRs identified to date are all intergenic elements, whereas the approximately 16 maternally methylated ICRs are situated at the promoters of genes [7]. The dependence of allele-specific patterns of gene expression on these ICRs has been demonstrated by experiments where the ICR has been deleted or the differential methylation pattern lost, resulting in an epigenotype switch. In the case of the *Dlk1-Dio3* locus on chromosome 12, the deletion of the maternally inherited unmethylated copy of the intergenic DMR (the ICR at this locus) results in the paternalization of the maternal chromosome, derepressing the normally paternally expressed protein-coding genes and silencing the maternally expressed non-coding RNAs, Figure 7.2B. Similarly, loss of methylation on the paternally inherited copy causes maternalization of the paternal chromosome, repression of the protein-coding genes and activation of the normally repressed non-coding RNAs, Figure 7.2C [9,10].

Secondary or somatic DMRs are promoter regions that do not acquire their differential methylation in the gametes, but develop parental-origin-specific methylation after fertilization during development. The acquisition of post-fertilization differential methylation requires the gametic ICR and is thought to reinforce imprinted gene expression. However, it is important to note that not all imprinted genes acquire a somatic DMR. Hence imprinted gene promoters can harbour gamete-derived differential methylation, post-fertilization somatic differential methylation, or be unmethylated at both parental alleles.

Histone modifications

A histone octamer of two H2A-H2B dimers and an H3-H4 tetramer acts as a scaffold around which DNA is wrapped into nucleosomes. Covalent modifications to key residues in the histone tails, especially those of H3 and H4, alters the interaction between adjacent nucleosomes and/or between histones and the DNA, changing the three-dimensional chromatin structure on both a global and a local scale. The resulting higher-order chromatin structures can be divided into tightly packed heterochromatic, transcriptionally silent domains, and euchromatic domains where the DNA is more accessible to nuclear protein complexes and thus the potential for transcription is maintained.

The dynamic modulation of histone modifications plays an important role in transcriptional control. Efforts have been made to divide histone modifications into repressive and active marks according to how

Figure 7.2 Imprinting control at the chromosome 12 *Dlk1-Dio3* domain (A) The imprinting control region of the *Dlk1-Dio3* locus is an intergenic germline differentially methylated region (IG-DMR) which is methylated on the paternally inherited chromosome and unmethylated on the maternally inherited chromosome. This results in the expression of the non-coding RNAs *Gtl2*, *anti-Rtl1*, *Mirg* and several clusters of microRNAs and SnoRNAs from the maternally inherited chromosome, whereas the protein coding genes *Dlk1*, *Rtl1* and *Dio3* are expressed from the paternally inherited chromosome. (B) The deletion of the maternally inherited unmethylated copy of the IG-DMR results in the paternalization of the maternal chromosome, the gain of methylation at post-fertilization somatic DMRs, de-repression of the normally paternally expressed protein-coding genes and silencing of non-coding RNAs on the maternally inherited chromosome. The loss of maternally expressed microRNA which normally reduce levels of *Rtl1* in *trans* results in approximately a fourfold upregulation of *Rtl1*, whereas *Dlk1* and *Dio3* are expressed at a double dose, reflecting their loss of imprinting. (C) Loss of methylation on the paternally inherited copy causes maternalization of the paternal chromosome, the loss of methylation at somatic DMRs, repression of the protein-coding genes and activation of the normally repressed non-coding RNAs [9,10]. (Adapted from Lin *et al.* 2003 [9].) (See colour plate section.)

they correlate with levels of transcriptional activity. Acetylation neutralizes the negative charge of lysine (K), disrupting tight packaging of chromatin, and is consistently associated with transcriptional activation. In contrast, the effect of lysine methylation is dependent on the residue and the histone modified and has been associated with both transcriptional activation: H3K4, H3K36, H3K79 and transcriptional repression: H3K9, H3K27, H4K20 [11]. However, this is an oversimplification and increasing evidence demonstrates that the impact of a particular histone mark is dependent on its genomic context [12].

Apart from descriptive studies associating active histone modifications with active imprinted alleles and repressive histone modifications with alleles repressed by imprinting, little is known about the role of histone modifications in imprinting control. Recently, a homozygous germline mutation in the H3K4 demethylase AOF1/KDM1B has been shown to perturb the acquisition of some maternal methylation imprints during oogenesis. This indicates a requirement for histone demethylation in order to establish germline CpG methylation at some ICRs [13]. Conversely, repressive histone modifications have been shown to precede and recruit DNA methylation, as exemplified by pluripotency gene silencing during embryonic stem cell differentiation [14].

With the exception of an essential role for ICRs, it is possible that cellular memory of transcriptional states and chromatin organization is to some degree tissue specific, with histone modifications being of greater importance in the extraembryonic tissues, and DNA methylation in the somatic lineages [15]. Several studies have shown imprinting in placental tissues to depend on the allele-specific accumulation of repressive histone marks under the control of the ICR [15,16]. The recent identification of a 'tri-mark', comprising the trimethylation of histone 3 at lysine 4 and 9 and the trimethylation of histone 4 at lysine 20 all known ICRs reignites the debate on the role of histone modifications in imprinting control [17].

Non-coding RNA

Pioneering work in yeast identified RNA-mediated mechanisms crucial for the control of the targeting of epigenetic modifications [18] and the heritability of histone modifications during replication [19]. Although these studies are in fission yeast that lack DNA methylation, small RNAs have also been shown to direct DNA methylation and are required for the silencing of transposable elements and other repetitive regions in mammalian germ cell development [20].

At imprinted loci, small non-coding RNAs have been implicated in dosage control, such as at the *Dlk1-Dio3* locus, where maternally expressed micro RNAs control the dosage of the paternally expressed *Rtl1* gene post-transcriptionally, Figure 7.2A [9]. In contrast much work has focused on the role of long non-coding RNAs, sometimes called macro non-coding RNAs, which are clearly associated with imprinting control at some loci. Imprinted expression of ncRNAs is required for long-distance bidirectional *cis*-acting control of transcription at four known imprinted clusters: the *Igf2r* domain, the *Kcnq1* domain, the *Snrpn* locus and the *Gnas* cluster. Targeted truncation or deletion of these large non-coding transcripts results in derepression in *cis* of protein-coding genes normally repressed on that chromosome. This is exemplified by the *Igf2r* locus, where the ncRNA *Airn* accumulates in *cis* at the promoter of the *Slc22a3* gene in the placenta, recruiting the histone H3K9 methyltransferase G9a, resulting in the silencing of the paternally inherited allele of *Slc22a3* [22].

Epigenetic reprogramming

Two episodes of genome-wide epigenetic reprogramming that are required for viability and reproductive success occur during early life. The first occurs in the zygote, and the second in the developing primordial germ cells of the embryo. There is increasing evidence that this reprogramming may be sensitive to environmental conditions, resulting in more heterogeneous outcomes than previously thought. It has been postulated that modification of epigenetic reprogramming by a compromised uterine environment may contribute to developmental programming and transgenerational inheritance of abnormal epigenetic states. This in turn could contribute to disease susceptibility, resulting, for example, in metabolic disease. Although these theories are highly attractive, the supporting evidence remains slim.

Preimplantation epigenetic reprogramming

Fertilization initiates a wave of epigenetic reprogramming affecting both the maternal and the paternal pronuclei. This involves a global loss of DNA methylation and alterations in histone modifications and is essential for totipotency. Demethylation of the paternal genome occurs extremely rapidly following fertilization and is believed to be an active process. Furthermore, during preimplantation stages the nuclear exclusion of the maintenance methyl-transferase DNMT1 required to catalyse methylation onto the nascent strand at DNA replication, results in the passive demethylation of the maternal genome through cellular division [22]. Normally, and important for maintaining their heritability, germline methylation marks at imprinted loci are protected from genome-wide preimplantation reprogramming. Several factors have recently been identified that can contribute to this process, including the zinc finger protein ZFP57 [10].

Nutritional availability around conception can affect adult phenotype

The Dutch hunger winter of 1944–5 provides a unique opportunity to examine the impact of *in utero* undernourishment on individuals conceived during the famine, and whether the timing of famine exposure affects outcome. There is some evidence of increased susceptibility to cardiovascular disease and earlier disease onset in individuals exposed to famine around conception. Exposure at this point has been associated with slight changes in methylation of three DMRs in different imprinted clusters in blood samples of affected versus unaffected siblings. However, the functional significance of this is unclear, as blood methylation is notoriously variable and the studies did not examine any associated expression changes or effects on imprinting, or attempt any correlation with known phenotypic outcomes [23].

Animal studies of peri-implantation developmental programming

Kwong *et al.* studied the effects of a peri-implantation low-protein diet on the cardiovascular health of rats. They found a male-specific reduction in birth weight and early onset of hypertension [24]. This was associated with reduced cell numbers at the blastocyst stage and a 30% reduction in blastocyst expression of *H19*, an imprinted ncRNA of unknown function. At term there was a 10% reduction in expression of hepatic *H19* and the reciprocally imprinted insulin-like growth factor 2, *Igf2*. The essential amino acid methionine is the penultimate carbon donor to the methyltransferase enzymes. Reduced methionine levels due to a low-protein diet may impinge on methyl-group availability for biological processes, including epigenetic modifications. Although *H19* DMR methylation was slightly altered, it did not correlate with the observed expression change, indicating that it was not mechanistically responsible. The expression data were only semiquantitative and the sample number was low, and so the relevance of the observed subtle changes in *Igf2* and *H19* expression to the cardiovascular phenotype are unclear.

The advent of *in vitro* embryo culture has had a huge impact on reproductive medicine and developmental biology. However, associations were reported between the use of assisted reproductive technologies and the increased incidence of rare imprinting syndromes [25]. Although these studies are controversial, as they are limited by small sample numbers and potentially confounded by the underlying causes of infertility, they have prompted much investigation into the impact of preimplantation culture and whether this represents an epigenetically labile period. Various investigators have since identified associations between *in vitro* culture and alterations in the expression levels of imprinted genes, sometimes accompanied by methylation changes and loss of imprinting [26]. However, all of these studies are potentially confounded by the use of superovulation, which can itself perturb imprinting [27]. Thus the extent of any susceptibility of zygotic epigenetic reprogramming to environmental modulation remains incompletely understood.

The strongest evidence for environmental modulation of epigenetic status *in utero* has come from analysis of the metastable alleles *Agouti* and *Axin Fused* (see below). As DNA methylation at these loci is somewhat stochastic, it is thought that they may be particularly sensitive to alterations in cellular epigenetic machinery. Gestational methionine supplementation increases the degree of methylation in the offspring at both of these alleles, altering both transcription and phenotype [28]. Further proof that changes in the intrauterine environment can produce epigenetic alterations even into late gestation was provided by Maclennan *et al.* [29]. Bilateral uterine artery ligation on day 19 of pregnancy in rat dams resulted in intrauterine growth restriction in the pups owing to the combination of hypoxia and nutritional deprivation. This was associated with altered hepatic one-carbon metabolism, global DNA hypomethylation and increased histone

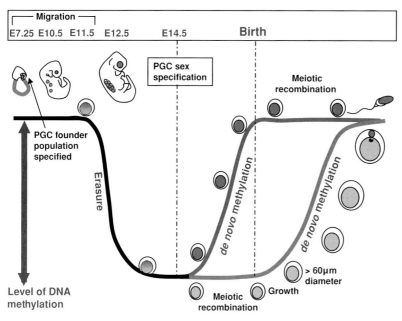

Figure 7.3 Reprogramming of DNA methylation at imprinting control regions during primordial germ cell development
Genome-wide epigenetic reprogramming, including at imprinting control regions during primordial germ cell (PGC) development is required for the viability of future offspring. This begins once the primordial germ cells (PGCs) have been specified at E7.25 and continues throughout the migration of the PGCs to the genital ridge. As the PGCs enter the ridge around E11.5 extensive demethylation occurs, including the dramatic and rapid erasure of methylation at imprinted loci. Following sex determination *de novo* methylation reinstates imprints according to PGC sex. In males this occurs between E14.5 and birth, concomitantly with *de novo* methylation of retrotransposon sequences in the G1 arrested PGCs. In the female germline methylation of maternally derived imprints occurs postnatally, concurrent with oocyte growth. Adapted from Bourc'his and Proudhon [76]. (See colour plate section.)

acetylation, although the transcriptional consequences of this were not assessed.

Epigenetic reprogramming in primordial germ cells

The existence of parental-origin-specific DMRs necessitates a process of epigenetic reprogramming during gamete development such that germ cells exhibit the appropriate maternally and paternally inherited epigenetic marks at ICRs to ensure the successful development of future offspring Figure 7.3. This begins once the primordial germ cells (PGCs) have been specified at E7.25 and continues throughout the migration of the PGCs to the genital ridge. A major wave of demethylation occurs around E11.5 and includes the dramatic and rapid erasure of methylation at imprinted loci [30]. Dogma dictates that this demethylation is complete, and that there is no epigenetic inheritance through meiosis. However, some repetitive elements, such as the retrotransposon-derived intracisternal A particles (IAPs) can partially escape this methylation reprogramming [31]. Analysis of the metastable alleles *Agouti* and *Axin Fused* has provided proof of principle that, at least at these IAP element-containing loci, alterations in the nutritional environment *in utero*, specifically changes in methionine availability, can result in epigenetic modulation of phenotype that is heritable to at least one further generation [32].

Following sex determination, *de novo* methylation reinstates imprints according to PGC sex. In males this occurs between E14.5 and birth concomitantly with *de novo* methylation of retrotransposon sequences in the G1-arrested PGCs. In the female germline methylation of maternally derived imprints occurs postnatally, concurrent with oocyte growth Figure 7.3 [30].

As the erasure and, at least in males, reacquisition of imprints in PGCs occurs *in utero* it is potentially vulnerable to alterations in the uterine environment. Evidence is accumulating for transgenerational developmental programming effects in human populations and experimental animal models. Analysis of maternal transmission is problematic, as altered maternal metabolism may recapitulate the initial insult [33]. However, transgenerational developmental programming through the paternal line has been identified in at least two different animal models involving nutrient restriction and glucocorticoid treatment [34,35]. Interestingly, in the glucocorticoid treatment model there was no inheritance to a third generation, implying that PGC reprogramming in the absence of an intrauterine insult was normal [35].

Imprinted genes play key roles in prenatal growth and development, in postnatal metabolism and in the regulation of energy balance, rendering them candidates for developmental programming

In mammals the evolution of imprinting coincides with the evolution of placentation. Given the dramatic phenotypic differences in extraembryonic tissue development between androgenetic and parthenogenetic conceptuses, and the large number of genes that show placental-specific imprinting, it is hypothesized that imprinting may have evolved to facilitate placentation. This is supported by the altered placental development of may murine imprinted gene knockouts and transgenics Figure 7.4. The placenta controls nutrient supply to the fetus and is the site of fetomaternal interaction. Alterations in placental development can therefore have a dramatic effect on fetal growth and maternal metabolism and behaviour; indeed, placental insufficiency is a leading cause of IUGR in the developed world. There is also evidence that interactions between imprinted genes act together in the fetus to regulate growth, thereby altering fetal demand for maternal resources. Imprinted genes also play critical roles in the development and postnatal function of key metabolic axes Figure 7.4. These varied roles have largely been revealed by the careful analysis and comparison of mouse genetic models where the dosage of a particular gene or cluster has been altered. These experiments will be discussed below, in addition to the existing data investigating whether imprinted genes play key roles in developmental programming.

Imprinted genes directly modulate fetal growth and affect fetal nutrient supply through placental development and function

Some imprinted genes are indispensable for the formation of extraembryonic tissues. Loss of the maternally expressed *Ascl2* or paternally expressed *Peg10* genes causes mid-gestational lethality due to the failure of placental development. *Ascl2* is required for the normal differentiation of spongiotrophoblast giant cells. By E9.5–10.5 $Ascl2^{-/+}$ placentas have excessive trophoblast giant cells but lack spongiotrophoblast in the junctional zone, impeding further placental development and resulting in embryonic lethality [36]. $Peg10^{+/-}$ conceptuses have a similar phenotype, although they have a normal number of giant cells. Overexpression of *Ascl2* and lack of *Peg10* in parthenogenetic conceptuses most likely contributes to their failure to make a normal placenta, and no *Ascl2* but two expressed copies of *Peg10* may promote hyperplasia in androgenetic extraembryonic tissue.

Other imprinted genes, albeit not essential for the formation of the placenta, do play important roles in placental growth, patterning and function. Indeed, there are three imprinted solute transporters, the organic cation transporters *Slc22a1/Impt1*, *Slc22a2/3* and the type A amino acid transporter *Slc48a4*. Altered dosage of these imprinted genes may be expected to alter nutrient transfer to the embryo. However, *Slc22a3* deletion has no effect on growth of the conceptus, presumably owing to transporter redundancy or placental plasticity [37].

Many imprinted genes play a direct role in fetal growth in addition to their role in placental development, which complicates the interpretation of genetic models. For example, *Grb10* expressed from the maternally inherited chromosome negatively regulates fetal growth through an *Igf2*-independent pathway [38]. Loss of *Grb10* produces embryo overgrowth accompanied by placental hyperplasia owing to an expansion in the labyrinthine zone where *Grb10* is expressed, resulting in increased placental efficiency which supports the increased fetal growth [39]. Conversely, loss of paternal expression of *Dlk1* results in placental growth restriction accompanied by dwarfism due to the combined effects of *Dlk1* in the placenta and embryo. Genetic models examining the roles of other imprinted

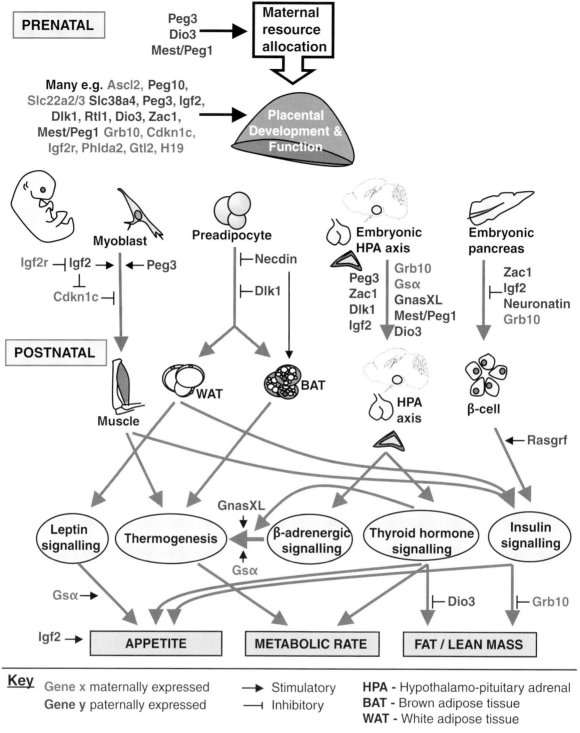

Figure 7.4 Imprinted genes act at multiple levels during prenatal and postnatal life to control energy homoeostasis
The correct dosage of imprinted gene expression is vital for energy homoeostasis during both prenatal and postnatal life. Imprinted genes play critical roles in the control of nutrient supply to the fetus through effects on maternal metabolism and energy partitioning, placentation and placental function. There is also evidence that interactions between imprinted genes act together in the fetus to regulate growth, thus altering fetal demand for maternal resources. Imprinted genes play key roles in the growth and differentiation of metabolic organs, with lasting implications for adult metabolism. Furthermore, there is increasing evidence that imprinted genes modulate the function of key adult metabolic pathways. These varied roles have largely been revealed by the careful analysis and comparison of mouse genetic models *in vivo* where the dosage of a particular gene or cluster has been altered. The terms maternally and paternally expressed refer to imprinted gene expression from maternally inherited and paternally inherited chromosomes. Adapted from Charalambous et al. [40]. (See colour plate section.)

genes, such as *Igf2*, *Igf2r*, *Zac1*, *Peg3* and *Peg1*, have uncovered similar intricate relationships [40]. Data from Varrault *et al.* suggest that the *Zac1* gene product can regulate the expression of *Igf2*, *H19*, *Cdkn1c*, *Grb10* and *Dlk1*, implying that *Zac1* may act as a nodal gene in a network of imprinted genes, coordinating the function of the two major growth pathways [41].

The complex coordinated regulation of the fetoplacental interface by imprinted genes is exemplified by *Igf2*. *Igf2* is a growth factor crucial for normal placental and fetal growth and development. It is expressed from the paternally inherited allele in all expressing fetal tissues (except the choroid plexus and leptomeninges of the brain where it is biallelically expressed) and is functionally antagonized by the reciprocally imprinted *Igf2r*. The *Igf2* gene has multiple alternative promoters. The fetal-type promoter controls expression in the fetus, the placental glycogen cells of the junctional zone and the labyrinthine zone endothelium, whereas the placental-specific promoter, P0, drives expression in the labyrinthine trophoblast, contributing 10% of placental *Igf2* [42]. The specific deletion of *Igf2P0* has allowed an elegant genetic dissection of the relative roles of *Igf2* in the fetus and placenta.

The *Igf2* knockout exhibits dwarfism. The size of the fetus and placenta are reduced to 50% and 60%, respectively, and the labyrinthine layer is disproportionately reduced [43]. Loss of *Igf2P0* reduces placental size by a similar amount (to 65%); however, it is proportional across all layers and is associated with increased barrier thickness and reduced passive diffusion of nutrients to the fetus. Compensatory upregulation of solute transporters allows the small placenta to support normal fetal growth up to E16. However, this increase in placental transfer efficiency cannot be sustained, resulting in 25% fetal growth restriction at birth [44]. Such placental functional adaptation is only possible if fetal IGF2 levels are normal [45]. Loss of fetal *Igf2* directly impairs fetal growth, reducing the demand for nutrients. Consequently, the placenta does not upregulate solute transporters and placental efficiency is unchanged. Postnatally, *Igf2P0*$^{+/-}$ animals catch up to normal size, whereas *Igf2*$^{+/-}$ animals remain dwarfed. Neither mutant has any reported postnatal metabolic phenotypes. Conversely, *Igf2r* knockouts show a 30% increase in birth weight and are neonatal lethal [46].

This clear demonstration of the importance of *Igf2* in placental and fetal growth led to the hypothesis that reduced *Igf2* expression in the placenta, the fetus or both may be implicated in the pathogenesis of IUGR. This has since been hotly debated, with a variety of studies finding that *Igf2* expression in placenta and/or cord blood correlates positively with birth weight, although others find no such relationship [47,48]. The discrepancies may partly be due to a failure to take into account the impact of changes in the levels of the circulating non-imprinted binding proteins that alter IGF2 bioavailability and which have been found in numerous studies to be associated with birth weight [48]. Fetal hepatic expression of these binding proteins is known to be exquisitely sensitive to fetal nutrition and insulin signalling. Undernourishment *in utero* results in increased expression of IGFBP1 and IGFBP2, which would be expected to sequester IGF2, thus matching fetal growth to nutritional status.

The *Phlda2/Ipl* imprinted gene expressed from the maternally inherited chromosome acts to restrain placental growth, whereas the reciprocally imprinted *Mest* gene promotes it [40]. Apostolidou *et al.* screened 200 human placentae by qPCR for *PHLDA2*, *IGF2*, *IGF2R* and *MEST*. Only *PHLDA2* significantly correlated (negatively) with birth weight, but no association was found between *PHLDA2* and placental weight. The change in *PHLDA2* expression was not associated with a loss of imprinting [49].

McMinn *et al.* assessed the transcriptome of a small sample of human IUGR and normal placentas. In contrast to the aforementioned study, they observed increased expression of *PHLDA2* and decreased expression of *MEST*, *MEG3*, *GATM*, *GNAS* and *ZAC1* in IUGR placentas. They observed no methylation changes at the *PHLDA2* or *MEST* ICRs, nor was the spatial distribution of PHLDA2 expression changed. Imprinted genes constituted 7% of their observed expression changes, a significantly higher proportion than would be expected, potentially implicating imprinted genes as a class in playing a key role in IUGR. However, some of these changes could not be confirmed with qPCR and others were not validated [50]. Furthermore, morphological adaptations occur in small placentas in an effort to sustain fetal growth [43] and thus the observed expression changes may be indirect, reflecting the altered morphology.

Imprinted genes affect fetal nutrient supply through alteration of maternal behaviour and resource mobilization

In addition to controlling nutrient transfer to the fetus the placenta is a highly active endocrine tissue, secreting factors that alter maternal metabolism and behaviour. Thus the developing fetus signals to its mother its demand for resources through the placenta, but there is also an independent drive for maternal weight gain during pregnancy. Imprinted genes have also been shown to play direct roles in both of these processes. Analysis of the *Peg3* imprinted gene mutant demonstrates the elegance of this complex interplay between fetal demand for maternal resources and maternal supply.

Peg3, which encodes a zinc-finger protein expressed from the paternally inherited chromosome, has been implicated in the control of β-catenin signalling and in the regulation of apoptosis through the p53 and TNF pathways. $Peg3^{+/-}$ pups had small placentas, reduced by 25% of wild type [51]. This was associated with failure of their wild-type mothers to gain weight appropriately during pregnancy, indicating that in the absence of conceptus *Peg3* expression, fetoplacental signalling of demand for maternal resources is impaired. However, $Peg3^{+/-}$ mothers bearing wild-type or maternal transmission pups also fail to lay down sufficient body weight reserves during pregnancy, resulting in growth restriction of their wild-type pups. This indicates that maternal *Peg3* is required to increase food intake appropriately in response to the placental endocrine signals of wild-type pups [52].

Imprinted genes affect the *in utero* development of key metabolic organs and the function of postnatal metabolic axes

Imprinted genes play critical roles in the development of key metabolic organs, with obvious consequences for postnatal metabolism. This includes organs involved in the control of homoeostatic metabolic axes such as the brain and neuroendocrine and endocrine organs, including the pituitary, adrenal and pancreas, as well as tissues such as muscle, white adipose tissue and liver which are critical for energy storage and utilization. Perturbations in the development and postnatal function of these tissues is thought to contribute to the metabolic sequelae of IUGR.

Imprinted genes are clearly important for the growth and development of muscle, as many genetic models show severe and contrasting phenotypes

Skeletal muscle is an important metabolic organ, being the prime site of glucose disposal. Lean mass is therefore a major determinant of insulin sensitivity, and altered development of skeletal muscle can have significant implications for postnatal glucose tolerance.

The ovine Callipyge mutation lies within the imprinted *Dlk1-Dio3* locus and causes hyperplasia of certain hindlimb muscles. Overexpression of *Dlk1* and *Rtl1* during muscle development in mutants has been implicated, although the relative contribution of these two paternally expressed genes remains to be resolved [53]. Similarly, maternal transmission of a null allele for *Grb10* results in increased lean mass, reduced adiposity and improved glucose tolerance, potentially through the disinhibition of insulin and IGF1 signalling [54]. The loss of *Peg3* has the opposite effect, reducing lean mass [55]. Although the mechanism involved is uncertain, *Peg3* has been implicated in the control of muscle stem cell number, cachexia and regeneration [56].

In the mouse, postnatal *Igf2* expression is normally restricted to the choroid plexus and leptomeninges of the brain, where it is not imprinted and where it has been implicated in the control of energy balance [57]. However, overexpression of *Igf2* postnatally, either in a transgenic model or due to a naturally occurring regulatory mutation, in the White Breed pig, results in increased muscle mass and, probably secondarily, reduced white adipose tissue [40].

White adipose tissue is an important endocrine organ in addition to being an energy store. Imprinted genes are critical for white adipose tissue development

Many imprinted gene models show altered adiposity. Some imprinted genes have a direct effect on adipose tissue development, whereas in other cases effects on white adipose tissue are secondary to changes in energy balance or insulin/IGF1 signalling, as discussed above [40,54,55].

Dlk1 (also known as *preadipocyte factor 1, Pref1*) is expressed at high levels in preadipocytes. Downregulation of *Dlk1* expression correlates with adipocyte differentiation *in vitro*, whereas over expression of

Dlk1 prevents differentiation in culture [58]. Loss of paternal expression of *Dlk1* is associated with increased adiposity [59], and overexpression of *Dlk1* under the control of its endogenous promoter results in reduced adipose tissue and lean, hypermetabolic animals [60]. However, it is unclear whether this is due to altered adipose development or secondary to the loss of *Dlk1* expression in other key metabolic tissues.

Jiminez Chillaron *et al.* demonstrated reduced *Dlk1* expression in the expanded adipose tissue of 4-month-old mice undernourished *in utero* [34]. However, *Dlk1* expression in adipose tissue during caloric restriction 3 weeks postpartum shows the opposite trend [61], indicating that these gene expression changes may reflect adiposity differences, or facilitate adipose tissue adaptation to altered metabolism. *Mest/Peg1* shows the opposite expression pattern to *Dlk1* as it is expressed in mature adipocytes and is associated with adipocyte expansion. *Mest* is upregulated during positive energy balance and downregulated in fasting [61]. Differential expression analysis of the adipose tissue of genetically identical inbred mice with differing levels of adiposity revealed upregulation of *Mest* in the adipose tissue of mice which subsequently developed obesity [62]. However, analysis of the metabolic phenotype of the paternal transmission of a *Mest* null allele is required to properly delineate the role of this gene in white adipose tissue.

Imprinted genes are critical for brown adipose tissue development

Brown adipose tissue (BAT) plays a vital role in the survival of small mammals such as rodents through non-shivering thermogenesis (NST). NST is especially critical during the neonatal stage, but plays a role throughout adulthood in the regulation of energy balance and temperature. The existence of functional brown adipose tissue in adult humans has recently been confirmed, with clear implications for metabolic health and disease.

The paternally expressed genes *Necdin* (involved in the imprinted Prader–Willi syndrome) and *Dlk1* and *Dio3* of the human chromosome 14 imprinted cluster are critical to the control of brown adipose development. *Necdin* interacts with the E2F transcription factors to block induction of $PPAR\gamma_1$. This is associated with increased *Dlk1* expression and the inhibition of BAT differentiation. Insulin/IGF1 suppresses *Necdin* and *Dlk1* expression, facilitating BAT maturation [63]. A murine regulatory mutation causing partial loss of imprinting of the *Dlk1–Dio3* cluster results in central hypothyrodism and the failure to downregulate BAT *Dlk1* expression during the transition to independent life. This impairs differentiation and thermogenic gene expression in BAT, with long-term consequences for thyroid axis function and energy balance [64]. These data emphasize the importance of correct imprinted gene dosage in prenatal and early postnatal life to determine appropriate functional 'set points' of metabolic axes.

Imprinted genes are crucial for the proliferation and maturation of the insulin secreting β-cells of the pancreas

Pancreatic sensitivity to blood levels of glucose, insulin, IGF1 and other hormones is critically important for metabolic health. The pancreas is a plastic organ and insulin resistance, associated either with pregnancy or with metabolic disease such as obesity, results in β-cell expansion and increased insulin secretion. Different individuals have different capacities for such β-cell adaptation, but once a ceiling of insulin production is reached, or β-cell failure commences, glucose intolerance ensues. Early life events may play a role in determining the capacity of adult pancreatic plasticity. Several animal models have demonstrated altered pancreatic development following *in utero* deprivation [34,65]. Imprinted genes play key roles in pancreatic development and maturation and may be involved in the pathogenesis of these defects.

IGF2 plays a critical role in pancreatic maturation. In rodents this takes place in the second week of postnatal life, but in humans occurs largely during the third trimester *in utero*. In the mouse the reduction in the level of circulating IGF2 postnatally is associated with a wave of apoptosis and the cessation of β-cell proliferation and gain of insulin sensitivity [66]. β-cell overexpression of *Igf2* results in islet hypertrophy and hyperinsulinaemia in early life; however, insulin release demonstrates reduced sensitivity to glucose and type 2 diabetes develops with age [67].

Martín *et al.* 2005 looked at the IGF axis and pancreatic function in rats that had been protein restricted *in utero* during the last week of gestation

[65]. These rats have a phenotype similar to that of local *Igf2* overexpression [67]. However, pancreatic *Igf2* mRNA expression was reduced and there was no change in hepatic expression or serum levels of IGF2, whereas pancreatic IGF1, IGF1R and IGFBP2 levels were increased. This was associated with an increase in the *in vitro* pancreatic mitogenic response to IGFs.

Two other imprinted genes modulate IGF signalling and are involved in pancreatic maturation. *Rasgrf* is required for the mitogenic response to IGF1 and *Rasgrf1$^{+/-}$* animals exhibit islet hyperplasia, reduced insulin output and impaired glucose tolerance. Conversely, the maternally expressed *Grb10* acts as a negative regulator of IGF1 and insulin signalling and is expressed in the developing pancreas. Ectopic overexpression of *Grb10* results in acinar atrophy [40].

Neuronatin is a paternally expressed imprinted gene widely expressed in neural and neuroendocrine tissues, skeletal muscle and the developing pancreas. NEURONATIN was found to act downstream of BETA2/NeuroD1, a basic helix–loop–helix transcription factor important in the development of the nervous system and the endocrine pancreas. siRNA knockdown of *Neuronatin* resulted in reduced glucose-stimulated insulin expression in an insulinoma cell line *in vitro* [68]. Waterland and Garza investigated the role of neonatal nutrition on pancreatic maturation by altering rat litter size during lactation [69]. Both overnourished and undernourished animals had impaired pancreatic islet glucose-stimulated insulin secretion. Expression of *Neuronatin* was found to be significantly reduced in the overnourished individuals. Given the phenotypic similarities with the aforementioned *in vitro* siRNA knockdown, reduced *Neuronatin* expression may have contributed to the insulin secretory defects.

Zac1 knockout mice show an IUGR phenotype and, as noted above, microarray meta-analysis suggested that *Zac1* is a key transcriptional modulator in a network of imprinted gene expression [41]. The *Zac1* locus is strongly associated with the rare disorder transient neonatal diabetes (TNDM). TNDM is thought to be caused by altered β-cell development and is characterized by severe growth retardation at birth, accompanied by hyperglycaemia requiring insulin treatment. Interestingly, although TNDM patients usually recover by 6 months of age, 40% develop type II diabetes later in life.

Imprinted genes in brain development and the central control of metabolic axes

In utero stress or deprivation and negative early life events have been associated in both humans and animals with lasting changes in behaviour, emotionality and increased susceptibility to various psychiatric diseases [70]. There is striking expression of imprinted genes in many areas critical for motivation, emotion and reward, such as the brainstem monoaminergic nuclei, the amygdala, nucleus accumbens and ventral tegmental area. Although there has been much speculation on the possible role of imprinted genes in these areas and hence in psychiatric illness, direct evidence of this is limited. However, *Dlk1* and *Grb10* have recently been shown to be involved in the development of the midbrain dopaminergic population, and loss of *Magel2* is associated with defects in serotonergic signalling [71,72].

There is increasing evidence that IUGR alters the central regulation of homoeostatic axes such as those involved in the control of blood volume, stress susceptibility and energy balance [73]. Many imprinted genes show high expression in key components of the hypothalamopituitary axis, and although several imprinted genetic mouse models show altered 'set points' of metabolic axes, to our knowledge there are currently no data linking IUGR with changes in the central nervous system expression of imprinted genes.

The *Gnas* imprinted locus on the mouse distal chromosome 2 has a complex pattern of gene expression, with parental-allele specific transcripts produced from four different initiation sites splicing onto the second exon of the gene. At this locus, *XLαs* and *Gsα* are reciprocally imprinted transcripts expressed from the paternal and maternal chromosomes, respectively. *XLαs* and *Gsα* code for alternative stimulatory G-protein isoforms and play opposing roles in the central regulation of energy expenditure and thermogenesis. Loss of the neuroendocrine tissue-specific G-protein *XLαs* impairs suckling, resulting in significant perinatal mortality. However, surviving individuals are lean and hypermetabolic owing to increased sympathetic drive [74]. In contrast, loss of *Gsα* expression results in obesity, glucose intolerance and multihormone resistance. *Gsα* has a tissue-specific pattern of imprinting, and the metabolic phenotype is more severe on maternal transmission of the null allele. Recent data indicate that *Gsα* is imprinted in the hypothalamic paraventricular nucleus, and loss of Gsα expression from the maternally inherited allele in the central nervous system is responsible for the obesity and glucose intolerance phenotype. Although altered paraventricular development cannot be excluded, it is likely that this is due to altered melanocortin 4 receptor signalling in the hypothalamic control of sympathetic output and energy expenditure [75].

Loss of *Peg3*, as described above, causes intrauterine growth retardation and impaired suckling and thermogenesis, which increases perinatal mortality and causes developmental delay [51]. Surviving animals show altered central control of energy balance. $Peg3^{+/-}$ animals lacking the paternally inherited expressed allele eat less and gain less weight postnatally, despite increased expression of the orexigenic (appetite stimulating) and reduced expression of the anorexigenic (appetite suppressing) neuropeptides of the hypothalamus. Puberty is delayed, as $Peg3^{+/-}$ animals require a higher level of adiposity to enter puberty, again indicative of a hypothalamic defect. Although the expression of the orexigenic and anorexigenic hypothalamic neuropeptides normalizes, food intake remains low throughout life, indicating that a lower 'set point' is reached. However, despite their lower appetite $Peg3^{+/-}$ animals develop obesity due to reduced energy expenditure. Core body temperature is significantly reduced and $Peg3^{+/-}$ animals are unable to respond to a cold challenge [55]. However, response to a peripheral norepinephrine challenge is normal, indicating that the defect in thermogenesis is due to impaired hypothalamic control of sympathetic output, reminiscent of the Gsα mutants. This is probably due to the altered function of the hypothalamic paraventricular nucleus where *Peg3* is strongly expressed. *Peg3* is also strongly expressed in adult muscle and has been reported to be expressed in adult white adipose tissue, two tissues of great importance in energy balance and feeding behaviour. The impact of *Peg3* on the development and function of these organs is unknown.

Recent work has demonstrated how the genes of the imprinted *Dlk1–Dio3* locus on mouse chromosome 12 locus act coordinately to regulate thyroid axis function and development during both embryonic and adult life. Thyroid hormones are crucial for the regulation of metabolic rate and thermogenesis, and also affect embryonic growth and development. High placental expression from mid-gestation of type III iodothyronine deiodinase (DIO3), converts bioactive T_3 and its precursor T_4 to inactive metabolites, acting as a barrier to maternal thyroid hormone. As thyroid hormones attenuate cellular proliferation and promote differentiation, it is thought that placental Dio3 expression acts to ensure the adequate growth of the fetus and to prevent premature cellular differentiation. A regulatory mutation causing partial loss of imprinting of the *Dlk1–Dio3* cluster in mice doubles the level at placental Dio3 activity, causing embryonic hypothyroidism. These animals exhibit postnatal central hypothyroidism and impaired brown adipose tissue development due (at least in part) to overexpression of *Dlk1*. As adults these animals have impaired release of thyroid hormone from their BAT, resulting in lifelong hypothyroidism, obesity and glucose intolerance [64].

Conclusion

The monoallelic restriction of imprinted gene expression is fundamentally dependent on epigenetic modifications that allow the transcriptional machinery to distinguish between the two parentally inherited chromosomes. The reprogramming of these marks in the developing gametes takes place at least partially *in utero*. It has therefore been hypothesized that this may render imprinted genes particularly vulnerable to intrauterine environmental insults that may impinge on the epigenetic machinery. It is known that imprinted genes such as *Igf2* play critical roles in determining both the fetal demand for and placental supply of nutrients. Genetic models have demonstrated the existence of networks of imprinted genes that act together to regulate development. Such experiments have also clearly shown how dosage sensitive imprinted genes are: alterations in expression level can have deleterious effects on *in utero* development and postnatal metabolism. Consequently, it has been suggested that altered expression of imprinted genes as a result of changes in the intrauterine environment may contribute to adult-onset diseases. It has also been suggested that the reprogramming of developing PGCs *in utero* may be affected by the uterine environment. Alterations in epigenetic reprogramming at imprinted loci have been postulated to play a role in transgenerational developmental programming.

The data currently available on the role of imprinted genes in developmental programming is insufficient to draw a firm conclusion as to whether imprinted genes are a key class of developmentally regulated genes contributing to adult onset disease. Most studies have been hampered by low sample size, but there is emerging evidence that imprinted genes such as *Phlda2* may be involved in altered placental development associated with IUGR. However, untangling cause and effect in such a morphologically plastic tissue is challenging [43].

Where there is some evidence of altered expression of imprinted genes in developmental programming, in the majority of cases this is not associated with substantial relaxation of imprinting, and does not consistently correlate with changes in DNA methylation. Therefore, any such changes in imprinted gene expression are likely to have been brought about by a transcription-factor-mediated mechanism. There is currently no evidence to suggest that imprinted genes are more vulnerable to environmental changes than other classes of genes. Indeed, it is conceptually possible that the converse is true: given the dependence of imprinted gene expression on epigenetic modifications, these may be more tightly safeguarded in the face of environmental perturbations during development. It is to be hoped that further research will expand our understanding of the mechanistic processes involved in developmental programming, such that we may attempt informed and effective clinical interventions to minimize the negative impact of fetal growth restriction.

References

1. **Hales CN, Barker DJ**. The thrifty phenotype hypothesis. *Br Med Bull* 2001; **60**: 5–20.
2. **Bird A**. Perceptions of epigenetics. *Nature* 2007; **447**: 396–8.
3. **Barton SC, Surani MA, Norris ML**. Role of paternal and maternal genomes in mouse development. *Nature* 1984; **311**: 374–6.
4. **McGrath J, Solter D**. Inability of mouse blastomere nuclei transferred to enucleated zygotes to support development in vitro. *Science* 1984; **226**: 1317–19.
5. **Surani MA, Barton SC, Norris ML**. Development of reconstituted mouse eggs suggests imprinting of the genome during gametogenesis. *Nature* 1984; **308**: 548–50.
6. **Beechey CV, Cattanach BM, Blake A, Peters J**. *MRC Harwell, Oxfordshire World Wide Web Site – Mouse Imprinting Data and References*, <http://wwwharmrcacuk/research/genomic_imprinting/> (2008).
7. **Edwards CA, Ferguson-Smith AC**. Mechanisms regulating imprinted genes in clusters. *Curr Opin Cell Biol* 2007; **19**: 281–9.
8. **Lister R** *et al*. Human DNA methylomes at base resolution show widespread epigenomic differences. *Nature* 2009; **462**: 315–22.
9. **Lin S** *et al*. Asymmetric regulation of imprinting on the maternal and paternal chromosomes at the Dlk1-Gtl2 imprinted cluster on mouse chromosome 12. *Nature Genet* 2003; **35**: 97–102.
10. **Li X** *et al*. A maternal-zygotic effect gene, Zfp57, maintains both maternal and paternal imprints. *Dev Cell* 2008; **15**: 547–57.
11. **Li B, Carey M, Workman JL**. The role of chromatin during transcription. *Cell* 2007; **128**: 707–19.
12. **Vakoc CR, Mandat SA, Olenchock BA, Blobel GA**. Histone H3 lysine 9 methylation and HP1gamma are associated with transcription elongation through mammalian chromatin. *Mol Cell* 2005; **19**: 381–91.
13. **Ciccone DN** *et al*. KDM1B is a histone H3K4 demethylase required to establish maternal genomic imprints. *Nature* 2009; **461**: 415–18.
14. **Kouzarides T**. Chromatin modifications and their function. *Cell* 2007; **128**: 693–705.
15. **Lewis A** *et al*. Imprinting on distal chromosome 7 in the placenta involves repressive histone methylation independent of DNA methylation. *Nature Genet* 2004; **36**: 1291–5.
16. **Pandey RR** *et al*. Kcnq1ot1 antisense noncoding RNA mediates lineage-specific transcriptional silencing through chromatin-level regulation. *Mol Cell* 2008; **32**: 232–46.
17. **McEwen KR, Ferguson-Smith AC**. Distinguishing epigenetic marks of developmental and imprinting regulation. *Epigenet Chromatin* 2010; **3**: 2.
18. **Martienssen RA, Zaratiegui M, Goto DB**. RNA interference and heterochromatin in the fission yeast *Schizosaccharomyces pombe*. *Trends Genet* 2005; **21**: 450–6.
19. **Kloc A, Zaratiegui M, Nora E, Martienssen R**. RNA interference guides histone modification during the S phase of chromosomal replication. *Curr Biol* 2008; **18**: 490–5.
20. **Kuramochi-Miyagawa S** *et al*. (DNA methylation of retrotransposon genes is regulated by Piwi family members MILI and MIWI2 in murine fetal testes. *Genes Dev* 2008; **22**: 908–17.
21. **Nagano T** *et al*. The Air noncoding RNA epigenetically silences transcription by targeting G9a to chromatin. *Science* 2008; **322**: 1717–20.

22. **Morgan HD**, **Santos F**, **Green K**, **Dean W**, **Reik W**. Epigenetic reprogramming in mammals. *Hum Mol Genet* 2005; **14** Spec No 1: R47–58.

23. **Heijmans BT** *et al*. Persistent epigenetic differences associated with prenatal exposure to famine in humans. *Proc Natl Acad Sci USA* 2008; **105**: 17046–9.

24. **Kwong WY**, **Wild AE**, **Roberts P**, **Willis AC**, **Fleming T**. Maternal undernutrition during the preimplantation period of rat development causes blastocyst abnormalities and programming of postnatal hypertension. *Development* 2000; **127**: 4195–202.

25. **Niemitz EL**, **Feinberg A**. Epigenetics and assisted reproductive technology: a call for investigation. *Am J Hum Genet* 2004; **74**: 599–609.

26. **Khosla S**, **Dean W**, **Brown D**, **Reik W**, **Feil R**. Culture of preimplantation mouse embryos affects fetal development and the expression of imprinted genes. *Biol Reprod* 2001; **64**: 918–26.

27. **Sato A**, **Otsu E**, **Negishi H**, **Utsunomiya T**, **Arima T**. Aberrant DNA methylation of imprinted loci in superovulated oocytes. *Hum Reprod* 2007; **22**: 26–35.

28. **Waterland RA** *et al*. Maternal methyl supplements increase offspring DNA methylation at Axin Fused. *Genesis* 2006; **44**: 401–6.

29. **MacLennan NK** *et al*. Uteroplacental insufficiency alters DNA methylation, one-carbon metabolism, and histone acetylation in IUGR rats. *Physiol Genomics* 2004; **18**: 43–50.

30. **Sasaki H**, **Matsui Y**. Epigenetic events in mammalian germ-cell development: reprogramming and beyond. *Nature Rev Genet* 2008; **9**: 129–40.

31. **Lane N** *et al*. Resistance of IAPs to methylation reprogramming may provide a mechanism for epigenetic inheritance in the mouse. *Genesis* 2003; **35**: 88–93.

32. **Rakyan VK** *et al*. Transgenerational inheritance of epigenetic states at the murine Axin(Fu) allele occurs after maternal and paternal transmission. *Proc Natl Acad Sci USA* 2003; **100**: 2538–43.

33. **Thamotharan M** *et al*. Transgenerational inheritance of the insulin-resistant phenotype in embryo-transferred intrauterine growth-restricted adult female rat offspring. *Am J Physiol Endocrinol Metab* 2007; **292**: E1270–9.

34. **Jimenez-Chillaron JC** *et al*. Intergenerational transmission of glucose intolerance and obesity by in utero undernutrition in mice. *Diabetes* 2009; **58**: 460–8.

35. **Drake AJ**, **Walker BR**, **Seckl JR**. Intergenerational consequences of fetal programming by in utero exposure to glucocorticoids in rats. *Am J Physiol Regul Integr Comp Physiol* 2005; **288**: R34–8.

36. **Guillemot F**, **Nagy A**, **Auerbach A**, **Rossant J**, **Joyner AL**. Essential role of Mash-2 in extraembryonic development. *Nature* 1994; **371**: 333–6.

37. **Zwart R**, **Verhaagh S**, **de Jong J**, **Lyon M**, **Barlow D**. Genetic analysis of the organic cation transporter genes Orct2/Slc22a2 and Orct3/Slc22a3 reduces the critical region for the t haplotype mutant t(w73) to 200 kb. *Mamm Genome* 2001; **12**: 734–40.

38. **Charalambous M** *et al*. Disruption of the imprinted Grb10 gene leads to disproportionate overgrowth by an Igf2-independent mechanism. *Proc Natl Acad Sci USA* 2003; **100**: 8292–7.

39. **Charalambous M** *et al*. Maternally-inherited Grb10 reduces placental size and efficiency. *Dev Biol* 2010; **337**: 1–8.

40. **Charalambous M**, **da Rocha ST**, **Ferguson-Smith AC**. Genomic imprinting, growth control and the allocation of nutritional resources: consequences for postnatal life. *Curr Opin Endocrinol Diabetes Obes* 2007; **14**: 3–12.

41. **Varrault A** *et al*. Zac1 regulates an imprinted gene network critically involved in the control of embryonic growth. *Dev Cell* 2006; **11**: 711–22.

42. **Moore T** *et al*. Multiple imprinted sense and antisense transcripts, differential methylation and tandem repeats in a putative imprinting control region upstream of mouse Igf2. *Proc Natl Acad Sci USA* 1997; **94**: 12509–14.

43. **Coan M** *et al*. Disproportional effects of *Igf2* knockout on placental morphology and diffusional exchange characteristics in the mouse. *J Physiol* 2008; **586**: 5023–32.

44. **Constancia M** *et al*. Placental-specific IGF-II is a major modulator of placental and fetal growth. *Nature* 2002; **417**: 945–8.

45. **Constancia M** *et al*. Adaptation of nutrient supply to fetal demand in the mouse involves interaction between the Igf2 gene and placental transporter systems. *Proc Natl Acad Sci USA* 2005; **102**: 19219–24.

46. **Barlow DP**, **Stoger R**, **Herrmann BG**, **Saito K**, **Schweifer N**. The mouse insulin-like growth factor type-2 receptor is imprinted and closely linked to the Tme locus. *Nature* 1991; **349**: 84–7.

47. **Giudice LC** *et al*. Insulin-like growth factors and their binding proteins in the term and preterm human fetus and neonate with normal and extremes of intrauterine growth. *J Clin Endocrinol Metab* 1995; **80**: 1548–55.

48. **Hung TY** *et al*. Relationship between umbilical cord blood insulin-like growth factors and anthropometry

in term newborns. *Acta Paediatr Taiwan* 2008; **49**: 19–23.

49. **Apostolidou S** *et al.* Elevated placental expression of the imprinted PHLDA2 gene is associated with low birth weight. *J Mol Med* 2007; **85**: 379–87.

50. **McMinn J** *et al.* Unbalanced placental expression of imprinted genes in human intrauterine growth restriction. *Placenta* 2006; **27**: 540–9.

51. **Li L** *et al.* Regulation of maternal behavior and offspring growth by paternally expressed Peg3. *Science* 1999; **284**: 330–3.

52. **Curley JP**, **Barton S**, **Surani A**, **Keverne EB**. Coadaptation in mother and infant regulated by a paternally expressed imprinted gene. *Proc Biol Sci* 2004; **271**: 1303–9.

53. **Fleming-Waddell JN** *et al.* Effect of DLK1 and RTL1 but not MEG3 or MEG8 on muscle gene expression in Callipyge lambs. *PLoS One* 2009; **4**: e7399.

54. **Smith FM** *et al.* Mice with a disruption of the imprinted Grb10 gene exhibit altered body composition, glucose homeostasis, and insulin signaling during postnatal life. *Mol Cell Biol* 2007; **27**: 5871–86.

55. **Curley J** *et al.* Increased body fat in mice with a targeted mutation of the paternally expressed imprinted gene Peg3. *Faseb J* 2005; **19**: 1302–4.

56. **Moresi V** *et al.* Tumor necrosis factor-alpha inhibition of skeletal muscle regeneration is mediated by a caspase-dependent stem cell response. *Stem Cells* 2008; **26**: 997–1008.

57. **Charalambous M** *et al.* An enhancer element at the Igf2/H19 locus drives gene expression in both imprinted and non-imprinted tissues. *Dev Biol* 2004; **271**: 488–97.

58. **Smas CM**, **Sul HS**. Pref-1, a protein containing EGF-like repeats, inhibits adipocyte differentiation. *Cell* 1993; **73**: 725–34.

59. **Moon YS** *et al.* Mice lacking paternally expressed Pref-1/Dlk1 display growth retardation and accelerated adiposity. *Mol Cell Biol* 2002; **22**: 5585–92.

60. **da Rocha ST** *et al.* Gene dosage effects of the imprinted delta-like homologue 1 (dlk1/pref1) in development: implications for the evolution of imprinting. *PLoS Genet* 2009; **5**: e1000392.

61. **Isganaitis E** *et al.* Accelerated postnatal growth increases lipogenic gene expression and adipocyte size in low-birth weight mice. *Diabetes* 2009; **58**: 1192–200.

62. **Koza RA** *et al.* Changes in gene expression foreshadow diet-induced obesity in genetically identical mice. *PLoS Genet* 2006; **2**: e81.

63. **Tseng YH** *et al.* Prediction of preadipocyte differentiation by gene expression reveals role of insulin receptor substrates and necdin. *Nature Cell Biol* 2005; **7**: 601–11.

64. **Charalambous M** *et al.* Dlk1 and Dio3 synergise to modulate the postnatal thyroid hormone axis. Submitted

65. **Martin MA** *et al.* Protein-caloric food restriction affects insulin-like growth factor system in fetal Wistar rat. *Endocrinology* 2005; **146**: 1364–71.

66. **Hill DJ** *et al.* Increased and persistent circulating insulin-like growth factor II in neonatal transgenic mice suppresses developmental apoptosis in the pancreatic islets. *Endocrinology* 2000; **141**: 1151–7.

67. **Devedjian JC** *et al.* Transgenic mice overexpressing insulin-like growth factor-II in beta cells develop type 2 diabetes. *J Clin Invest* 2000; **105**: 731–40.

68. **Chu K**, **Tsai MJ**. Neuronatin, a downstream target of BETA2/NeuroD1 in the pancreas, is involved in glucose-mediated insulin secretion. *Diabetes* 2005; **54**: 1064–73.

69. **Waterland RA**, **Garza C**. Early postnatal nutrition determines adult pancreatic glucose-responsive insulin secretion and islet gene expression in rats. *J Nutr* 2002; **132**: 357–64.

70. **Beydoun H**, **Saftlas AF**. Physical and mental health outcomes of prenatal maternal stress in human and animal studies: a review of recent evidence. *Paediatr Perinat Epidemiol* 2008; **22**: 438–66.

71. **Bauer M** *et al.* Delta-like 1 participates in the specification of ventral midbrain progenitor derived dopaminergic neurons. *J Neurochem* 2008; **104**: 1101–15.

72. **Kozlov SV** *et al.* The imprinted gene Magel2 regulates normal circadian output. *Nature Genet* 2007; **39**: 1266–72.

73. **Murgatroyd C** *et al.* Dynamic DNA methylation programs persistent adverse effects of early-life stress. *Nature Neurosci* 2009; **12**: 1559–66.

74. **Xie T** *et al.* The alternative stimulatory G protein alpha-subunit XLalphas is a critical regulator of energy and glucose metabolism and sympathetic nerve activity in adult mice. *J Biol Chem* 2006; **281**: 18989–99.

75. **Chen M** *et al.* Central nervous system imprinting of the G protein G(s)alpha and its role in metabolic regulation. *Cell Metab* 2009; **9**: 548–55.

76. **Bourc'his D**, **Proudhon C**. Sexual dimorphism in parental imprint ontogeny and contribution to embryonic development. *Mol Cell Endocrinol* 2008; **282**(1–2): 87–94.

Discussion

MAGNUS: I am curious about the influence on brain growth. There are some studies indicating that children of old fathers have an increased risk of neurodevelopmental disorders. Are there any possible imprinting mechanisms at work here?

FERGUSON SMITH: Yes. Two of the earliest imprinted disorders that were identified are the two reciprocally imprinted neurological disorders, Prader–Willi and Angelman syndromes. Angelman syndrome is a severe mental retardation defined by inappropriate bouts of laughter, ataxic gait and absence of speech. Prader–Willi syndrome is an obesity disorder. People are born hypertonic but subsequently they cannot control their appetite. This is caused by a cluster of imprinted genes that are predominantly imprinted in the brain. In another context, but also relevant for the brain, we have been working on the paternally expressed imprinted gene *DLK1*, which seems to be important in adult neurogenesis. Its dosage is important for regulating postnatal neurogenic processes. There is another cluster of imprinted genes on chromosome 2, the *GSα* cluster, that is important in regulating metabolism via central signalling processes. I think the brain is a very interesting place from an imprinting perspective, and we need to be watching this space in terms of epigenetic control of programming in brain. Because it is predominantly a post-mitotic tissue, it may have evolved more unusual mechanisms of epigenetic control that allow the control of dosage in particular cell types at particular times.

CRITCHLEY: From a clinical perspective, a year or two ago there was this association between children born by ART and an increased incidence of Angelman syndrome. What is the story here now?

FERGUSON SMITH: These imprinted syndromes have a very low frequency in the general population. In ART both Beckwith–Wiedemann and Angelman syndrome have increased incidences. This could be happening in one of two ways: in the female germline it could be through perturbations in the epigenetic states that are driven through the hormonal regimen used in ART, or another hypothesis suggests that defects could occur during preimplantation embryo culture. There is considerable interest in this area and prospective studies are in progress to look at this more closely.

TYCKO: The feeling is that this is recruitment of oocytes that might not otherwise make it. It is an oocyte phenomenon. The cases of Beckwith–Wiedemann syndrome that are ART-associated are those with loss of methylation of the *KvDMR1* imprinting control region (as if the imprint was not correctly executed in female gametogenesis). We never see cases of the other type of Beckwith–Wiedemann syndrome which involve gain of methylation upstream of *H19* on the maternal allele (a mistake that might occur at a postzygotic stage).

FLEMING: There are mouse studies which have shown that in healthy eggs and embryos culturing can disturb imprinted genes.

MOFFETT: In humans, do these imprinted genes show any appreciable polymorphism?

FERGUSON SMITH: For imprinted genes there is both intra- and interspecific polymorphism. For example, in humans the *IGF2R* gene is imprinted in some individuals and not in others. And as an example of interspecific polymorphism, in mouse several imprinted genes that are imprinted specifically in the placenta do not show imprinting in human. I think this means that imprinting may be more adaptable and variable than we originally thought.

Chapter 8

Trophoblast invasion and uterine artery remodelling in primates

Robert Pijnenborg, Lisbeth Vercruysse and Anthony M. Carter

Introduction: implantation and placentation in primates

Trophoblast invasion and remodelling of the uterine spiral arteries during pregnancy show marked species differences. Not only do invasion pathways (interstitial and endovascular) differ, but also the depth (decidual and myometrial) and extent (central, lateral and marginal) of the invasion. An additional important feature is the degree of decidual erosion, as this may lead to a shift in the position of spiral artery outlets and subsequent necrosis of superficial areas of decidua [1]. We consider trophoblast invasion to be a key process, not only for the elaboration of the placental bed, but also for shaping the basal plate of the placenta itself, which may have important consequences for maternal blood supply to the intervillous space. Ultimately the existing interspecies differences can only properly be understood within an evolutionary context, and such a perspective will undoubtedly also broaden our insights into human placentation. In this chapter we will try to identify some trends of placental evolution in primates by comparing placental organization between those with haemochorial placentas, with special attention to the great apes. Interspecies differences may also be relevant for understanding the impact of placental defects on fetal development in species that differ in normally having either shallow or deep trophoblast invasion.

Although New World monkeys (Platyrrhini, e.g. squirrel monkeys and marmosets), Old World monkeys (Cercopithecoidea, e.g. rhesus monkeys and baboons) and great apes (Hominoidea: gibbons, orangutans, gorillas, chimpanzees and humans) share a haemochorial type of placentation, they differ in some aspects of placental organization. New World monkey placentas are said to be 'pseudo-labyrinthine' because of the presence of intervillous connections ('trabeculae') [2], whereas the placental villi of Old World monkeys and great apes are entirely free-floating [3]. This difference has important repercussions for the intervillous flow (see next section). Although the organization of Old World monkey and great ape placentas is basically similar, the two groups differ in the depth of blastocyst implantation, which is superficial and interstitial, respectively. Trophoblast invasion during further placental development has been studied in a few Old World monkey species [4–7], but no published reports exist for great apes. Because trophoblast invasion in Old World monkeys seems to be mostly restricted to the decidua, it is relevant to investigate this phenomenon in great apes, the more so as a few published case reports indicate the occasional occurrence of pre-eclampsia – a disease believed to be linked to impaired trophoblast invasion [8,9] – in the gorilla and the chimpanzee [10–12]. Trophoblast invasion is a highly regulated process which is thought to be at least partially controlled by maternal cells such as uterine natural killer (uNK) cells. An early report on candidate gene families that might contribute to phenotypic differences between human and chimpanzee [13] highlighted the killer inhibitory receptors that are expressed on NK cells. This theme was further developed in a recent study of shifting uNK–trophoblast interactions in primate evolution [14]. These findings inspired one of us (AMC) to explore the availability of preserved pregnant uteri of great apes in museum collections [15] for investigating trophoblast invasion patterns. The most important source of material turned out to be the Museum für Naturkunde in Berlin, where J. P. Hill's [3] original collection of primate uteri is kept. Thanks to the cooperation of the curator, Dr Peter Giere, we were able to sample placental and placental bed tissues from preserved

The Placenta and Human Developmental Programming, ed. Graham J. Burton, David J. P. Barker, Ashley Moffett and Kent Thornburg. Published by Cambridge University Press. © Cambridge University Press 2011.

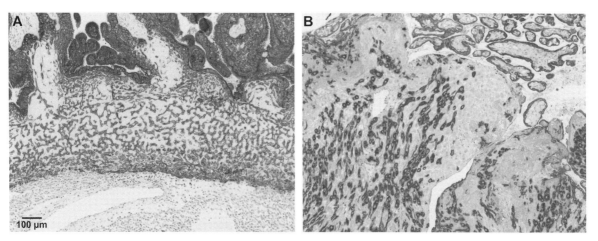

Figure 8.1 (A) Basal plate of a baboon placenta, showing an intact cytotrophoblastic shell and virtual absence of invaded interstitial trophoblast in the underlying decidua. (B) Basal plate of a human placenta, showing numerous interstitially invading trophoblasts in the decidua.

gorilla, chimpanzee, orangutan and gibbon specimens for histological evaluation. Because of the possible relevance to placental pathologies and pregnancy complications we focused particularly on trophoblast invasion and uterine spiral artery remodelling [16]. An important caveat of this study is that very few specimens could be examined, and that therefore only tentative conclusions can be drawn.

Maternal placental perfusion, trophoblast invasion and decidual erosion

In order to allow efficient oxygen and nutrient exchange, rapid bulk transport of maternal blood is directed towards the fetal side of the placenta, from where it returns more slowly towards the maternal side. In rodent labyrinthine placentas this rapid bulk transport occurs typically via maternal arterial channels, which are transplacental elongations of uterine spiral arteries. In primates with haemochorial placentas, similar maternal arterial channels are formed in the pseudo-labyrinthine placentas of tarsiers and New World monkeys [3,17,18]. In contrast, both the Old World monkeys and the Hominoidea (humans and great apes) possess free-floating villi and transplacental maternal channels are absent. Instead, the maternal blood enters the intervillous space as free 'spurts' from spiral artery outlets at the basal plate [19,20]. Such bulk transport to the fetal side of the placenta by intervillous spurts would not be as effective in the pseudo-labyrinthine placentas of the New World monkeys, as the intervillous connections may hinder free flow, but this potential problem is obviously circumvented by the formation of transplacental channels.

Another important difference between Old World and New World primates is that 'extravillous' trophoblast seems to be absent in the latter [3], although it has to be admitted that little information is available in the literature. Hill [3] pointed out that the maternal side of the placenta in New World monkeys consists of a syncytial mass, possibly directly derived from the 'primitive syncytium' which originates in the wall of the implanted blastocyst. Mononuclear cytotrophoblasts proliferate at the inner (fetal) side and grow into this syncytial mass, providing stem cells for villous development. In contrast to the Old World monkeys, great apes and humans, however, this proliferating cytotrophoblast does not penetrate the primitive syncytium to reach the decidua, and therefore neither extravillous columns nor a cytotrophoblastic shell are formed. In Old World monkeys a cytotrophoblastic shell is formed but remains intact throughout pregnancy, showing a clear demarcation from the underlying decidua where there is little interstitial invasion by extravillous trophoblast (Figure 8.1A) [7]. In early human pregnancy the cytotrophoblastic shell disintegrates as interstitial invasion of extravillous trophoblast starts into the decidua and progresses into the inner myometrium (Figure 8.1B). The eroding action of the shell on the decidua must be significantly

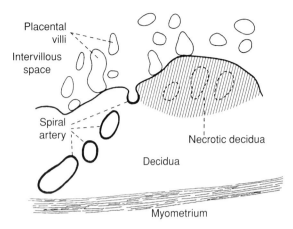

Figure 8.2 Diagram showing successive cross-sections of a decidual spiral artery in the human, opening into the intervillous space through a gap created halfway through the decidual segment. This figure was previously published by R. Pijnenborg in *Trophoblast Research* 11: 232, 1998 [1].

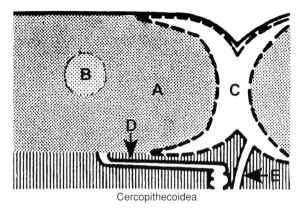

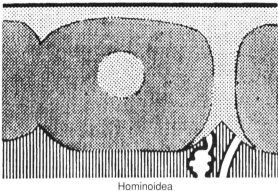

Figure 8.3 Diagrams of cross-sectioned placentas of Cercopithecoidea (Old World monkeys, top) and Hominoidea (great apes and humans, bottom), showing the position of spiral artery outlets, directed towards the centre of a placental lobule and an interlobular region respectively. A. Lobular region. B. Lobular centre. C. Interlobular region. D. Spiral artery. E. Vein. Modified from Gruenwald, *Am J Anat* 136: **145**; 1973 [18].

enhanced by this interstitial invasion, which may contribute by breaching the spiral artery walls to allow their opening into the intervillous space. In great apes the fate of the cytotrophoblastic shell has not so far been investigated.

In early anatomical studies of delivered human placentas, spiral artery outlets were assumed to be directed towards the centres of the placental lobuli [21,22], and it was therefore thought that the relatively villus-free lobular centres were created by the intralobular flow [22]. This was contradicted by Gruenwald's findings that in term human placentas most spiral artery outlets are situated in interlobular regions [23]. He made similar observations in great apes – although the situation was not very clear in the gibbons because of a poor definition of the lobular outlines – whereas in Old World monkeys spiral artery outlets were clearly directed to the lobular centres [17,18]. Expansion of the placental site in the course of pregnancy leads to an oblique stretching of the spiral arteries, especially underneath the more lateral placental areas. In the human, secondary outlets are created in these outstretched vessels [24,25], presumably by the eroding action of the interstitially invading trophoblast (Figure 8.2). The very limited interstitial invasion in Old World monkeys may be the reason why this shift of spiral artery outlets from intralobular to interlobular does not occur in such animals (Figure 8.3). Because a similar shift of arterial outlets occurs in both chimpanzees and gorillas [18], it was to be expected that extensive interstitial trophoblast invasion would also occur in these species. This was confirmed by our chimpanzee and gorilla specimens, which showed marked interstitial invasion of the decidua and the inner myometrium just as in the human [16]. This was not the case in gibbons, which may account for the ambiguous position of spiral artery outlets reported for the latter species [18]. There is a hint of restricted interstitial invasion in an orangutan specimen, but these results need to be substantiated.

Endovascular trophoblast invasion and spiral artery remodelling

There are no reports about a possible endovascular trophoblast invasion of spiral arteries in New World monkeys, which at first sight is not to be expected because of the alleged absence of extravillous trophoblast. One may consider whether the transplacental maternal channels, which communicate with the spiral arteries, might provide a possible pathway for endovascular invasion by placenta-derived cytotrophoblast, as happens in rodents such as the rat [26,27]. To our knowledge, however, the presence of endovascular trophoblast has never been reported in spiral arteries of New World monkeys.

Among the Old World monkeys endovascular invasion of spiral arteries has so far only been described in detail in rhesus monkeys and baboons [4,5]. Endovascular invasion starts much earlier in rhesus monkeys than in the human [28]. There is some uncertainty about the actual depth of invasion in this particular species [6], but in baboons the decidua-restricted endovascular invasion proposed by Ramsey was confirmed in a later study [7]. Little attention has been paid so far to early (decidua-associated) remodelling steps that might precede trophoblast invasion. As Ramsey pointed out that the decidua is much less developed in such species, decidua-associated remodelling might be limited, and not include the deeper non-invaded myometrial segments. In both rhesus monkeys and baboons the vascular smooth muscle of the invaded arteries is replaced by fibrinoid with intramurally embedded trophoblasts, which in these species must have an almost exclusive endovascular origin because of the paucity of interstitial trophoblasts.

Deep invasion and associated remodelling of spiral arteries have intensively been studied in the human [29–31], particularly because of their probable relevance in pre-eclampsia, where impaired invasion and vascular remodelling have been described [8,9]. Not only invasion depth *per se* has to be considered, but also the fact that there is a greater lateral extension of the placenta than in other primates [17], implying that more spiral arteries are incorporated in the placental bed which equally undergo deep invasion. A similar lateral extension of the placenta has been described in great apes [17], but the depth and spread of invasion have not so far been studied in these species. We can now confirm that in both the chimpanzee and the gorilla deep endovascular trophoblast invasion occurs in the decidua as well as the inner myometrium, similar to the human (Figure 8.4) [16]. Incorporation of this trophoblast in the vessel wall is also associated with a loss of vascular smooth muscle and deposition of fibrinoid. Because only a very limited number of often fragmented specimens could be examined, the

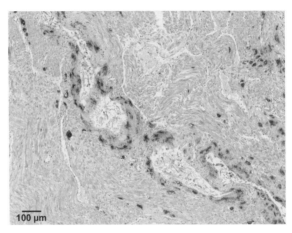

Figure 8.4 Section of a myometrial spiral artery in the placental bed of a chimpanzee. The original vascular smooth muscle has been replaced by trophoblast (cytokeratin immunostaining, black) embedded in fibrinoid material. Scattered interstitially invaded trophoblasts are also present.

exact pattern in time and lateral extent of invasion could not be elucidated. Nevertheless, there is a hint that, at least in the chimpanzee, endovascular invasion may be initiated later than in the human, as there was a striking absence of endovascular trophoblast in a specimen corresponding to a 12-week human pregnancy (according to fetal size) that was sampled at the centre of the placental bed [16]. Because the early endovascular migration in the human seems to coincide with rising placental oxygen [32], a later intravascular appearance of trophoblast in chimpanzees would imply a later onset of increased oxygen supply to the placenta. Furthermore, the degree of the early decidua-associated remodelling may be different in chimpanzees, as in a later specimen (corresponding to an 18-week human pregnancy according to fetal size) intraluminal trophoblasts were present in deep myometrial spiral arteries, which still showed a virtually intact vascular smooth muscle and elastic membrane. In the gibbon, endovascular invasion of the terminal decidual parts of a spiral artery was observed (Figure 8.5), but no trophoblast could be detected in the deeper decidua or the inner myometrium. This is in line with the limited interstitial invasion described in the previous section.

Potential advantages of increased invasion depth

Deep trophoblast invasion as described in the human has evolved in at least some of the great apes as well.

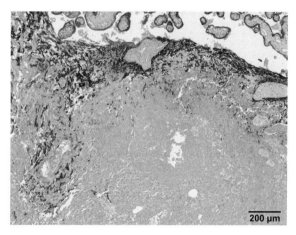

Figure 8.5 Trophoblast invasion (cytokeratin immunostaining, black) of a distal spiral artery segment close to the basal plate of a gibbon. Only little interstitial invasion has occurred into the decidual stroma immediately underneath the basal plate.

This raises questions about the advantage of this extra step for placental function and fetal development. It has been proposed that deeper invasion and the associated widening of a longer stretch of spiral arteries may have compensated for vena cava compression in hominid primates which had evolved bipedalism [33]. This idea is contradicted by the deep invasion in chimpanzees and gorillas, which are mainly knuckle-walkers, certainly during their pregnancy. Great apes and humans show more extensive brain development than other primates, and the question has been asked whether a more invasive placenta with a more reliable oxygen and nutrient supply might have been a contributory factor [34]. Because chimpanzees and gorillas have a similar deep invasion as humans, it is clear that this feature cannot explain the superior brain development in our own species as claimed by Robillard and colleagues [35]. Indeed, Martin pointed out that the very extensive brain development in humans occurs mainly postnatally, and may therefore not be an immediate result of deep placentation [34]. Conversely, a later onset of endovascular trophoblast invasion compared to the human, as suggested by our youngest chimpanzee specimen, might affect fetal brain development. Another possible advantage of deep invasion is to provide extra nutritional support for larger fetuses, which may be related to the previous advantage, as body size is also related to brain size [34]. There is no doubt that the Hominoidea superfamily includes the biggest of all existing primates, with the exception of gibbons, which lack deep invasion.

Pre-eclampsia in the human is associated with a lack of deep invasion of the majority of spiral arteries [8,9], although the primary cause of the disease is still unknown. In that respect it is interesting that cases of pre-eclampsia have been described in both the gorilla [10,11] and the chimpanzee [12], which both have deep invasion in normal pregnancy. Unfortunately, no placental bed biopsies were collected in these pathological cases. In pre-eclamptic human pregnancies not only the invasion depth is compromised [8], but also the lateral extent of invasion and spiral artery remodelling [36,32]. Also, in cases of pre-eclampsia in great apes an impaired lateral spread of this invasion may be a critical factor, but unfortunately no further information is available.

Because early-onset cases of pre-eclampsia in particular are associated with intrauterine growth restriction of the fetus, impaired spiral artery invasion and associated placental malfunctioning contribute also to the health risk of these babies in later life. This does not necessarily imply that failed trophoblast invasion *per se* may be the primary factor. It has been well established that trophoblast-associated remodelling of the spiral arteries is preceded by a decidua-associated remodelling as part of endometrial decidualization, starting around the time of implantation [31]. Optimal placental development depends on an intricate dialogue between trophoblast and surrounding maternal tissues, to alleviate the potential conflict of interest between mother and fetus in the course of pregnancy [37]. Decidualization in the human is more extensive than in rhesus monkeys and baboons, and has been related to the deeper trophoblast invasion in our own species [4]. One could speculate that a relatively early onset of decidualization, starting around the time of implantation in the human, is an essential preparation for an interstitial rather than a superficial type of implantation as occurs in rhesus monkeys and baboons. Interstitial implantation occurs in all great apes [38,39], including the gibbons, but no information is available about early decidual changes during the luteal phase of the cycle in these species. Efficiency of decidualization may be an important factor with possible long-term effects on fetuses developing in a suboptimal intrauterine environment.

Conclusion

Placental evolution has apparently followed diverging pathways in different primate groups, the closest similarity being found between Old World monkeys and great apes. However, deep invasion to include the inner (junctional zone) myometrium has only been observed so far in humans, chimpanzees and gorillas. Failure of deep invasion and impaired spiral artery remodelling have been shown to be associated with pre-eclampsia in the human. A few cases of pre-eclampsia have been described in chimpanzees and gorillas, but no information is available about a possibly impaired invasion depth or defects in spiral artery remodelling in these animals. It is to be expected that invasion-related defects must interfere with a regular oxygen and nutrient supply to the developing fetus, and may therefore have an impact on the quality of later life.

References

1. **Pijnenborg R.** The human decidua as a passage-way for trophoblast invasion. *Trophoblast Res* 1998; **11**: 229–41.
2. **Wislocki GB.** Remarks on the placentation of a platyrrhine monkey (*Ateles geoffroyi*). *Am J Anat* 1925; **36**: 467–87.
3. **Hill JP.** The developmental history of the Primates. *Phil Trans Roy Soc Lond* 1932; B221: 45–178.
4. **Ramsey EM, Houston ML, Harris JWS.** Interactions of the trophoblast and maternal tissues in three closely related primate species. *Am J Obstet Gynecol* 1976; **124**: 647–52.
5. **Enders AC, Lantz KC, Schlafke S.** Preference of invasive cytotrophoblast for maternal vessels in early implantation in the macaque. *Acta Anat (Basel)* 1996; **155**: 145–62.
6. **Blankenship TN, Enders AC, King BF.** Trophoblastic invasion and the development of uteroplacental arteries in the macaque: immunohistochemical localization of cytokeratins, desmin, type IV collagen, laminin and fibronectin. *Cell Tissue Res* 1993; **272**: 227–36.
7. **Pijnenborg R, D'Hooghe TD, Vercruysse L, Bambra C.** Evaluation of trophoblast invasion in placental bed biopsies of the baboon, with immunohistochemical localization of cytokeratin, fibronectin, and laminin. *J Med Primatol* 1996; **25**: 272–81.
8. **Brosens IA, Robertson WB, Dixon HG.** The role of spiral arteries in the pathogenesis of preeclampsia. In: Wynn RM, ed. *Obstetrics and Gynecology Annual*. New York: Appleton-Century-Crofts, 1972; 177–91.
9. **Robertson WB, Brosens I, Dixon G.** Uteroplacental vascular pathology. *Eur J Obstet Gynecol Reprod Biol* 1975; **5**: 47–65.
10. **Baird JN Jr.** Eclampsia in a lowland gorilla. *Am J Obstet Gynecol* 1981; **141**: 345–6.
11. **Thornton JG, Onwude JL.** Convulsions in pregnancy in related gorillas. *Am J Obstet Gynecol* 1992; **167**: 240–1.
12. **Stout C, Lemmon WB.** Glomerular capillary endothelial swelling in a pregnant chimpanzee. *Am J Obstet Gynecol* 1969; **105**: 212–15.
13. **Varki A, Altheide TK.** Comparing the human and chimpanzee genomes: searching for needles in a haystack. *Genome Res* 2005; **15**: 1746–58.
14. **Abi-Rached L, Parham P.** Natural selection drives recurrent formation of activating killer cell immunoglobulin-like receptor and Ly49 from inhibitory homologues. *J Exp Med* 2005; **201**: 1319–32.
15. **Carter AM.** Sources for comparative studies of placentation I. Embryological collections. *Placenta* 2008; **29**: 95–8.
16. **Pijnenborg R, Vercruysse L, Carter AM.** Deep trophoblast invasion and spiral artery remodelling in the placental bed of the chimpanzee. In preparation.
17. **Gruenwald P.** Expansion of placental site and maternal blood supply of primate placentas. *Anat Rec* 1972; **173**: 189–204.
18. **Gruenwald P.** Lobular structure of haemochorial primate placentas, and its relation to maternal vessels. *Am J Anat* 1973; **136**: 133–52.
19. **Ramsey EM.** Circulation in the maternal placenta of the rhesus monkey and man, with observations on the marginal lakes. *Am J Anat* 1956; **98**: 159–90.
20. **Ramsey EM, Donner MW.** *Placental Vasculature and Circulation*. Stuttgart: Georg Thieme, 1980.
21. **Wilkin** Contribution à l'étude de la circulation placentaire d'origine foetale. *Gynécol Obstét* 1954; **53**: 239–63.
22. **Reynolds SRM.** Formation of fetal cotyledons in the hemochorial placenta. A theoretical consideration of the functional implications of such an arrangement. *Am J Obstet Gynecol* 1966; **94**: 425–39.
23. **Gruenwald** The lobular architecture of the human placenta. *Bull Johns Hopkins Hosp* 1966; **119**: 172–90.
24. **Harris JWS, Ramsey EM.** The morphology of human uteroplacental vasculature. *Contrib Embryol* 1966; **38**: 43–58.
25. **Pijnenborg R, Dixon G, Robertson WB, Brosens I.** Trophoblastic invasion of human decidua from 8 to 18 weeks of pregnancy. *Placenta* 1980; **1**: 3–19.
26. **Caluwaerts S, Vercruysse L, Luyten C, Pijnenborg R.** Endovascular trophoblast invasion and associated

structural changes in uterine spiral arteries of the pregnant rat. *Placenta* 2005; **26**: 574.

27. **Vercruysse L, Caluwaerts S, Luyten C, Pijnenborg R.** Interstitial trophoblast invasion in the decidua and mesometrial triangle during the last third of pregnancy in the rat. *Placenta* 2006; **27**: 22–33.

28. **Blankenship TN, Enders AC.** Modification of uterine vasculature during pregnancy in macaques. *Microsc Res Tech* 2003; **60**: 390–401.

29. **Brosens I, Robertson WB, Dixon HG.** The physiological response of the vessels of the placental bed to normal pregnancy. *J Pathol Bacteriol* 1967; **93**: 569–79.

30. **Pijnenborg R, Bland JM, Robertson WB, Brosens I.** Uteroplacental arterial changes related to interstitial trophoblast migration in early human pregnancy. *Placenta* 1983; **4**: 397–414.

31. **Pijnenborg R, Vercruysse L, Hanssens M.** The uterine spiral arteries in human pregnancy: facts and controversies. *Placenta* 2006; **27**: 939–58.

32. **Pijnenborg R, Vercruysse L, Hanssens M, Brosens I.** Endovascular trophoblast and preeclampsia: a reassessment. *Pregnancy Hypertension* 2010; **1** (in press).

33. **Rockwell LC, Vargas E, Moore LG.** Human physiological adaptation to pregnancy: inter-and intraspecific perspectives. *Am J Hum Biol* 2003; **15**: 330–41.

34. **Martin RD.** Evolution of placentation in primates: implications of mammalian phylogeny. *Evol Biol* 2008; **35**: 125–45.

35. **Robillard PY, Dekker GA, Hulsey TC.** Evolutionary adaptations to preeclampsia/eclampsia in humans: low fecundability rate, loss of oestrus, prohibitions of incest and systematic polyandry. *Am J Reprod Immunol* 2002; **47**: 104–11.

36. **Brosens IA.** The uteroplacental vessels at term: the distribution and extent of the physiological changes. *Trophoblast Res* 1988; **3**: 61–7.

37. **Haig D.** Altercation of generations: genetic conflicts of pregnancy. *Am J Reprod Immunol* 1996; **35**: 226–32.

38. **Strahl H.** Primaten-Placentan. In: Hubrecht AAW, Strahl H, Keibel F eds. *Menschenaffen (Anthropomorphae) Studien über Entwicklung und Schädelbau*. Wiesbaden: CW Kreidel's Verlag, 1903; 415–91.

39. **Luckett WP.** Comparative development and evolution of the placenta in primates. *Contrib Primatol* 1974; **3**: 142–234.

Discussion

BURTON: One theory as to why deep invasion might occur is the conflict of the role of the spiral arteries in restricting blood loss at menstruation and providing free blood flow during pregnancy. To what extent do these higher apes also menstruate, and is there any correlation with the degree of invasion?

PIJNENBORG: The great apes do menstruate, as do Old World monkeys. The famous textbook of mammalian reproduction by S. A. Asdell (*Patterns of Mammalian Reproduction*, 2nd ed. Cornell University Press, 1964) is a good source of all these sorts of data. Menstruation has been described in most of the monkeys and apes studied, but the amount of menstrual bleeding may differ a great deal in different species. I am not sure about the New World monkeys. Bats also menstruate.

Moore: The speculation you were making about bipedalism – is the point that pre-eclampsia may be restricted to humans because we are the only bipedal primate? This was a point we made some years ago (Rockwell LC, Vargas E, Moore LG. Human physiological adaptation to pregnancy: inter- and intraspecific perspectives. *Am J Human Biology* 2003; 15: 330–341) based on the idea that bipeds are reliant upon spiral artery conversion (and other factors lowering uteroplacental vascular resistance) late in pregnancy to raise blood flow because cardiac output is declining, at least in some postures. Bipedalism imposes a burden on humans that great apes don't have.

HUNT: The question of why trophoblasts invade deeply in some species and not others is an entrancing one, but I am also interested in the mechanisms that are in place to facilitate invasion. I know you have done some work on various cytokines and interactions with cells in the decidua. Can you comment on this?

PIJNENBORG: I am glad you picked that up. I am sure that decidualization is an important factor. It goes back to Elizabeth Ramsey's 1976 paper (*Am J Obstet Gynecol* 1976; 124: 647–2) in which she compared the species she had available. She made the comment that in rhesus monkeys or baboons the decidua is much less differentiated and developed than in humans. She tried to explain this by suggesting that decidualization is needed as a support for deep invasion. The late Bill Robertson used to say that decidua and trophoblast are not enemies, they are good friends. This was completely different from the current thinking in the 1970s, which was mainly based on mouse data, that decidua restricts trophoblast invasion. However, R. L. Jones *et al.*

(*Endocrinology* 2006; 147: 724–2) published data showing that decidual cells in the human do produce factors such as activins that stimulate trophoblast cells to produce metalloproteinases. I am convinced that the event of trophoblast invasion is a question of trophoblast–maternal cell interactions. The inner myometrium is now increasingly considered to be a special compartment, undergoing changes in early pregnancy which are linked to the decidualization process. The muscle walls of the spiral arteries in the inner myometrium show a marked disorganization preceding trophoblast invasion. Some obstetricians like to call this part of the myometrium the junctional zone. In contrast to what happens in the decidua, the early vascular changes in the junctional zone myometrium cannot be affected by uterine NK cells because they are not present in that area, but macrophages may have a role to play.

ROBERT BOYD: Trying to relate these fascinating data to programming, two thoughts enter my mind. If such closely related species have such significant differences, placental shape *per se* seems unlikely to be related to programming, when we get programming in species such as mice that are so far away. Second, gestational length is of interest. If you move away from the tropics to areas where food is seasonal, you are likely to get a different sort of programming for those with a long gestational period compared with those with a short gestational period.

CRITCHLEY: I accept that these are very few specimens, and they are old and we don't know how they were fixed. Classic antibodies that we might use to look for the presence of NK cells or macrophages might be difficult to use. But even at a histological level, did you see an immune cell infiltrate close to the spiral arteries?

PIJNENBORG: Yes: small cells with dark nuclei.

MOFFETT: Can you do a phloxine tartrazine stain to identify uterine NK cells?

PIJNENBORG: I am not sure. This staining may show the typical granules of uterine NK cells, provided they have been preserved in this archival material.

CRITCHLEY: They will always be present where there is marked decidualization.

THORNBURG: I'd like to come back to the topic of placental efficiency. Do you think it would be possible in these primate models to have a bigger, wider spiral artery to make up for the fact that there are fewer of them? In other words, is it possible that a placenta could work better with fewer spiral arteries depending on how they are placed in the placenta, to make it more efficient?

PIJNENBORG: Great apes have numerous spiral arteries. There are indeed fewer spiral arteries in the Old World monkeys such as baboons. Whether or not this relates to placental efficiency is unclear, but all these placentas must be efficient enough to allow survival of the species.

THORNBURG: What biological constraints are placed on a placenta to become efficient? One might be the way it can remodel its arteries; another might be how many of these units it can make.

SIBLEY: I'd like to throw the geometry of the blood flow into the mix. Where you have the more human-like model with multivillous pool flow, one seems to have spiral arteries going in parallel with the channels, which might lead to a classical concurrent or countercurrent flow system.

PIJNENBORG: Yes, this is the situation in rodents.

MOORE: Further exploration of New World monkeys might be informative. They span such a large range, from miniature marmosets to howler monkeys. Some of these questions of geometry might be answered.

PIJNENBORG: I would love to see more studies on them. There's nothing in the literature. Placental blood flow in New World monkeys is in a sense comparable with the rat and mouse placentae. Spiral arteries are prolonged as transplacental channels and there are countercurrent flow systems. This is completely different to Old World monkeys and great apes.

SIBLEY: It would be interesting to know the fetal: placental weight ratio in New World monkeys. In the countercurrent models in the rodents the fetal: placental weight ratio was much greater than in the human with the multivillous system. Are there data on this?

PIJNENBORG: I don't think that information exists.

MOFFETT: This erosion of the terminal ends of the spiral arteries seems to be a particularly human feature. Is that related to this layer of superficial necrosis seen just beneath the placenta?

PIJNENBORG: Basically this layer is a mixture of fibrinoid material and fibrin. I have not been able to see continuity between this layer and a spiral artery in any specimen that I have looked at.

MOFFETT: Is it responsible for the erosion of the arteries?

PIJNENBORG: No. Erosion of the arteries is below this superficial necrotic layer. The trophoblasts move deeper and breach the walls of spiral arteries. Superficial necrosis of the overlying tissues may be secondary to the invasive process, which also causes shifting of the arterial outlets. In my view, the superficial fibrinoid layer is just dead material left over.

BARKER: Robert, you made an interesting point about length of gestation. It is true that a given maternal/fetal placental phenotype will predict things differently if the baby was born before 40 weeks or after. Breast cancer is associated with being born at 40 weeks or later; so too is polycystic ovary disease. In Finland, between 1924 and 1944, the length of gestation increased by one week, from 39 to 40 weeks. Robert Pijnenborg, you said that the use of the word invasion was optional, and while it was customary for placentologists to think of invasion it might in fact be seduction by the mother, who puts out things that entice so-called 'invasion'.

PIJNENBORG: Invasion is not just the inherent property of cells such as cancer cells, it is the result of interactions between fetal cells and maternal cells.

BARKER: It is more 'movement' than 'invasion' then.

LOKE: The picture I see is that nutrients in the maternal blood can only be effective to fetal growth if they can get through to the fetus. The placenta can affect this passage in two ways. One is what we have already discussed: there could be down regulation of certain molecules such as transporters, which make it less efficient to transport whatever there is in the maternal blood. The second way is that placental cells do not invade deep enough to convert the spiral arteries. In my mind, there are two mechanisms by which the placenta can affect fetal growth, regardless of maternal nutrition.

MOFFETT: I'd reverse them and put the latter first, upstream of the other.

ROBERT Boyd: That might be a specialty bias. If you have more transporters you can cope with less invasion.

TYCKO: It may be that maternal cells attract invasion, but it is clear from the imprinting data that maternal alleles are attempting to repel invasion. The data I showed were for a maternally expressed imprinting gene which is reducing the ability of trophoblast cells to invade the decidua. If decidual cells are attracting invasion, they may be doing so because they are instructed to do so by paternal alleles in those decidual cells.

PIJNENBORG: Are you talking about humans or mice?

TYCKO: We can only do the experiments in mice, but your data suggest to me it would be interesting to study the evolution of imprinting, even among primates. Some of the genes that are dictating this invasiveness are likely to be imprinted genes.

PIJNENBORG: I am a bit unhappy about the continuous use of rodent models for humans, because there may be a very different sort of balance struck. The whole idea of decidual barrier was developed in mice, for example, and it doesn't reflect what happens in humans.

TYCKO: The details will be different, but the concept of a process that drains maternal resources more efficiently or protects maternal resources will be the same.

JAUNIAUX: Do you have any data on monkey obstetrics? We know that we have 70% miscarriages in early human conception but as far as I know there is very little information on miscarriage in monkey. Miscarriages are found to be associated with a profound defect in all trophoblastic biological activities including trophoblastic invasion.

PIJNENBORG: Unfortunately not. That is an interesting point.

JACKSON: When does the placenta know when to stop?

PIJNENBORG: It is not the first time that this question has been asked. No one knows. There are two end points of invasion. The first end point is the fusing together of interstitially invading trophoblasts to form multinuclear giant cells, which are no longer invasive. How this is controlled is not known. There are indications that some cytokines may stimulate more rapid fusion to giant cells. The other end point is what we observe in endovascular trophoblast invasion: these cells remain mononuclear but end up by being buried in fibrinoid.

JACKSON: Most placentas achieve a size and level of invasion that is appropriate for what the fetus is going to require, before they have had a chance for the fetus to speak to them to tell them what it requires. Is this just chance, or is there some form of messaging that enables the placenta to know that it has done enough?

ROBERT BOYD: The message would come from the mother, reflecting the environment, rather than the fetus.

CRITCHLEY: Those of us in clinical practice will be aware that we are now seeing increased incidence of abnormal placentation in women. I am talking about women where the invasion is too much, and the placenta buries into the myometrium. This happens because we have an increasing caesarean section rate, so women have a scar on the uterus. They no longer have a normal decidua, which calls a halt. Do we have an opportunity to look at the problem in the other direction? None of us in our single hospitals will have enough patients, but if collectively we have the opportunity to look at the uterus where the placenta has been abnormal, this might be informative.

HUNT: I have a comment on the mouse versus human controversy. There is a rat strain that is unlike the usual rodent model where there is huge invasion right up through the smooth muscle. This rat model might show what portion of the process by which the trophoblast moves has a genetic or programmed basis.

Chapter 9

The role of the maternal immune response in fetal programming

Ashley Moffett

Nutrients and oxygen must cross the placenta into the fetal circulation for normal growth and development of the fetus. The supply of oxygen and nutrients in the intervillous space is dependent upon an increasing uterine blood supply throughout pregnancy. Placental trophoblast cells achieve this by invading the uterus and converting the walls of the arteries into high-conductance vessels. It is therefore the primary role of extravillous cytotrophoblast to regulate maternal blood flow into the intervillous space by their invasion and transformation of decidual arteries. Disturbance of this process is the primary defect in major disorders of pregnancy such as pre-eclampsia and fetal growth restriction. The decidua plays an important role in regulation of placental invasion. Too much fetal intrusion would benefit fetal nourishment while compromising the mother. Too little invasion could put the fetus at a disadvantage. Although suboptimal invasion may not lead to overt clinical conditions during pregnancy, the deprivation of nutrient supply that results could contribute to fetal programming. The role of the uterine immune system in the regulation of trophoblast transformation of spiral arteries will be discussed in this chapter.

Trophoblast and the maternofetal boundary

Two weeks after attachment of the blastocyst, the inner cytotrophoblast layer of the trophectoderm proliferates into buds, which then protrude through the overlying syncytium. Subsequently this cytotrophoblast layer will have two cell fates: villous trophoblast and extravillous trophoblast. These two trophoblast subpopulations make different contributions in establishing efficient transport between the maternal and fetal circulations. Villous trophoblast covers the chorionic villi providing the barrier through which metabolic exchange between mother and fetus occurs, and it interacts with maternal blood in the intervillous space. Extravillous trophoblast cells (EVT) are the cells that invade the decidua, myometrium and spiral arteries, and include interstitial and endovascular trophoblast and placental bed giant cells [1,2].

Initially one population of EVT (interstitial) migrates into the decidua and come to surround the maternal spiral arteries. Where this occurs the muscular walls of the arteries are destroyed and the endothelial cells swell. The second subtype of EVT (endovascular) invades into these altered arteries. Endovascular trophoblast cells migrate down the luminal walls of these arteries and eventually replace the maternal endothelial cells. In healthy pregnancies EVT invade as far as the inner third of the myometrium, where they fuse to form placental bed giant cells. The consequence of these adaptations is to convert the arteries from muscular vessels to low-resistance channels that are incapable of responding to vasoactive stimuli.

Clinical implications of abnormal trophobast invasion

The health of both the fetus and the mother is contingent upon regulating the extent of trophoblast invasion. Disruption of the normal balance between the itinerant trophoblast cells and the uterine tissues they colonize during placentation results in various clinical problems [3]. These conditions give an insight into how the delineation of the territorial boundary between two individuals is achieved. Trophoblast-cell penetration of the uterine epithelium and invasion into the uterine wall and arteries is potentially highly dangerous, particularly in humans.

The Placenta and Human Developmental Programming, ed. Graham J. Burton, David J. P. Barker, Ashley Moffett and Kent Thornburg. Published by Cambridge University Press. © Cambridge University Press 2011.

Uncontrolled trophoblast-cell invasion is seen when decidua is deficient or absent, as in tubal pregnancy or when the placental genome implants on scar tissue from a previous caesarian section, a condition known as placenta creta [4]. Without medical intervention, these conditions result in maternal death from haemorrhage. At the opposite extreme, excessive restraint of trophoblast cells by the decidua may result in pregnancies in which trophoblast-cell invasion into the arteries and uterine wall is inadequate. In this case, the territorial boundary has moved in favour of the mother and the blood supply to the fetus becomes poor. The major problems that result from such reduced blood supply are fetal prematurity, fetal growth restriction, stillbirth and pre-eclampsia [5]. Early studies in which trophoblast cells were transplanted to ectopic sites in mice and pigs demonstrated the inherent invasive proclivities of trophoblast cells [6]. The decidua has been considered to behave as a procrustean bed, forcing conformity on its guests whatever their shape or size [7].

Pre-eclampsia is a specific disorder of human pregnancy that is associated with a significant maternal mortality and morbidity. The disease develops in stages, the maternal systemic illness being a late manifestation of the placental stress responses that occur as a result of the reduced uteroplacental blood flow [5]. Genetic studies of pre-eclampsia have implicated a contribution from both maternal and paternal genotypes. A family history in maternal relatives is associated with a fourfold risk [8]. Despite this, no concordance is found in monozygotic twins [9]. In population-based studies a contribution of both paternal and maternal factors in the development of the disease is apparent [10,11]. That immune recognition is involved is suggested by the increased risk in first pregnancies and in multiparae, but only after changing partners [12]. How a change of partner by the women affects the risk is still controversial and will require studies of very large cohorts such as those from Norway [13]. Significantly, the frequency of pre-eclampsia is ~30% in pregnancies resulting from oocyte donation, where the fetus shares neither haploid genome with the mother [14].

Why should such a devastating disease be maintained despite the strong selective pressures for reproductive success? The answer probably lies in the delicate negotiation between trophoblast-cell invasion and decidua that is required during every human pregnancy. The few mothers who die from pre-eclampsia can be viewed as an evolutionary consequence, or indeed sacrifice, owing to the need to control the aggressive behaviour of human trophoblast cells.

Immunology of pregnancy

Why should the immune system be involved in the regulation of trophoblast invasion and successful placentation? The placental genome, being a hybrid between paternal and maternal genomes, has the potential to express paternal antigens and thus could potentially be recognized as non-self by the maternal immune system. This has always been considered a problem that needs to be evaded, because in transplantation such recognition of non-self results in rejection of the grafted tissue [15]. If similar mechanisms operate in the uterus, then reproductive success must rely on some form of immunological accommodation of the mother to placental tissue.

Traditionally, when immunologists have thought about the immune system during pregnancy they have focused on the acquisition of maternal tolerance to the allogeneic fetus [16]. This view is too simplistic, because it does not take into consideration the anatomical fact that it is the maternal relationship with the placenta rather than with the fetus that holds the key to our understanding of the 'immunological paradox' of pregnancy. In particular, the focus should be on the intermingling of placental and maternal cells in the uterine wall, as this is where direct tissue contact occurs during placentation. Failure to distinguish between the local uterine immune response to the placenta and the systemic immune response to fetal cells (that usually cross to the mother at delivery) has led to a great deal of confusion. It is the close apposition of placental and uterine tissues that creates a particular immunological dilemma.

Medawar drew the analogy between the fetus and an allograft following the discovery that immune responses to polymorphic MHC molecules resulted in rejection of allografts [15]. Although this paper has since driven the direction of research into the immunology of pregnancy, it has been difficult to define whether any classical, alloreactive, adaptive immune responses ever do occur in pregnancy. That reproductive failure in humans ever results from histoincompatibility between mother and fetus has been even more difficult to demonstrate. There are several difficulties in answering these questions that are often not appreciated. At delivery fetal cells can indeed gain access to the maternal circulation and initiate the production of anti-HLA antibodies (the original source of anti-HLA sera in tissue typing) and even cytotoxic T cells restricted to paternal HLA. Importantly, though, there is no detectable effect on the success of the pregnancy whether these antifetal adaptive immune responses are present or absent. Second, the strategies used to form an efficient placenta that has optimal access to maternal nutrients are remarkably different even between closely related species [3,17]. The mouse is the most valuable model for understanding the immune system *in vivo*, but its reproductive system differs in many fundamental ways from that of humans, including the short gestation and lack of deep placental invasion into the uterus [18]. Third, the anatomy of placentation needs to be considered, specifically how the immune response will differ when blood is in contact with the trophoblast exchange area, or at the site of trophoblast contact with the local uterine immune response and in the regional draining lymph nodes.

Trophoblast defines the boundary between the mother and fetus and has many unusual properties [19]. It is extraembryonic and invasive, and expresses endogenous retrovirus products, oncofetal proteins and imprinted genes. In addition, the DNA in trophoblast is relatively unmethylated [20]. All these properties could have some relevance in interaction with the maternal immune system in terms of pathogen-associated molecular pattern (PAMP) recognition. Of most relevance to immunologists is the expression of MHC and MHC-like genes by trophoblast cells. These would be the potential ligands for immune receptors on uterine NK cells, T and B lymphocytes and myelomonocytic cells. Human trophoblast cells have been studied most extensively and display a unique and intriguing array of HLA class I molecules, the functions of which may hold the key to the successful temporary coexistence of two individuals.

Trophoblast HLA molecules

The major MHC molecules in humans are HLA class I and class II molecules. No trophoblast cells express HLA class II molecules even after stimulation with IFNγ. The various trophoblast populations differ both with respect to their expression of HLA class I molecules and in which maternal immune cells they encounter. The subdivisions result in different interfaces existing between mother and placenta. Villous syncytiotrophoblast lines the intervillous space and is thus in contact with immune cells in the maternal blood. This syncytium does not express any HLA class I antigens and thus is presumably immunologically inert to T cell-mediated responses. Indeed, there is no evidence that the mother makes antibodies or T-cell responses to villous trophoblast in either normal or pathological pregnancies. Furthermore, the ultimate fetal barrier, the villous syncytiotrophoblast, is a continuous syncytium that is likely to resist cytolytic attack by complement or cytotoxic lymphocytes. The absence of HLA class I does not cause cytolytic activation of blood NK cells, which are generally inhibited by MHC molecules expressed on a cell's surface. This could be because of a lack of activating ligands and also protection of the surface by high levels of glycosylation.

Although the fetus proper expresses the full array of its paternally inherited HLA genes, the EVT that invade the decidua only express a select few of the MHC antigens at their disposal [21]. The two main classic HLA class I antigens, HLA-A or HLA-B, which are expressed by the majority of adult somatic cells, are not transcribed. Only three HLA class I subtypes have been detected on certain subsets of human trophoblast. These are HLA-C, HLA-G and HLA-E. A report that HLA-F is present awaits confirmation [22]. Neither HLA-G nor HLA-E shows any appreciable polymorphism. HLA-C is a classical MHC class I molecule: its expression is widespread on somatic cells, but HLA-C is expressed at about 10% of the levels of HLA-A and HLA-B and is less polymorphic. Trophoblast has been shown to express mRNA and protein from both the maternal and paternal alleles of HLA-C [21,23 and unpublished].

The lack of highly polymorphic HLA-A and -B molecules will preclude strong anti-HLA class I T cell alloreactions that could damage fetal trophoblast. Indeed, viruses such as HIV, with elaborate immunoevasion strategies, also express HLA-C, while downmodulating -A and -B. It is nevertheless still possible that the polymorphic HLA-C or minor/trophoblast-specific proteins could initiate alloresponses. Furthermore, lymphatics are present in the decidual mucosal lining, but the events occurring in draining lymph nodes have not been well characterized during pregnancy [24]. Many overlapping mechanisms have been described to explain how these potential T-cell responses are avoided [25]. These include IDO production (which results in tryptophan depletion), the presence of FAS ligand and of Treg cells [26,27]. Sequences encoding the transcription factor Foxp3, associated with regulatory T cells, are generally found in placental vertebrates, but not in most non-placental and lower organisms. It was therefore proposed that regulatory T cells may have been a prerequisite for the evolution of placentation [28].

The HLA-G locus is non-polymorphic, but transcripts encoding a number of potential isoforms generated by alternative splicing, including a putative soluble variant, have been identified. Whether any appreciable protein products of these isoforms are ever expressed is still uncertain. HLA-G appears to be unusual in that high molecular weight forms that are β_2-m-associated homodimers can be detected at the cell surface [29,30]. Another novel feature among class I molecules is the nature of the carbohydrate structure. HLA-G carries N-acetyl galactosamine when expressed by trophoblast, but this is not a feature of HLA-G transfectants [31]. Despite many studies, the function of HLA-G in pregnancy is still not known. Adults homozygous for a highly truncated 'null' allele of HLA-G have been identified [32]. The birth of these individuals suggests that the expression of full-length membrane-associated HLA-G in trophoblast is not essential for pregnancy. Recent studies suggesting that the decidual immune responses may be all polarized towards tolerogenic rather than immunogenic responses as a result of the interaction of HLA-G with an innate immune receptor (LILRB1) on decidual myelomonocytic cells are of particular interest [30]. Because HLA-G, uniquely among HLA class I molecules, forms β_2-m-associated homodimers, which have higher avidity to LILRB1 than other conventional HLA-I monomeric molecules, the placenta itself could be driving the local immune response in the uterus to prevent fetal damage [29].

The third HLA antigen to be discovered on trophoblast was HLA-E, which, like HLA-G, is a nonclassic class I molecule with limited sequence polymorphism [33]. The cell surface expression of HLA-E is dependent on its binding of signal peptides cleaved from other class I molecules. Signal peptides from both HLA-G and HLA-C have been shown to bind to HLA-E. Potentially even the null allele of HLA-G could provide peptide and thus cause an upregulation of HLA-E expression.

Decidualization

The human non-pregnant uterine mucosa (endometrium) is unlike other mucosal surfaces in that it undergoes cycles of breakdown and regeneration throughout reproductive life [34]. Following ovulation the oestrogen-primed endometrium undergoes differentiation in the secretory phase. It is during this phase that preparations for implantation can be seen, primarily in the glands and stroma of the functionalis. The glands begin their secretion into the uterine cavity and the stromal cells, especially around the spiral arteries, change from fibroblastic to epithelioid in appearance. This morphological change is known as decidualization, and in humans it will commence even in the absence of fertilization. The secretory phase is also characterized by the appearance in the endometrium of a large number of leukocytes [35].

Because of the close correlation between the invasion of trophoblast cells and the extent of decidualization, it was argued that the decidual tissue has a permissive influence that favours trophoblast-cell invasion into the uterus. The alternative view was that the decidua provides a defensive riposte to the highly invasive trophoblast cells. It now seems likely that both ideas are correct and that the decidua allows orderly access of trophoblast cells to the maternal nutrient supply by achieving the right balance between under- and overinvasion. A compromise is reached between mother and fetus and a boundary is drawn. Leukocytic infiltration persists and is a particularly striking feature of the decidua basalis, where the placenta implants. Therefore, the invading trophoblast will potentially interact not only with the glands and stroma but also with maternal immune cells. It is reasonable to suppose that cells of the immune system might act at the uterine interface to regulate the extent of trophoblast invasion. If this were the case, we would expect to find maternal leukocytes at the site of placentation specialized for recognition of fetal antigens expressed by EVT. In fact, just such a population of lymphocytes, the uterine natural killer (uNK) cells, is found in the decidua of early pregnancy [36].

Uterine mucosal lymphocytes

Immunohistology using an antibody against CD45 (leukocyte common antigen) has shown that during the first trimester of pregnancy up to 40% of cells in the decidua are leukocytes. Detailed phenotyping of this population has revealed a unique composition of immune cells, which is unlike that seen in other tissues or mucosal surfaces [1].

Only a few B cells can be detected in endometrium and decidua in small lymphoid aggregates found in the basal layer. Their number does not vary during the menstrual cycle or pregnancy. Uterine B cells can respond to antigenic challenge; in the event of intrauterine infection germinal centres form and plasma cells can be found, but these are not features of normal pregnancy [37].

Endometrial macrophages express the CD14 macrophage differentiation antigen and class II MHC antigens [38]. *In vitro*, decidual macrophages are adherent and phagocytic. Immunohistology reveals that approximately 20% of leukocytes in endometrium are macrophages. This percentage is relatively invariant throughout the menstrual cycle and also in pregnancy. However, the distribution of macrophages in the pregnant uterus is striking. There is enrichment of these cells at the implantation site; in the dicidua parietalis, where no trophoblast invasion occurs, macrophages are sparse. Decidual macrophages, albeit not as numerous as uNK, are found in close association with the invading trophoblast. They can bind to placental HLA-G via the LILRB1 receptor (see above) [30].

T cells account for around 10–25% of leukocytes in decidua. They are found scattered throughout the stroma, within epithelia and also in the basal lymphoid aggregates. As with B cells, their number is static during the menstrual cycle and they do not accumulate at the implantation site.

Uterine NK cells (uNK cells)

The most distinctive feature of the uterine mucosa during reproductive life is the presence of a large population of NK cells. Up to 70% of decidual leukocytes are CD56$^+$ [39]. Ultrastructurally a typical uterine CD56$^+$ cell is a large lymphocyte with a reniform, eccentric nucleus and short cytoplasmic projections. They also possess varying numbers of membrane-bound cytoplasmic granules that contain cytolytic molecules such as granzyme and perforin [40]. In early pregnancy decidual NK cells make up around 30% of the stromal compartment. They are particularly populous at the implantation site, where they are found scattered throughout the stroma and thickly clustered around glands and spiral arteries. Their high-level infiltration puts them in close apposition to the invading trophoblast. The stimulus for the presence of so many NK cells in the uterus is incompletely understood. In humans uNK cells are present before implantation and in non-pregnant cycles where the number of uNK cells changes during the course of the menstrual cycle [41]. CD56$^+$ cells are rare in the proliferative phase but increase in number from the mid-secretory phase onwards, an increase that is sustained into early pregnancy.

As yet there is no convincing demonstration of a fundamental role for uNK cells in human reproduction [42]. They may be important in the renewal, differentiation and breakdown of the endometrium during the menstrual cycle and pregnancy. The close encirclement of spiral arteries by uNK cells suggests that they may influence mucosal vascularization. Endometrial NK cells have been shown by *in situ* hybridization to express mRNA for VEGF-C, PGF and Ang-2 in the secretory phase, all factors that could influence arterial stability [43]. NKG5, an NK-cell derived soluble factor, was isolated from decidua and is mitogenic for endothelial cells [44]. Secretion of a variety of other angiogenic factors has also been found from decidual NK cells [45,46]. Future investigations should address what influence these NK cells might have on processes such as decidualization, angiogenesis and menstrual breakdown.

NK cell receptors

NK activity is controlled by the integration of both activating and inhibitory signals. One of the most important families of NK receptors is the killer immunoglobulin receptors (KIR) [47]. KIR are Ig-superfamily (Ig-SF) receptors, which are expressed mainly by NK cells but also by certain T-cell subsets. These genes are classified by the number of extracellullar Ig-like domains they encode (2D or 3D) and by whether they have a long or short cytoplasmic domain (L or S). In general, the long-tailed KIR are inhibitory receptors, whereas the short-tailed KIRs transduce activating signals to the NK cell by associating with the adaptor protein DAP12. Certain loci have been found in all human haplotypes that have been genotyped or sequenced so far [48]. They are KIR3DL3, KIR3DL2, and KIR2DL4. These loci form a framework with regions of variable gene content in between. Unrelated individuals are unlikely to share the same KIR genotype owing to haplotypic differences in

KIR gene number and also intragenic polymorphism [49].

The uNK cells express surface KIR receptors, which are known to bind to HLA-C and HLA-G on trophoblast. The role of HLA-G and its receptor KIR2DL4 is still controversial [50]. It has been proposed that KIR2DL4 couples with soluble HLA-G and makes its way to Rab5+ve endosomes in a dynamin-dependent process. This results in secretion of inflammatory and angiogenic factors such as TNF, IL-1β, IFNγ and IL-8. The suggestion is that HLA-G interaction with KIR2DL4 on uNK promotes vascularization of the maternal decidua. Other data are consistent with decidual NK cells regulating development of reproductive tissue, by a range of cytokines, without necessarily supporting a role for KIR2DL4 [45]. It is also not clear how the HLA-G reaches the endosomes, as soluble forms of HLA-G have been difficult to find *in vivo* [30]. It will be of interest to see whether the HLA-G monomer or dimer binds to KIR2DL4. Little of the HLA-G monomer appears to bind to LILRB1, so if the monomer binds to KIR2DL4 instead this would be a mechanism whereby a single trophoblast molecule can signal differentially to the two major maternal leukocytes in the decidua.

KIR and HLA-C

HLA-C is the only known polymorphic HLA class I molecule on trophoblast, as HLA-E and -G are almost invariant. Because both HLA-C and KIR are polymorphic, the combinations of these ligands and receptors on mother and fetus will potentially differ in each pregnancy. The KIR family of receptors is highly diverse [47,51]. KIR haplotypes in different individuals may vary for gene number, in the sequence of their alleles and in the proportion of genes encoding activating or inhibitory receptors. The various KIR haplotypes fall into two basic types, a distinction which provides a useful basis for epidemiological study. KIR A haplotypes are generally shorter and mostly posses inhibitory KIR, with long cytoplasmic tails, associated with ITIM motifs. B haplotypes, on the other hand, generally encompass several activating, as well as inhibitory, KIR. This stratification helps to simplify analysis of this aspect of immune interaction in pregnancy. The maternal genotype may be AA (no activating KIR), AB (generally a small number of activating KIR) or BB (generally multiple activating KIR).

Similarly, considering the trophoblast, the highly polymorphic HLA-C ligands may be grouped according to those parts of their sequences known to interact with KIR. The HLA-C1 group comprise an asparagine residue at position 80. Such molecules interact with inhibitory KIR receptors 2DL2 and 2DL3, as well as, potentially, 2DS2. The HLA-C2 group has a lysine residue at position 80. These class I molecules interact with 2DL1 or the activating counterpart, 2DS1. Some exceptions to these rules have been reported: some alleles of KIR2DL2 and 2DL3 can also interact with some HLA-C2 molecules. KIR2DL2, generally exhibited strong interactions with HLA-C1 as well as with the few exceptional HLA-C2 molecules that it bound [52].

Maternal KIR and fetal HLA-C genes and reproductive success

HLA-C+ trophoblast cells intermingle with KIR+ uNK cells at the site of placentation where EVT transform the uterine arteries. Although variations in the extent of trophoblast invasion will obviously be influenced by multiple biological processes, the KIR– HLA-C axis is still the only known polymorphic interaction between maternal and fetal cells. This has led to the hypothesis that KIR– HLA-C interaction is critical to regulation of placentation [16] Support for this has come from two sources. First, there is genetic epidemiological evidence showing an association of preeclampsia with certain combinations of maternal KIR with fetal HLA-C. The strongest association was with increased frequency of maternal KIR corresponding to the AA genotype in combination with HLA-C2 groups in the fetus [54,55]. Similar findings come from a small cohort of women with three or more recurrent miscarriages where both male and females had an increased frequency of HLA-C2 groups [56]. Although the frequency of the KIR AA genotype was similar to that of controls in the males, it was increased in the female partners. The KIR genes on the B haplotype that were most reduced in frequency are located in the telomeric region where the locus for KIR2DS1 is found. This is plausible biologically, as 2DS1 is the activating KIR for HLA-C2 groups and would provide the balance needed to counteract the strong inhibitory signal the NK receives via KIR2DL1 and HLA-C2.

A second compelling set of arguments comes from consideration of cell biology and biochemistry. Uterine NK cells, as the dominant lymphocytes in the decidua in early pregnancy, are ideally poised to mediate invasion of the trophoblast. Moreover, KIR that bind HLA-C are increased in expression on uNK cells compared to blood NK cells [57]. In other words, the uterine KIR repertoire is skewed towards HLA-C recognition. Tetrameric HLA-C constructs could be demonstrated to bind to KIR2D on uNK cells [57]. Engagement of KIR on uNK cells altered cytokine production, although conventional NK cytokines such as IFNγ are only produced in small amounts by uNK cells. Of additional interest is the conformation of HLA-C at the trophoblast cell surface. Whereas HLA-C seems to be unstable on somatic cells, with unfolded conformers found in addition to the $β_2$-associated forms, on trophoblast HLA-C is all in this stable form [58]. How this affects KIR recognition is an interesting question. Taken together, these findings are consistent with the biological importance of the KIR–HLA-C interaction in fetal growth regulation.

Polymorphic genes and reproduction

The hypothesis that such an important biological process as reproduction is controlled by a chance encounter between maternal receptors and receptor ligands that are both highly variable is at first counterintuitive. A key factor is this discussion is that both KIR and HLA haplotypes are under strong diversifying selection from resistance to infection [59]. The diversity of KIR and HLA haplotypes in different populations suggests they are under strong selective pressure, and large genetic studies have clearly implicated a role for KIR in infectious disease [60]. The clearest disease associations come from studies involving large numbers of patients (1500+) with viral disease. In hepatitis C, homozygosity for both inhibitory KIR2DL3 and its ligand, HLA-C of the C1 group, is associated with resolution of infection, whereas possession of the allelic KIR2DL2 shows no advantage [61]. Functional evidence explaining this differential NK cell response is shown by analysis of the NK-inhibited subset of NK cells in an individual. This response is larger, IFNγ production is more rapid, and there is greater degranulation in HLA-C1 homozygous than in HLA-C2 homozygous individuals [62].

In HIV-infected individuals a reduced progression to AIDS is found in patients with KIR3DS1 in combination with HLA-Bw4 allotypes that have an isoleucine at position 80 (HLA-Bw4–80I) [63]. Again, functional studies support this genetic association and show that KIR3DS1-expressing NK cells can inhibit HIV replication in CD4+ T cells from individuals with this genotype [64]. In addition, the influence of inhibitory KIR3DL1 with HLA-B has shown that distinct allelic combinations also affect disease progression. KIR3DL1 alleles known to be highly inhibitory were protective [65]. This may seem contradictory, considering the beneficial effect of the activating KIR3DS1, but can be explained by considering the role of strong inhibitory responses in educating NK cells to be fully responsive during development [66–68]. In individuals with KIR AA genotypes who have no activating KIR there will be strong inhibitory signals during the development of NK cells, allowing highly responsive effector responses of mature NK cells. These will be beneficial for responses to infectious agents, but in contrast will be detrimental for reproduction [59]. It is likely that the so-far unexplained complexity of KIR and HLA reflects the intersection of two important biological functions, namely response to infection and regulation of trophoblast invasion. In the case of infection, selection may result from evolutionary pressure to periodically set the balance between over-and underactive NK responses.

HLA-C was thought to be of little importance compared to HLA-A and -B because of its reduced expression levels, although it is a key receptor for KIR2D molecules [69]. It appeared in primates relatively recently in evolution; it is present only in chimpanzees and humans, and makes a variable appearance on orangutan MHC haplotypes [47]. Whether the late appearance of KIR: HLA-C2 in primate evolution relates to the deeper forms of placentation required for upright gait and a large brain is an interesting idea [3]. As regards fetal development, balancing selection for KIR and HLA-C genotypes may be necessary to set the birth weight between two extremes in the face of changing environmental conditions, such as levels of maternal nutrition.

Conclusion

Pregnancy is a unique situation which has intrigued immunologists for more than half a century. The growing fetus requires maternal blood for its nutrition, but direct interaction of maternal leukocytes with allogeneic fetal cells would be unfavourable. Instead, the fetus surrounds itself with a specialized extraembryonic tissue with which it can tap into the maternal blood without provoking a detrimental immune response. The cells of the placenta have an unusual surface phenotype, especially in terms of the complement of MHC proteins they present.

The mother prepares for and responds to pregnancy with both systemic and local adaptations. Locally within the uterus the highly invasive nature of trophoblast is kept in check by a specialized layer – the decidua – but exactly how the decidua functions in this capacity is incompletely understood. What is known is that the immunological response to the trophoblast is not like that seen in transplantation, and that the populations of maternal leukocytes present in the uterus are unique.

Much about decidual NK cells also remains mysterious. They are very numerous at the site of trophoblast invasion and can be seen in close proximity to trophoblast. However, although they are potentially cytolytic, they do not kill trophoblast in normal pregnancy. Decidual NK cells have been shown to have receptors that could respond to trophoblast MHC. However, how decidual NK cells respond to trophoblast and what purpose this serves are unanswered questions [41]. Decidual NK cells may exert control over trophoblast through the production of cytokines, or they may be necessary for maintenance of the decidua.

Hopefully future studies will result in a more integrated understanding of how the different cell types in the decidua interact with each other and with trophoblast to create the correct environment for implantation to be a success from both maternal and fetal perspectives. Overall, the conclusion is that the interaction between maternal immune cells and the trophoblast may have a physiological function in regulating the development of the placenta, rather than the idea that there is a maternal immunological defence against her allogeneic fetus. This would be a new perspective on an old problem.

References

1. **Loke YW**, **King A**. *Human Implantation: Cell Biology and Immunology*. Cambridge: Cambridge University Press, 1995.
2. **Pijnenborg R**, **Vercruysse L**, **Hanssens M**, **van Assche A**. Incomplete trophoblast invasion: the evidence. In: *Pre-eclampsia*. Critchley H, MacLean A, Poston L, Walker J, eds. London: RCOG Press, 2003; 15–26.
3. **Moffett A**, **Loke C**. Immunology of placentation in eutherian mammals. *Nature Rev Immunol* 2006; **6**(8): 584–94.
4. **Khong TY**, **Robertson WB**. Placenta creta and placenta praevia creta. *Placenta* 1987; **8**: 399–409.
5. **Burton GJ**, **Woods AW**, **Jauniaux E**, **Kingdom JC**. Rheological and physiological consequences of conversion of the maternal spiral arteries for uteroplacental blood flow during human pregnancy. *Placenta* 2009; **30**: 473–82.
6. **Kirby DRS**. The invasiveness of trophoblast. In: *The Early Conceptus, Normal and Abnormal*. Park WW, ed. *Proceedings of Symposium*. Queen's College, Dundee, 1965; 68–73.
7. **McClaren A**. Maternal factors in nidation. In: *The Early Conceptus, Normal and Abnormal*. Park WW, ed. *Proceedings of Symposium*. Queen's College, Dundee, 1965; 27–33.
8. **Cincotta RB**, **Brennecke SP**. Family history of pre-eclampsia as a predictor for pre-eclampsia in primigravidas. *Int J Gynaecol Obstet* 1998; **60**: 23–7.
9. **Treloar SA**, **Cooper DW**, **Brennecke SP**, **Grehan MM**, **Martin NG**. An Australian twin study of the genetic basis of preeclampsia and eclampsia. *Am J Obstet Gynecol* 2001; **184**: 374–81.
10. **Esplin MS**, **Fausett MB**, **Fraser A** et al. Paternal and maternal components of the predisposition to preeclampsia. *N Engl J Med* 2001; **344**: 867–72.
11. **Lie RT**. Intergenerational exchange and perinatal risks: a note on interpretation of generational recurrence risks. *Paediatr Perinatol Epidemiol* 2007; 21 Suppl **1**: 13–18.
12. **Trupin LS**, **Simon LP**, **Eskenazi B**. Change in paternity: a risk factor for preeclampsia in multiparas. *Epidemiology* 1996; 7: 240–4.
13. **Skjaerven R**, **Vatten LJ**, **Wilcox AJ** et al. Recurrence of pre-eclampsia across generations: exploring fetal and maternal genetic components in a population based cohort. *Br Med J* 2005; **331**: 877.
14. **Salha O**, **Sharma V**, **Dada T** et al. The influence of donated gametes on the incidence of hypertensive disorders of pregnancy. *Hum Reprod* 1999; **14**(9): 2268–73.
15. **Medawar PB**. Some immunological and endocrinological problems raised by the evolution of viviparity in vertebrates. In: *Society for Experimental Biology*. New York: Academic Press, 1953; 320–38.
16. **Trowsdale J**, **Betz AG**. Mother's little helpers: mechanisms of maternal-fetal tolerance. *Nature Immunol* 2006; **7**(3): 241–6.
17. **Steven DH** (ed) *Comparative Placentation*. New York: Academic Press, 1975.
18. **Georgiades P**, **Ferguson-Smith AC**, **Burton GJ**. Comparative developmental anatomy of the murine and human definitive placentae. *Placenta* 2002; **23**: 3–19.

19. **Johnson M**. Origins of pluriblast and trophoblast in the eutherian conceptus. *Reprod. Fertil. Dev* 1996; **8**: 699–709.

20. **Blackburn DG**. Reconstructing the evolution of viviparity and placentation. *J. Theoret. Biol* 1998; **192**: 183–90.

21. **Apps R**, **Murphy SP**, **Fernando R** et al. Human leucocyte antigen (HLA) expression of primary trophoblast cells and placental cell lines, determined using single antigen beads to characterize allotype specificities of anti-HLA antibodies. *Immunology* 2009; **127**: 26–39.

22. **Ishitani A**, **Sageshima N**, **Lee N** et al. Protein expression and peptide binding suggest unique and interacting functional roles for HLA-E, F, and G in maternal-placental immune recognition. *J Immunol* 2003; **171**: 1376–84.

23. **King A**, **Allan DS**, **Bowen M** et al. HLA-E is expressed on trophoblast and interacts with CD94/NKG2 receptors on decidual NK cells. *Eur J Immunol* 2000; **30**: 1623–31.

24. **Red-Horse K**, **Rivera J**, **Schanz A** et al. Cytotrophoblast induction of arterial apoptosis and lymphangiogenesis in an *in vivo* model of human placentation. *J Clin Invest* 2006; **116**: 2643–52.

25. **Mellor AL**, **Munn DH**. Creating immune privilege: active local suppression that benefits friends, but protects foes. *Nature Rev Immunol* 2008; **8**: 74–80.

26. **Tilburgs T**, **Roelen DL**, **Van Der Mast BJ** et al. Evidence for a selective migration of fetus-specific CD4+CD25 bright regulatory T cells from the peripheral blood to the decidua in human pregnancy. *J Immunol* 2008; **180**(5737): 45.

27. **Mellor AL**, **Sivakumar J**, **Chandler P** et al. Prevention of T cell-driven complement activation and inflammation by tryptophan catabolism during pregnancy. *Nature Immunol* 2001; **2**: 64–8.

28. **Aluvihare VR**, **Betz AG**. The role of regulatory T cells in alloantigen tolerance. *Immunol Rev* 2006; **212**: 330–43.

29. **Boyson JE**, **Erskine R**, **Whitman MC** et al. Disulfide bond-mediated dimerization of HLA-G on the cell surface. *Proc Natl Acad Sci USA* 2002; **99**: 16180–5.

30. **Apps R**, **Gardner L**, **Sharkey AM**, **Holmes N** and **Moffett A**. A homodimeric complex of HLA-G on normal trophoblast cells modulates antigen-presenting cells via LILRB1. *Eur J Immunol* 2007; **37**: 1924–37

31. **McMaster M**, **Zhou Y**, **Shorter S** et al. HLA-G isoforms produced by placental cytotrophoblasts and found in amniotic fluid are due to unusual glycosylation. *J Immunol* 1998; **160**: 5922–8.

32. **Ober C**, **Aldrich C**, **Rosinsky B** et al. HLA-G1 protein expression is not essential for fetal survival. *Placenta* 1998; **19**: 127–32.

33. **King A**, **Burrows TD**, **Hiby SE** et al. Surface expression of HLA-C antigen by human extravillous trophoblast. *Placenta* 2000; **21**: 376–87.

34. **Finn C**. Menstruation: A nonadaptive consequence of uterine evolution. *Q Rev Biol* 1998; **73**: 163–73.

35. **Bulmer JN**, **Lash GE**. Human uterine natural killer cells: a reappraisal. *Mol Immunol* 2005; **42**: 511–21.

36. **Moffett-King A**. Natural killer cells and pregnancy. *Nature Rev Immunol* 2002; **2**: 656–63.

37. **More IAR**. The normal human endometrium. In: *Haines and Taylor's Obstetrical and Gynaecological Pathology*. Edinburgh: Churchill Livingstone, 1987; **1**: 302–19.

38. **Gardner L**, **Moffett A**. Dendritic cells in the human decidua. *Biol Reprod* 2003; **69**: 1438–46.

39. **Bulmer JN**, **Morrison L**, **Longfellow M**, **Ritson A**, **Pace D**. Granulated lymphocytes in human endometrium: histochemical and immunohistochemical studies. *Hum Reprod* 1991; **6**: 791–8.

40. **King A**, **Wooding P**, **Gardner L**, **Loke YW**. Expression of perforin, granzyme A and TIA-1 by human uterine CD56+ NK cells implies they are activated and capable of effector functions. *Hum Reprod* 1993; **8**: 2061–7.

41. **King A**, **Wellings V**, **Gardner L**, **Loke YW**. Immunocytochemical characterization of the unusual large granular lymphocytes in human endometrium throughout the menstrual cycle. *Hum Immunol* 1989; **24**: 195–205.

42. **Lash GE**, **Robson SC**, **Bulmer JN**. Review: functional role of uterine natural killer (uNK) cells in human early pregnancy decidua. *Placenta* 2010; **31** (Suppl 1): S87–S92.

43. **Li XF**, **Charnock-Jones DS**, **Zhang E** et al. Angiogenic growth factor messenger ribonucleic acids in uterine natural killer cells. *J Clin Endocrinol Metab* 2001; **86**: 1823–34.

44. **Langer N**, **Beach D**, **Lindenbaum ES**. Novel hyperactive mitogen to endothelial cells: human decidual NKG5. *Am J Reprod Immunol* 1999; **42**: 263–72.

45. **Hanna J**, **Goldman-Wohl D**, **Hamani Y** et al. Decidual NK cells regulate key developmental processes at the human fetal-maternal interface. *Nature Med* 2006; **12**(9): 1065–74.

46. **Lash GE**, **Schiessl B**, **Kirkley M** et al. Expression of angiogenic growth factors by uterine natural killer cells during early pregnancy. *J Leukoc Biol* 2006; **80**: 572–80.

47. **Parham P**. MHC class I molecules and KIRs in human history, health and survival. *Nature Rev Immunol* 2005; **5**(3): 201–14.

48. **Vilches C, Parham P**. KIR: diverse, rapidly evolving receptors of innate and adaptive immunity. *Annu Rev Immunol* 2002; **20**: 217–51.

49. **Shilling HG, Guethlein LA, Cheng NW** et al. Allelic polymorphism synergizes with variable gene content to individualize human KIR genotype. *J Immunol* 2002; **168**: 2307–15.

50. **Rajagopalan S, Bryceson YT, Kuppusamy SP** et al. Activation of NK cells by an endocytosed receptor for soluble HLA-G. *PLoS Biol* 2006; **4**(1): e9.

51. **Carrington M, Noman P**. *The KIR Gene Cluster*. Bethesda, MD: National Center for Biotechnology Information, 2003.

52. **Moesta AK, Norman PJ, Yawata M** et al. Synergistic polymorphism at two positions distal to the ligand-binding site makes KIR2DL2 a stronger receptor for HLA-C than KIR2DL3. *J Immunol* 2008; **180**(6): 3969–79.

53. **Trowsdale J, Moffett A**. NK receptor interactions with MHC class I molecules in pregnancy. *Semin Immunol* 2008; **20**: 317–32.

54. **Hiby SE, Walker JJ, O'Shaughnessy KM** et al. Combinations of maternal KIR and fetal HLA-C genes influence the risk of preeclampsia and reproductive success. *J Exp Med* 2004; **200**: 957–865.

55. **Hiby SE, Apps R, Sharkey AM** et al. Material activating KIR protect against human reproductive failure mediated by fetal HLA-C2. *J Clin Invest* 2010; (in press).

56. **Hiby SE, Regan L, Lo W** et al. Association of maternal killer-cell immunoglobulin-like receptors and parental HLA-C genotypes with recurrent miscarriage. *Hum Reprod* 2008; **23**: 972–6.

57. **Sharkey AM, Gardner L, Hiby S** et al. Killer Ig-like receptor expression in uterine NK cells is biased toward recognition of HLA-C and alters with gestational age. *J Immunol* 2008; **181**: 39–46.

58. **Apps R, Gardner L, Hiby SE, Sharkey AM, Moffett A**. Conformation of human leucocyte antigen-C molecules at the surface of human trophoblast cells. *Immunology* 2008; **124**: 322–28.

59. **Parham P**. The genetic and evolutionary balances in human NK cell receptor diversity. *Semin Immunol* 2008; **20**: 311–16.

60. **Khakoo SI, Carrington M**. KIR and disease: a model system or system of models? *Immunol. Rev* 2006; **214**: 186–201.

61. **Khakoo SI, Thio CL, Martin MP** et al. HLA and NK cell inhibitory receptor genes in resolving hepatitis C virus infection. *Science* 2004; **305**(5685): 872–4.

62. **Ahlenstiel G, Martin MP, Gao X, Carrington M, Rehermann B**. Distinct KIR/HLA compound genotypes affect the kinetics of human antiviral natural killer cell responses. *J Clin Invest* 2008; **118**: 1017–26.

63. **Martin MP, Gao X, Lee JH** et al. Epistatic interaction between KIR3DS1 and HLA-B delays the progression to AIDS. *Nature Genet* 2002; **31**: 429–34.

64. **Alter G, Martin MP, Teigen N** et al. Differential natural killer cell-mediated inhibition of HIV-1 replication based on distinct KIR/HLA subtypes. *J Exp Med* 2007; **204**(12): 3027–36.

65. **Martin MP, Qi Y, Gao X** et al. Innate partnership of HLA-B and KIR3DL1 subtypes against HIV-1. *Nature Genet* 2007; **39**(6): 733–40.

66. **Kim S, Poursine-Laurent J, Truscott SM** et al. Licensing of natural killer cells by host major histocompatibility complex class I molecules. *Nature* 2005; **436**(7051): 709–13.

67. **Kim S, Sunwoo JB, Yang L** et al. HLA alleles determine differences in human natural killer cell responsiveness and potency. *Proc Natl Acad Sci USA* 2008; **105**(8): 3053–8.

68. **Anfossi N, Andre P, Guia S** et al. Human NK cell education by inhibitory receptors for MHC class I. *Immunity* 2006; **25**: 331–42.

69. **Colonna M, Borsellino G, Falco M, Ferrara GB, Strominger JL**. HLA-C is the inhibitory ligand that determines dominant resistance to lysis by NK1- and NK2-specific natural killer cells. *Proc Natl Acad Sci USA* 1993; **90**(24): 12000–4.

Discussion

HUNT: I think your studies on HLA-C are the most exciting in the literature on the class I genes that have been explored, particularly in relation to the NK cell. I was interested in your identification of the KIR B genes that have a protective effect. Perhaps the uterine NK cell is a special target for this system of interaction between HLA-C and the KIR B haplotype. Do you have information on how cancer cells or other normal cells might use similar systems?

MOFFETT: Do you mean how do peripheral blood NK cells use the system? KIR are expressed in NK cells, and the NK cells in the uterus are biased towards KIR that recognise HLA-C. They have far higher levels of expression for these KIR that bind HLA-C than any other NK cells. Furthermore, the levels of HLA-C on trophoblasts are far higher than on any other cells. We have an antibody specific for

HLA-C so we can determine this. I think it has particular biological significance related to placentation. But of course KIR are immune system genes, as are HLA. There are other studies that clearly show that KIR and HLA-C or HLA-B are important in the defence against HIV and other viral infections. Thus, these polymorphisms are also important in defence against infection. These are the ultimate Darwinian gene roles in reproduction and immune defence.

HUNT: So this is a unique system for learning how the HLA genes and KIRs have organ-specific effects.

MOFFETT: I don't think there is any evidence that they have a role in tumours yet. In transplantation they are mainly important in bone marrow transplantation and may not be so relevant in responses to solid allografts.

JACKSON: Do the 20% of macrophages play any role in the interaction at all?

MOFFETT: Yes, but we have no evidence that this macrophage response varies in different pregnancies. The macrophages have an important receptor for HLA-G (LILRBI), and it is probably this interaction that is deviating these cells away from making immune responses that are damaging to the embryo. Neither the LILRBI nor the HLA-G is appreciably polymorphic so this seems to occur in the same way in every pregnancy.

THORNBURG: Which environmental factors help regulate the expression of HLA-C?

MOFFETT: I don't know any environmental factors that do. There is an interesting SNP found in the genome-wide association in HIV which means that certain individuals have higher levels of HLA-C than others (Fellay J. *et al.* A whole-genome association study of major determinants for host control of HIV-1. Science, 2007;**317**:944–7). This is associated with progression in HIV. Thus, levels of HLA-C may be another important variable.

THORNBURG: Something must regulate it. How might this fit with programming? It is so important for pre-eclampsia.

MOFFETT: It is not known why HLA-C and not the other classical HLA class 1 molecules, HLA-B + A is expressed by extravillous trophoblast cells.

MOORE: This association with pre-eclampsia, does it differ in early versus late onset disease?

MOFFETT: We haven't studied this yet. We expect it would, but we haven't teased out the database enough.

LOKE: If I may try to answer Ken Thornburg's question, Ashley Moffett has shown that the variation is there to determine birth weight. The general dogma is that the evolutionary drive for the generation of MHC diversity is infection. Reproduction is also a powerful drive, and the two go together. There is no point in being able to survive infections if you can't breed properly. The generation of HLA-C and KIR diversity is for infection as well as for reproduction. The two should go together.

LAMPL: The graph you showed is quite central to much of our discussion. Earlier evidence that documents selection of birth weight (Karn MN and Penrose LS, 1951, Birth weight and gestational time in relation to maternal age, parity and infant survival, Ann Eugen 16:147–164), identified that optimal survival is associated with neonates who have greater than the average birth weight. There are many processes that may contribute to, or drive birth weight. What is important from an evolutionary perspective is birth weight, and these observations may reflect such selection.

SMITH: How would you relate the genetic associations with pre-eclampsia that you have described to the well recognised associations between both oocyte donation and nulliparity and the risk of pre-eclampsia?

MOFFETT: In Oocyte donation the risk of pre-eclampsia is 30%. I think you could predict that if you are a surrogate mother it is probably not a good idea to be AA and have an HLA-C2 embryo. The first thing that one might suggest is that sperm donors are not C2 homozygous. The "dangerous" males might be these C2 homozygous ones. The first pregnancy issue is also interesting. It was always said that NK cells act immediately and have no memory; they are quite unlike T cells. But recently a paper in Nature [Ugolini *et al.* Immunology: Natural killer cells remember. Nature. 2009;**457**:544–545] showed that NK cells do remember (this may not be like T cell memory, but it seems that there is a clonal expansion). So in the pregnancy there could be a clonal expansion of the KIR that would recognise that HLA-C group of that baby. These NK cells might proliferate more readily next time, but NK memory may not last very long. So it would rather fit with partner specificity fading away after several years. There is a sort of short term 'memory'.

MOORE: The Penrose observations remind us that birth weight is important (Karn MN, Penrose LS. Birth weight and gestational time in relation to maternal age, parity and infant survival. Eugenics. 1957;**16**:147–64). It is also important to recognise that while in a modern hospital care centre, babies who are growth restricted are probably not more likely to die; this is not true throughout our evolutionary history. From an evolutionary perspective, the mortality that matters is mortality before the end of the reproductive span. Heart disease and cancer generally occur later in life after one's reproduction has been completed. Thus such kinds of mortality are in some respects selectively neutral, but programming gives us an inter-generational link. A baby who is small at birth today might not be at increased risk of mortality, but it will be at increased risk later in life from these other factors. This emphasises that evolution has not stopped, and that the selective effect of birth weight is preserved if not accentuated in the modern setting.

BARKER: It is sometimes forgotten that the pattern of chronic disease is incredibly unstable. 100 years ago coronary heart disease was rare, and then it rose. Asthma suddenly appeared, as did breast cancer. Diabetes is on the rise. It is not as if infectious disease has been replaced by some relatively constant set of diseases; the pattern is amazingly unstable. Behind this, perhaps, is that early developmental processes are very sensitive and small perturbations can set the embryo off on one path or another. We don't know whether there are paths of human development that are optimal for a great range of disorders, or whether there is a lot of trading off between one disease and another.

CETIN: One of the main factors that influences short-term outcomes in growth of babies is gestational age when they are born. We are discussing many issues here, but it seems not so much dependent on how they were when they were born, but their gestational age when they were born. We should also not forget that there are some nice immunological data looking at postpartum depression. The incidence is strictly related to the place where you are living. In Japan it is very low. In Italy it is much higher, and in the USA it is even higher. There is a nice direct relationship to the amount of fish that is sold in that year! [Hibbeln JR, Nieminen LRG, Blasbalg TL, *et al*. Healthy intakes of n-3 and n-6 fatty acids: estimations considering worldwide diversity. Am J Clin Nutr 2006;83: 1483S-1493S]. It is an incredible relationship. We are touching on a number of issues, but it is complex, and maternal nutrition is important, not just in pregnancy.

MAGNUS: Do women with the AA haplotypes have increased risk of cardiovascular disease?

MOFFETT: I have been asking the people who look at these KIR polymorphisms whether they have thought of addressing this question. People with KIR AA (there is a much higher frequency in Africans) have NK cells that are much more responsive. If you are AA you are much better at responding to infection and this might mean more "inflammatory". Very simply, KIR AA is good for infection and bad for reproduction.

ROBERT BOYD: What is the relationship of the birth weight cohorts to overall longevity, as opposed to specific disease states?

BARKER: There is no literature on this. It's a very good point. We have preliminary data showing that different maternal/placental/fetal/child phenotypes are linked to longevity.

Chapter 10

Clinical causes and aspects of placental insufficiency

Irene Cetin and Emanuela Taricco

Introduction

Placental function is a major factor in determining the successful outcome of pregnancy, that is, the birth of a normal baby. Placental insufficiency is a complication of pregnancy in which the placenta cannot carry enough oxygen and nutrients to the growing fetus. Intrauterine growth is indeed the result of genetic potential, nutritional status and endocrine regulation that will determine the growth trajectory during intrauterine life [1]. Changes in placental function that influence the availability and utilization of substrate to the fetus may alter its growth trajectory.

Clinical relevance of intrauterine growth restriction (IUGR)

Intrauterine growth restriction (IUGR) occurs when the fetus fails to achieve its full growth potential. IUGR is diagnosed in approximately 7–15% of all pregnancies [2–3] and is the most important cause of perinatal mortality and morbidity [3,4]. IUGR can affect pregnancy outcome (fetal death, preterm delivery, maternal pre-eclampsia) [5,6] as well as perinatal and neonatal outcome (respiratory distress, cerebral damage, necrotising enterocolitis etc.) [7–9]. Birth weight and gestational age at delivery are both factors that determine the outcome of IUGR pregnancies, and the survival rate of IUGR fetuses increases with advancing gestational age [2]. Moreover, in the last two decades many studies have established link between low birth weight and the predisposition to develop metabolic syndrome in adult life (hypertension, obesity, diabetes and cardiac damage) [5–6].

In spite of its clinical relevance, there is still no internationally recognized definition of IUGR. The term fetal growth restriction (FGR) is often used instead of intrauterine growth restriction, and the two can be used interchangeably. Small for gestational age (SGA) is usually a neonatal term to define a birth weight of less than the 10th percentile for gestational age. Although many studies define IUGR as an infant falling below the 10th, 5th or 3rd centile for gestational age, this definition does not necessarily reflect a condition arising from placental insufficiency. The difference between IUGR and SGA is that in the former there has been a clinical demonstration of a shift in the intrauterine growth curve occurring *in utero* and identified by ultrasound.

Since 1993 we have defined IUGR by birth weight below the 10th percentile for the reference population associated with the *in utero* demonstration of a longitudinal decrease in the growth of the abdominal circumference, with a correct ultrasound datation before 20 weeks of gestation [10,11].

Ultrasound measurement of the fetal abdominal circumference at the level of the fetal liver seems to hold the best accuracy and is currently considered the single best indicator of intrauterine fetal growth abnormalities in the second half of pregnancy [12]. The rationale for this measurement is that it corresponds most closely with the size of the fetal liver; moreover, it includes the measurement of fetal subcutaneous fat, an indicator of fetal body composition.

Because most fetal weight is gained during the second half of gestation, when a large and exponential deposition of fat tissue occurs [13], the clinical diagnosis of IUGR is typically made then, when the placental:fetal ratio progressively decreases [14].

However, the pathogenesis of IUGR is probably multifactorial and not yet perfectly known. There are some known causes, or risk factors, for the development of IUGR (Table 10.1), such as

---

The Placenta and Human Developmental Programming, ed. Graham J. Burton, David J. P. Barker, Ashley Moffett and Kent Thornburg. Published by Cambridge University Press. © Cambridge University Press 2011.

Table 10.1 Known causes of intrauterine growth restriction

Maternal
- Pre-existing diseases: hypertension, diabetes, autoimmune, cardiovascular, renal and oral diseases, thrombophilia
- Malnutrition
- Race
- Low socio-economic status
- Toxic exposures

Fetal
- Infections: CMV, rubella, malaria
- Structural anomalies
- Chromosomal or genetic diseases
- Placental abnormalities
- Multiple pregnancies

maternal habits (smoking, drugs), pre-existing maternal diseases (autoimmune, endocrine, cardiovascular, hypertensive), acquired blood borne infections (CMV, rubella, malaria), and abnormal placental position. Moreover, the potential for fetal growth can be altered in the presence of fetal abnormalities, such as chromosomal defects or malformations, or multiple gestations. However, in most situations no known cause can be detected. Although in the past these cases were defined as idiopathic, they share a common placental phenotype, called for many years 'placental insufficiency' [15]. In these cases, the recurrence of IUGR in a subsequent pregnancy ranges between 10% and 20% [16,17], a figure that indicates the presence of a specific maternal continuing risk factor, but not in all cases. The finding of an abnormal uterine artery Doppler is related to an increased risk of recurrence of IUGR during the next pregnancy [16]. Placental insufficiency may be caused by an impairment in the maternal circulation, fetal circulation or both by vascular thrombosis. Genetic defects, such as deficit of antithrombin III, protein C and protein S, and gain-of-function factor V and prothrombin mutations have been associated with an increased risk of venous thromboembolism and may have a role in pregnancy complications [18]. Recently, several reports have investigated a possible association between IUGR and maternal heritable causes of thrombophilia [19–21], but this relationship has not been consistently proved.

The role of paternal inheritance is also important, perhaps more than ever expected. Polymorphisms in genes that regulate how the placenta invades maternal tissues, differentiates and functions, and how the mother adapts to pregnancy have been identified as candidates that confer risk to pregnancy success. These polymorphisms may be relevant when occurring not only in the mother but also in the father, and therefore acting in the trophoblast [22].

Moreover, placental development is specifically regulated by imprinted genes that play a major role in the regulation of nutrient supply by the placenta. Dysregulation of imprinted genes has been implicated in the pathogenesis of IUGR. Mutations, epimutations and uniparental disomies affecting imprinted loci can cause placental impairment and fetal alterations both in mice and humans [23,24].

Placental weight and surface

Placental growth occurs mainly in the first half of gestation, whereas the fetal growth curve becomes exponential in the second half. These two different growth patterns may partially explain why IUGR occurs when the mass of the placenta per gram of fetus becomes critical, in a period of maximal fetal growth. Moreover, it suggests that the reduction in placental growth is already defined during the placentation process [25,26]. A strong relationship has been observed between placental weight and birth weight [27], and data arising from large cohort studies have shown that the combination of a large placenta and low birth weight is a strong independent risk factor for cardiovascular disease in adulthood [28].

Placentas from SGA pregnancies have been reported to be smaller than those from appropriate for gestational age (AGA) pregnancies, with a significantly higher placental:fetal weight (P/F) ratio in SGA than in AGA at comparable gestational ages [27]. When we examined the relationship between fetal and placental weights in IUGR, we observed a steeper relationship in IUGR with increasing severity [29]. Moreover, the placental:fetal weight ratio has been recently shown to be increased in IUGR fetuses in relation to IUGR severity. IUGR fetuses with placental weight ≥ 250 g exhibited a fetal:placental weight ratio significantly lower than that in AGA fetuses, suggesting that IUGR may be due to a reduction of placental function per gram of tissue [30]. The fetal:placental weight ratio may be the more important variable when alterations in fetal growth occur.

Besides placental weight, placental surface area and thickness determine its transfer capacities. In order to increase the surface for nutrient and oxygen exchange the placenta can increase the surface area of villi [31].

This has recently been shown by histomorphometric studies of placentas from idiopathic IUGR [32]. The placental surface area was estimated in the Helsinki Birth Cohort, which comprised 13 345 people born in the city during 1934–44. In this cohort, the prevalence of hypertension was significantly higher with low placental weight and surface area, but only in the offspring of mothers who were short or of low socio-economic status, and therefore presumably undernourished. On the contrary, in the offspring of mothers who were likely to have been the best nourished, being tall and middle class, hypertension was predicted by large placental size in relation to birth weight [33]. This is similar to what has been reported for pregnant women who were undergoing famine in Holland during the first trimester [34].

Transfer of oxygen and nutrients in placental insufficiency

The placenta is the key organ integrating maternal nutritional, metabolic, endocrine and vascular conditions and fetal requirements [35]. Fetal nutritional requirements may not be met in conditions that alter placental function, as the availability of oxygen and nutrients to the fetus depends on the interplay between placental transporters and metabolic activity. The causes of placental damage can be multiple and simultaneous (trombophilic and genetic factors can be involved), with complicated interactions between genes and the environment.

Implantation occurs in the first weeks of gestation when trophoblast cells invade the maternal decidua and the muscular wall of spiral arteries, reducing their resistance [36,37]. In pregnancies complicated by IUGR and/or maternal pre-eclampsia this invasion is limited to the decidual portion of the spiral arteries [38], leading to increased placental resistance to uterine blood flow. In the past, impaired trophoblast invasion was thought to result in an under-perfused placenta, leading to placental hypoxia and fetal undernutrition. Accumulating new data no longer support this view, suggesting instead that impaired invasion leads to localized increases and turbulence in blood flow (increased velocity and turbulence) rather than in a generalized decrease in blood flow [39].

Reduced transfer of oxygen and nutrients occurs as a result of inadequate uterine – umbilical blood flow exchange as well as reduced transport (i.e. for amino acids) and altered nutrient metabolism (i.e. for fatty acids). These changes occur together with changes in placental weight, surface and permeability, as well as in uterine/umbilical blood flow leading to uneven perfusion [10]. This placental phenotype is associated with changes in the fetal growth trajectory that reflect fetal adaptation processes, including reduced metabolic rate and endocrine adaptations (Figure 10.1) [40]. IUGR fetuses grow less and show decreased deposition of tissue, specifically fat, as a result of the reduced uptake of nutrients from the umbilical circulation.

The relationship between the maternal and fetal circulation in the placenta is crucial for efficient exchange of oxygen and nutrients. In the ideal situation, proportional flows of maternal and fetal blood lead to an even circulation, whereas in a situation of impaired placental perfusion, maternal and fetal blood flows face each other in different proportions at different cotyledons, leading to an uneven circulation and inefficient oxygen and nutrient exchange.

In humans, umbilical venous pO_2 is significantly related to, and lower than, that in the uterine vein, suggesting that the human placenta simulates a venous equilibrator [10]. In IUGR, the coefficient of uterine oxygen extraction is significantly reduced and uterine venous pO_2 levels and uterine–umbilical venous gradients are significantly increased compared to normal pregnancies. This finding suggests that the reduced pO_2 levels observed in the IUGR circulation may be due to reduced placental permeability and utilization, rather than to limited uterine blood flow. An open question is: what levels of oxygen are the trophoblast cells exposed to in IUGR?

Mitochondrial DNA is an indicator of cellular activity and has been shown to increase in conditions of nutritional deprivation. In IUGR, placental mitochondrial DNA content is significantly increased, and this rise is inversely related to umbilical venous pO_2 (Figure 10.2) [41]. This may imply that placental mitochondrial DNA increases as a compensatory mechanism to hypoxia and/or to the metabolic changes representing placental metabolic adaptations to the reduced nutrient availability. In contrast, it may well also be that mitochondrial function is damaged in IUGR trophoblast cells, representing in itself a limiting factor for the metabolic requirements.

Recently, umbilical oxygen uptake has been measured in human pregnancies by coupling the measure of umbilical vein blood flow by ultrasound prior

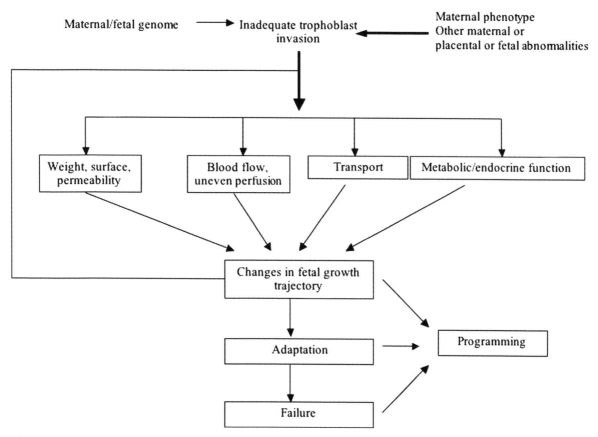

Figure 10.1 Placental phenotype of IUGR. changes in fetal growth trajectory that reflect fetal adaptation processes, including reduced metabolic rate and endocrine adaptations.

to delivery to the umbilical v-a O_2 content difference [42,43]. Reported values range between 0.24 and 0.29 mmol min kg in normal pregnancies, which is quite comparable to values obtained during intrauterine life in other species [25]. However, preliminary results in IUGR fetuses demonstrate significantly lower fetal oxygen uptake, even on per kg basis, i.e. approximately 50% of the values reported in normal fetuses, suggesting that metabolic rate is significantly decreased in IUGR [44].

Chronic fetal hypoxia and acidemia occur not only in IUGR fetuses. In pregestational diabetic pregnancies, fetal hypoxia and acidaemia have been reported to occur in relation to poor glycaemic control in the mother. Recently we have shown that fetuses from pregnancies with gestational diabetes have lower oxygen and higher lactate values in the umbilical vein. Although this finding may be due partly to changes in fetal metabolism as a result of hyperglycaemia and hyperinsulinaemia, the lack of differences between umbilical oxygen content and uptake seems also to suggest placental alterations occurring in these pregnancies [45].

Considerable progress has been made in recent years towards understanding the mechanisms that regulate placental handling of nutrients in human pregnancies. Most of the mechanism involved in placental uptake across the microvillous plasma membrane are well described for glucose as well as for amino acids. However, the extent to which nutrients are metabolized within the placenta and the mechanisms by which they are transported into the umbilical circulation across the basal membrane are not well understood. One of the most valuable models in reaching this objective has been the dually perfused human placenta, which allows investigation of both transport and metabolism, with gradients in both materno-fetal and feto-maternal directions.

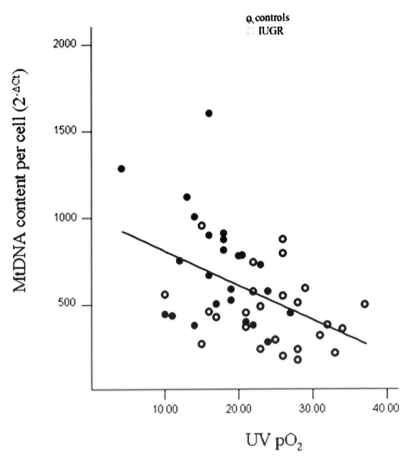

Figure 10.2 Relationship between placental mitochondrial DNA content and umbilical vein pO2 in normal (o) and IUGR fetuses (●). Reprinted with permission (Lattuada D, Placenta 2008) [41].

Placental transport of glucose, the most relevant nutrient during intrauterine life, depends on the maternofetal glucose concentration gradient. Maternofetal differences in glucose concentration increase during pregnancy due to a positive gradient in relation to increased fetal requirements [46]. In IUGR fetuses, the maternofetal glucose concentration difference is significantly increased in relation to the severity of IUGR, thus preserving a constant glucose placental transport [46]. Morover, studies infusing ^{13}C-glucose in IUGR pregnancies at the time of cordocentesis have not demonstrated a significant dilution of the tracer in the fetal circulation, suggesting that there is no compensatory glucogenesis in IUGR fetuses; the fetus is therefore dependent on the mother for the delivery of glucose [47].

Placental transport of amino acids is an active transport with fetal concentrations higher than the maternal. *In vitro* studies have shown that the placenta also determines the composition of amino acids transported to the fetus by interconversion of amino acids. Interorgan conversion cycles have been demonstrated between the placenta and the fetal liver for glutamate and glutamine as well as for glycine and serine [48]. Studies with stable isotopes infused into the mother before caesarean section or cordocentesis have further demonstrated that non-essential amino acids such as glycine and proline are produced in significant amounts in the fetal-placental unit, whereas the essential amino acids leucine and lysine are transferred in significant amounts from the maternal to the fetal circulation [49].

Recently, it has been suggested that the placenta acts as a 'nutrient sensor', regulating its transport function according to the ability of the maternal supply line to provide nutrients [50]. Placental amino acid uptake from the fetal circulation can also contribute to the regulatory function of the placenta. In human pregnancies, significant net placental uptake from the umbilical circulation has been demonstrated for glutamate and serine [51]. Recently, modification of fetal plasma amino acid composition has been demonstrated to occur *in vitro* through the activity of placental amino acid exchangers [52]. Studies of isolated perfused human placental cotyledons have shown that fetal boluses of alanine and leucine could stimulate the release of other amino acids from the placenta, suggesting a placental mechanism to modify the composition but not the total amount of amino acids in the fetal circulation.

In IUGR fetuses, umbilical venous amino acid concentrations have been reported to be significantly lower, with a decreased fetomaternal gradient, and this finding is independent of the severity of IUGR [53]. Studies using a maternal primed constant infusion of ^{13}C leucine have shown a decreased fetomaternal leucine enrichment ratio in relation to IUGR severity, suggesting the presence of increased fetal catabolism in IUGR [54].

The role of the placenta in the transfer of maternal fatty acids into the umbilical circulation has been investigated in some detail only in recent years. Studies show a modified fetal fatty acid profile that favours the availability of long-chain polyunsaturated fatty acids (LC-PUFA), which are needed for brain tissue accretion and membrane fluidity [55]. This enrichment of LC-PUFA during intrauterine life has been defined as a 'biomagnification process' and could be due to both placental and fetal metabolic processes. However, the degree to which the human fetus is capable of producing LC-PUFA by desaturation and elongation of their precursors is not clear, and therefore the placental supply of LC-PUFA is critical for the synthesis of structural lipids, and hence for normal fetal development, including, importantly, that of the brain [56]. Increasing PUFA concentrations have been shown to significantly increase the uptake of fatty acids in BeWo cells [57]. Recently the role of nuclear receptors (NR) like PPARs and LXRs in regulating placental uptake and synthesis of fatty acids has raised interest, as NR could represent the mediators of increased or decreased fatty acid placental uptake and metabolism by regulating the synthesis of genes such a CD36 and ABCA1 [58].

In IUGR fetuses significant changes have been reported in the ratios between the LC-PUFA docosahexaenoic acid and arachidonic acid and their precursors α linolenic acid and linoleic acid [59]. These differences might have an important role in the increased susceptibility to brain damage shown by IUGR fetuses.

Moreover, significant alterations in lipoprotein receptors and lipoprotein lipases, necessary for the fetal supply of maternally derived fatty acids, have been reported in IUGR placentas [60–63].

From adaptation to failure

Once diagnosed, IUGR is a condition that needs monitoring for the correct timing of delivery. However, it is quite difficult to establish a fixed protocol for this obstetric condition, as many factors need to be taken into account: maternal conditions, parents' expectations, gestational age, estimated fetal weight, and the evaluation of fetal well-being. In particular, the impact of prematurity on neonatal outcome needs to be weighed against the risks to the fetus of staying *in utero*. For the last 20 years IUGR fetuses have been monitored and classified for optimal timing of delivery by computerized cardiotocography, together with pulsed Doppler ultrasound for the study of the placental and fetal circulation to identify fetal damage better and earlier.

A classification of IUGR based on the pulsatility index (PI) of the umbilical artery and on computerized fetal heart rate (FHR) shows a progressive decrease in oxygen content and an increase in lactacidaemia with an abnormal umbilical artery PI and FHR [64].

Reduced umbilical venous oxygen tension and increased lactate concentration together reflect the fetal and placental metabolic switch to failure, which should be regarded as a terminal phase prior to intrauterine fetal death. The effect of early delivery, rather than delaying birth for as long as possible, was evaluated in the Growth Restriction Intervention Trial (GRIT), a multicenter controlled randomized trial performed in 13 countries [65]. Although no significant differences in mortality were observed when perinatal data were evaluated (intrauterine death was higher in the delayed group, but this was similar to neonatal death in the early group), the overall rate of severe disabilities at 2 years was higher in the early-delivered group in infants younger than 31 weeks at randomization [65].

A sequence of metabolic and cardiovascular changes are observed in the IUGR fetus prior to the development of overt hypoxia. Whereas amino acid concentrations are reduced in all groups of IUGR fetuses [53–63], the fraction of intrauterine protein catabolism increases progressively with increasing severity of IUGR [54]. When placental nutrient supply

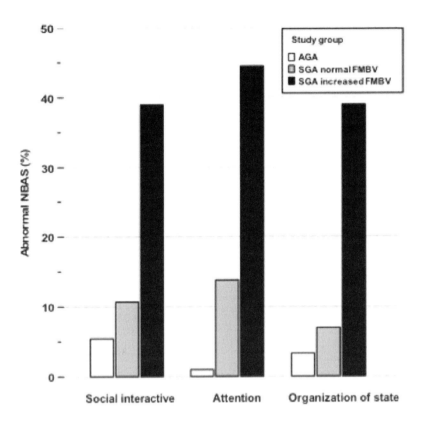

Figure 10.3 Increased brain perfusion discriminates SGA fetuses at risk for abnormal behaviour at 2 years of age, independently from umbilical arterial Dopplers. AGA, appropriate-for-gestational-age; SGA, small-for-gestational-age. FMBV, fractional moving blood volume; *NBAS*, Neonatal Behavioral Assessment Scale. *Reprinted with permission (Cruz-Martinaz R, American Journal of Obstetrics & Gynecology 2009).*

becomes insufficient, the fetus reduces its tissue deposition, increases the fraction of protein catabolism, and it probably reduces its oxygen consumption. At the same time, a redistribution of the blood flow to the fetal organs occurs, aimed at preserving brain perfusion and reducing the blood supply to the peripheral districts (brain-sparing effect) [66].

The temporal sequence of abnormal Doppler changes in the fetal circulation has been described in severe IUGR. Early changes occur in peripheral vessels (umbilical and middle cerebral arteries). Late changes (umbilical artery reverse flow and abnormal changes in the ductus venosus, aortic and pulmonary outflow tracts) are significantly associated with perinatal death [67].

As a follow up to the classification system developed in 1993 [64], three levels of pregressive severity have been identified in fetuses with a diagnosis of IUGR:

1. reduced placental nutrient transport (particularly for amino acids), no signs for reduced uteroplacental blood flow, no signs of hypoxia⇒initial adaptation;
2. reduced placental nutrient exchange (amino acids, fatty acids), reduced placental perfusion, oligohydramnios (reduced kidney function), brain sparing, reduced liver perfusion, initial signs of moderate hypoxia⇒initial organ failure;
3. reduced fetal nutrient utilization and increased catabolism, liver, heart and brain failure, severe hypoxia and hyperlactacidaemia⇒generalized failure leading to intrauterine or neonatal death.

Although the first and the third stages seem well characterized, new evidence challenges the view that cerebral vasodilation is an adaptive mechanism that preserves brain function. Recent data have shown that increased blood perfusion of the brain discriminates against SGA fetuses that are at risk for abnormal neuro-behaviour (Figure 10.3) [68]. Moreover, abnormal Doppler velocimetry in the umbilical artery has been shown to be predictive of necrotizing enterocolitis and mortality in the early neonatal period [69].

In summary, placental insufficiency is associated with a specific placental and fetal phenotype, with progressive levels of compromise leading to organ failure. The long-term outcome in relation to the different levels of severity needs to be further studied, as well as potential therapies for genes regulating nutrient placental transfer.

Acknowledgements

The contents of this review are largely due to the lifelong significant scientific contribution and constant mentorship of Giorgio Pardi, an outstanding teacher and scientist.

References

1. **Cetin I, Sparks JW**. Determinants of intrauterine growth. In: Hay WW, Thureen P, eds. *Neonatal Nutrition and Metabolism* 2nd edn. Cambridge: Cambridge University Press, 2005.
2. **Baschat AA**. Fetal responses to placental insufficiency: an update. *Br J Obstet Gynaecol* 2004; **111**: 1031–41.
3. **Alexander GR, Kogan M, Bader D** et al. US birth weight/gestational age-specific neonatal mortality: 1995–1997 rates for whites, hispanics, and blacks. *Pediatrics* 2003; **111**: e61–6.
4. **Baschat AA, Cosmi E, Bilardo CM** et al. Predictors of neonatal outcome in early-onset placental dysfunction. *Obstet Gynecol* 2007; **109**: 253–61.
5. **Prysak M, Lorenz RP, Kisly A**. Pregnancy outcome in nulliparous women 35 years and older. *Obstet Gynecol* 1995; **85**(1): 65–70.
6. **Visser GH, Sadovsky G, Nicolaides KH**. Antepartum heart rate patterns in small for gestational age third-trimester fetuses: correlations with blood gas values obtained at cordocentesis. *Am J Obstet Gynecol* 1990; **162**: 698–703.
7. **Berkowitz GS, Skovron ML, Lapinski RH, Berkowitz RL**. Delayed childbearing and the outcome of pregnancy. *N Engl J Med* 1990; **322** (10): 693–4.
8. **Bernstein IM, Horbar JD, Badger GJ, Ohlsson A, Golan**. A Morbidity and mortality among very-low-birth weight neonates with intrauterine growth restriction. *Am J Obstet Gynecol* 2000; **182**: 198–206.
9. **Gortner L, Wauer RR, Stock GJ** et al. Neonatal outcome in small for gestational age infants: do they really better? *J Perinat Med* 1999; **27**: 484–9.
10. **Pardi G, Cetin I**. Human fetal growth and organ development: fifty years of discoveries. *Am J Obstet Gynecol* 2006; **194**: 1088–99.
11. **Parazzini F, Cortinovis I, Botulus R, Fedele L**. Standard di peso alla nascita in Italia. *Ann Ost Gin Med Perinatol* CXII 1991; 203–46.
12. **Cetin I, Boito S, Radaelli T**. Evaluation of fetal growth and fetal well-being. *Semin Ultrasound CT MRI* 2008; **29**: 136–46.
13. **Enzi G, Zanardo V, Caretta F, Inelmen EM, Rubaltelli F**. Intrauterine growth and adipose tissue development. *Am J Clin Nutr* 1981; **34**: 1785–90.
14. **Molteni RA, Stys SJ, Battaglia FC**. Relationship of fetal and placental weight in human beings: fetal/placental weight ratios at various gestational ages and birth weight distributions. *J Reprod Med* 1978; **21**: 327–34.
15. **Sibley CP, Turner MA, Cetin I** et al. Placental phenotypes of intrauterine growth. *Pediatr Res* 2005; **58**: 827–32.
16. **Saemundsson Y, Svantesson H, Gudmundsson S**. Abnormal uterine artery Doppler in pregnancies suspected of a SGA fetus is related to increased risk of recurrence during next pregnancy. *Acta Obstet Gynecol Scand* 2009; **88**(7): 814–17.
17. **Berghella V**. Prevention of reccurent fetal growth restriction. *Obstet Gynecol* 2007; **110**: 904–12.
18. **Kupferminc MJ, Eldor A, Steinman N** et al. Increased frequency of genetic thrombophilia in women with complications of pregnancy. *N Engl J Med* 1999; **340**: 9–13.
19. **Infante-Rivard C, Rivard GE, Yotov WV** et al. Absence of association of thrombophilia polymorphism with intrauterine growth restriction. *N Engl J Med* 2002; **347**: 19–25.
20. **Franchi F, Cetin I, Todros T** et al. Intrauterine growth restriction and genetic predisposition to thrombophilia. *Haematologica* 2004; **89**: 444–9.
21. **Rodger M, Paidas M, McLintock C** et al. Inherited thrombophilia and pregnancy complications revisited. *Obstet Gynecol* 2008; **112**: 320–4.
22. **Roberts CT**. Review: complicated interactions between genes and the environment in placentation, pregnancy outcome and long term health. *Placenta* 2010; **31**: S47–S53.
23. **Diplas AI, Lambertini L, Lee MJ** et al. Differential expression of imprinted genes in normal and IUGR human placentas. *Epigenetics* 2009; **4**: 235–40.
24. **Grati FR, Miozzo M, Cassani B** et al. Fetal and placental chromosomal mosaicism revealed by QF-PCR in severe IUGR pregnancies. *Placenta* 2005; **26**: 10–18.
25. **Battaglia FC, Meschia G**. *An Introduction to Fetal Physiology*. London: Academic Press, 1986.

26. **Bell AW**, **Wilkening RB**, **Meschia G**. Some aspects of placental function in chronically heat-stressed ewes. *J Dev Physiol* 1987; **9**: 17–29.

27. **Heinonen S**, **Taipale P**, **Saarikoski S**. Weights of placentae from small-for-gestational age infants revisited. *Placenta* 2001; **22**: 399–404.

28. **Barker DJP**, **Bull AR**, **Osmond C** et al. Fetal and placental size and risk of hypertension in adult life. *Br Med J* 1990; **301**: 259–62.

29. **Pardi G**, **Marconi AM**, **Cetin I**. Placental-fetal interrelationship in IUGR fetuses – a review. *Trophoblast Res* 2002; **23**, S136–41.

30. **Marconi AM**, **Paolini CL**, **Zerbe G**, **Battaglia FC**. Lactacidemia in intrauterine growth restricted (IUGR) pregnancies: relationship to clinical severity, oxygenation and placental weight. *Pediatr Res* 2006; **59**: 570–4.

31. **Salafia CM**, **Charles AK**, **Maas EM**. Placenta and fetal growth restriction. *Clin Obstet Gynecol* 2006; **49**: 236–56.

32. **Biswas S**, **Ghosh SK**, **Chhabra S**. Surface area of chorionic villi of placentas: an index of intrauterine growth restriction of fetuses. *J Obstet Gynaecol Res* 2008; **34**: 487–93.

33. **Barker DJP**, **Thornburg KL**, **Osmond C**, **Kajante E**, **Eriksson**. The surface area of the placenta and hypertension in the offspring in later life. *Int. J. Dev. Biol* 2010; **54**: 525–30.

34. **Lumey LH**. Decreased birth weights in infants after maternal in utero exposure to the Dutch famine of 1944–1945. *Paediatr Perinatol Epidemiol* 1992; **6**(2): 240–53.

35. **Jansson T**, **Powell TL**. IFPA 2005 Award in Placentology Lecture. Human placental transport in altered fetal growth: does the placenta function as a nutrient sensor? a review. *Placenta* 2006; **27** Suppl A: S91–7.

36. **Brosens I**, **Robertson WB**, **Dixon**. The physiological response to vessels of the placental bed to normal pregnancy. *J Pathol Bacteriol* 1967; **93**, 569–79.

37. **Meekins JW**, **Pijnenborg R**, **Hanssens M**, **McFadyen IR**, **VanAsshe A**. A study of placental bed spiral arteries and trophoblast invasion in normal and severe pre-eclamptic pregnancies. *Br J Obstet Gynaecol* 1994; **101**: 669–74.

38. **Pijnenborg R**, **Bland JM**, **Robertson WB**, **Brosens I**. Uteroplacental arterial changes related to interstitial trophoblast migration in early human pregnancy. *Placenta* 1983; **4**: 397–413.

39. **Burton GJ**, **Woods AW**, **Jauniaux E**, **Kingdom JC**. Rheological and physiological consequences of conversion of the maternal spiral arteries for uteroplacental blood flow during human pregnancy. *Placenta* 2009; **30**: 473–82.

40. **Fowden AL**, **Forhead AJ**, **Coan PM**, **Burton GJ**. The placenta and intrauterine programming. *J Neuroendocrinol* 2008; **20**: 439–50.

41. **Lattuada D**, **Colleoni F**, **Martinelli A** et al. Higher mitochondrial DNA content in human IUGR placenta. *Placenta* 2008; **29**: 1029–33.

42. **Radaelli T**, **Boito S**, **Cozzi V** et al. Fetal oxygen consumption in term normal pregnancies. Society for Gynecologic Investigations, Los Angeles, CA, USA, March 23–26, 2005. *J Soc Gynecol Invest* 2005; Abstract **175**, 12.

43. **Acharya G**, **Sitras V**. Oxygen uptake of the human fetus at term. *Acta Obstet Gynecol Scand* 2009; **88**: 104–9.

44. **Radaelli T**, **Boito S**, **Taricco E** et al. Evaluation of Oxygen Uptake in IUGR Fetuses. Abstract #211 of the 55th Annual Meeting of the Society for Gynecologic Investigation, San Diego, CA, 2008.

45. **Taricco E**, **Radaelli T**, **Rossi G** et al. Effects of gestational diabetes on fetal oxygen and glucose levels in vivo. *Br J Obstet Gynaecol* 2009; **116**(13): 1729–35.

46. **Marconi AM**, **Paolini C**, **Buscaglia M** et al. The impact of gestational age and fetal growth on the maternal-fetal glucose concentration difference. *Obstet Gynecol* 1996; **87**: 937–42.

47. **Marconi AM**, **Cetin I**, **Davoli E** et al. An evaluation of fetal glucogenesis in intrauterine growth-retarded pregnancies. *Metabolism* 1993; **42**: 860–4.

48. **Cetin I**. Amino acid interconversions in the fetal-placental unit: the animal model and human studies in vivo. *Pediatr Res* 2001; **49**: 148–54.

49. **Cetin I**, **Marconi AM**, **Baggiani AM** et al. In vivo placental transport of glycine and leucine in human pregnancies. *Pediatr Res* 1995; **37**: 571–5.

50. **Paolini CL**, **Marconi AM**, **Ronzoni S** et al. Placental transport of leucine, phenylalanine, glycine, and proline in intrauterine growth-restricted pregnancies. *J Clin Endocrinol Metab* 2001; **86**: 5427–32.

51. **Cetin I**, **Nobile De Santis MS**, **Taricco E** et al. Maternal and fetal amino acid concentrations in normal pregnancies and in pregnancies with gestational diabetes mellitus. *Am J Obstet Gynecol* 2005; **192**: 610–17.

52. **Cleal JK**, **Brownbill P**, **Godfrey KM** et al. Modification of fetal plasma amino acid composition by placental amino acid exchangers in vitro. *J Physiol* 2007; **582**: 871–82.

53. **Cetin I**, **Ronzoni S**, **Marconi AM** et al. Maternal concentrations and fetal-maternal concentration differences of plasma amino acids in normal and intrauterine growth-restricted pregnancies. *Am J Obstet Gynecol* 1996; **174**(5): 1575–83.

54. **Marconi AM**, **Paolini CL**, **Stramare L** et al. Steady state maternal-fetal leucine enrichments in normal and intrauterine growth-restricted pregnancies. *Pediatr Res* 1999; **46**(1): 114–19.

55. **Cetin I**, **Koletzko B**. Long-chain ω-3 fatty acid supply in pregnancy and lactation. *Curr Opin Clin Nutr Metab Care* 2008; **11**: 297–302.

56. **Haggarty P**. Placental regulation of fatty acid delivery and its effect on fetal growth–a review. *Placenta* 2002; **23** Suppl A: S28–38.

57. **Tobin KA**, **Johnsen GM**, **Staff AC**, **Duttaroy AK**. Long-chain polyunsaturated fatty acid transport across human placentachoriocarcinoma (BeWo) cells. *Placenta* 2009; **30**: 41–7.

58. **Stefulj J**, **Panzenboeck U**, **Becker T** et al. Human endothelial cells of the placental barrier efficiently deliver cholesterol to the fetal circulation via ABCA1 and ABCG1. *Circ Res* 2009; **104**(5): 600–8.

59. **Cetin I**, **Giovannini N**, **Alvino G** et al. Intrauterine growth restriction is associated with changes in polyunsaturated fatty acid fetal-maternal relationships. *Pediatr Res* 2002; **52**: 750–5.

60. **Magnusson AL**, **Waterman IJ**, **Wennergren M**, **Jansson T**, **Powell TL**. Triglyceride hydrolase activities and expression of fatty acid binding proteins in the human placenta in pregnancies complicated by intrauterine growth restriction and diabetes. *J Clin Endocrinol Metab* 2004; **89**: 4607–14.

61. **Tabano S**, **Alvino G**, **Antonazzo P** et al. Placental LPL gene expression is increased in severe intrauterine growth restricted pregnancies. *Pediatr Res* 2006; **59**: 250–3.

62. **Wadsack C**, **Tabano S**, **Maier A** et al. Intrauterine growth restriction (IUGR) is associated with alterations in placental lipoprotein receptors and maternal lipoprotein composition. *Am J Physiol Endocrinol Metab* 2007; **292**: 476–84.

63. **Gauster M**, **Hiden U**, **Blaschitz A** et al. Dysregulation of placental endothelial lipase and lipoprotein lipase in intrauterine growth-restricted pregnancies. *J Clin Endocrinol Metab* 2007; **92**: 2256–63.

64. **Pardi G**, **Cetin I**, **Marconi AM** et al. Diagnostic value of blood sampling in fetuses with growth retardation. *N Engl J Med* 1993; 11; **328**(10): 728–9.

65. **Thornton JG**, **Hornbuckle J**, **Vail A**, **Spiegelhalter DJ**, **Levene M**; GRIT Study Group. Infant wellbeing at 2 years of age in the Growth Restriction Intervention Trial (GRIT): multicentred randomised controlled trial. *Lancet* 2004; **364**: 513–20.

66. **Cetin I**, **Boito S**, **Radaelli T**. Evaluation of fetal growth and fetal well-being. *Semin Ultrasound CT MRI* 2008; **29**(2): 136–46.

67. **Ferrazzi E**, **Bozzo M**, **Rigano S** et al. Temporal sequence of abnormal Doppler changes in the peripheral and central circulatory systems of the severely growth-restricted fetus. *Ultrasound Obstet Gynecol* 2002; **19**: 140–6.

68. **Cruz-Martinez R**, **Figueras F**, **Oros D** et al. Cerebral blood perfusion and neurobehavioral performance in full-term small-for-gestational-age fetuses. *Am J Obstet Gynecol* 2009; **201**(5): 474. e1–7.

69. **Bhatt AB**, **Tank PD**, **Barmade KB**, **Damania KR**. Abnormal Doppler flow velocimetry in the growth restricted foetus as a predictor for necrotising enterocolitis. *J Postgrad Med* 2002; **48**(3): 182–5.

Discussion

ROBERT BOYD: We see a lot of placental abnormalities. Do they have a causal role in fetal programming or are they placenta-restricted responses to whatever is programming the fetus?

CETIN: Programming is probably also going on at the placental level. It is probably also taking place at the fetal level, but we don't know exactly what is going on. The placenta is already damaged very early on in pregnancy. We have always said that the fetus has exponential growth after midgestation and that the placenta is programmed much more than the fetus at the beginning of pregnancy. But really we don't know what input the fetus receives earlier on. All the metabolic data start at 20 weeks, not earlier, although there are now some markers of pre-eclampsia from the first trimester. The mother is probably affected very early on.

TYCKO: It would be nice if we could start from your first thoughts of inadequate trophoblast invasion. Are there any data that distinguish this from the other possibility that the maternal vessels have a deficit that make them poorly susceptible to the proper transformation? How does one distinguish the primary pathogenesis being trophoblast versus maternal?

CETIN: We have debated this for many years. For amino acid transport systems, we knew that they were reduced, but we didn't know whether this was an adaptation. Two studies come to my mind that are relevant here. One is with the *Igf2* knockout mice which have changes in placental transport systems before the intrauterine growth restriction occurs. Similarly, in the rat model of growth restriction the placental changes occur first, in both the growth and transport systems of the placenta, before there is a decrease in fetal growth. These suggest that changes in the placenta are causing the growth restriction.

TYCKO: It seems that the pathogenesis of IUGR in humans is related to vascular insufficiency. Amino acid transport deficiency may occur, but still one has to explain the vascular insufficiency. It seems this could be due to either poor trophoblast function early on or poor receptivity of the maternal environment so those vessels don't transform.

POSTON: I think this is an important point. Work in pre-eclampsia suggests that women who get pre-eclampsia have underlying cardiovascular risk. Some people think that pre-eclampsia unmasks cardiovascular risk in women who are more prone to cardiovascular disease in later life. There is good evidence that they may have cardiovascular risk to start with. These are people with underlying dyslipidaemia and endothelial cell dysfunction. There is emerging evidence from the SCOPE cohort in Auckland and Adelaide that low birth weight of the mother is a strong predisposing factor to pre-eclampsia in first pregnancy. This really does suggest that underlying cardiovascular risk may be a predisposing factor. I have always thought that there has been too little emphasis on trying to understand the maternal endothelial role in this.

CETIN: There can be many causes of IUGR. Just the maternal BMI can predict a number of potential alterations of fetal growth. On the other hand, it is not the only explanation. The maternal phenotype doesn't account for all cases of pre-eclampsia or IUGR. There are probably some situations where the endothelial damage in the mother is the leading cause, or other situations such as anorexic mothers where there is nutritional deficiency. Most cases of IUGR involve absolutely normal mothers.

BURTON: Brosens' original findings showed that trophoblast invasion in pre-eclampsia is a mixed bag. Some cases have superficial invasion, whereas others have equally deep invasion, but it doesn't go into the arteries. So there may be something different in the arteries.

PIJNENBORG: I would suggest you replace the first box in your diagram (inadequate trophoblast invasion) by 'inadequate uteroplacental blood flow'. We don't know the causes of this inadequate flow. Trophoblast invasion is just one possible factor, but it may well be that increased flow precedes increased invasion. Trophoblast invasion and spiral artery remodelling is much more complex than we think, starting in the centre and then spreading to the periphery of the placental bed. We still don't understand the different steps of spiral artery remodelling and we still have no adequate tools to follow this process in the course of pregnancy.

SIBLEY: One comment on the blood flow issue: your data have shown that the uterine blood flow is more than sufficient for the IUGR fetus; what is happening is that the IUGR fetus is not able to metabolize amino acids efficiently. There are two effects of blood flow. Altered blood flow can lead to altered growth, but the fact that you get ischaemia–reperfusion injury early on could cause the primary damage to the placenta and then the growth restriction could follow this. When people write grants for my group they always start off by saying we are interested in IUGR because of neonatal morbidity and mortality, and because of later programming. I cross this out, and say that the evidence is that the relationship is between later programming and fetal size across the normal range. How much do you think understanding IUGR helps us understand programming?

BARKER: That is exactly the question I was going to ask. What are the negative experiences that a six-pound baby has had which are not shared with an eight-pound baby? And do you think studying this heterogeneous group of very small babies is going to illuminate our problem?

CETIN: It is going to illuminate the problem as much as Lorna Moore's data will. These extreme models don't necessarily reflect the range of what we consider to be normal situations. But what is normal?

BARKER: We want illumination of the usual experiences of babies with normal healthy mothers who end up at six pounds rather than eight. I want to comment on your interesting slide that showed that despite head sparing, certain IUGR babies have evidence of brain damage. Kent Thornburg and I have published a paper on stroke. Two kinds of babies are at risk of stroke. One has a small head circumference (the inference being that something went wrong with brain development). The other has a relatively large head in relation to the length of the body. This speaks to your point that in order to be 'brain spared' something negative has to happen, and this negative thing seems to set up stroke in later life.

GIUSSANI: I want to come back to this question of where the programming is happening, and change not only some of the boxes, but also the direction of some of the arrows in the last diagram you showed. The fetus is not a passive box at the end funnel: it can feed back to the placenta. Perhaps we can get some clues from the fetal oxygen uptake data you have shown. The short-term benefit under adverse nutrition conditions is to reduce growth trajectory. Then it is needing less oxygen to maintain that growth trajectory. It is matching and feeding back.

THORNBURG: I want to reinforce David Barker's point about whether studying these small babies is important for programming. He answered himself by saying that it is important to see how brain sparing works. I am interested in these small babies because they tell us something about the biological capability of a fetus in adapting to a poor environment.

CETIN: One other possibility is that these IUGR babies are those who have been unable to adapt. They have completely different metabolic reactions to similar damage at the beginning. Maybe those who don't become IUGR were able to adapt in a different way.

THORNBURG: In animal models we often use extreme cases in order to understand biological mechanisms, and you are seeing these in some of your studies. So they shouldn't be discounted.

JACKSON: Irene, you may be uniquely placed to address this question. In hypoxia–reperfusion injury in a number of organs or tissues, there is a specific protective effect by superfusion with glycine that isn't accounted for by glutathione or acting as an osmolyte. Is there any evidence in the context of the placenta that you have been talking about of this particular protective effect?

CETIN: I don't know of any data on this.

ROBERT BOYD: The reduced oxygen extraction may not be a permeability effect; it might result from altered geometry of the vasculature on the maternal side.

Chapter 11

Uterine blood flow as a determinant of fetoplacental development

Lorna G. Moore

We shall not cease from exploration
And the end of all our exploring
Will be to arrive where we started
And know the place for the first time. [1]

Introduction

In keeping with the theme of this book, this chapter explores a critical aspect of fetal programming; namely the determinants of uterine[1] blood flow and their role in fetoplacental development. This important topic has been the subject of investigation for more than 50 years. Early studies were begun by Donald Barron and his distinguished team in the 1950s in Peru, and at about the same time by John Lichty and co-workers in Colorado. These studies combined with later work done by our and other groups have firmly established that high altitude restricts fetal growth and therefore serves as a valuable study design for investigating the determinants of fetoplacental development. The lessons learned from these studies are reviewed here, including current work investigating the physiological and genetic mechanisms regulating uterine oxygen delivery. This chapter closes with a brief discussion of areas likely to be fruitful for future research.

Early studies

Normal fetoplacental development requires adequate and sustained tissue delivery of oxygen. Given the lower barometric pressure and the resultant decline in the partial pressure of oxygen, high altitude has served as a natural laboratory for studying the effects of hypoxia on fetoplacental development. Among the first such studies were those conducted by some of the 'greats' of fetal physiology – Donald Barron, André Hellegers, William Huckabee, Giacomo Meschia, James Metcalfe, Harry Prystowsky and others. With support from the Josiah Macy Foundation, this distinguished team showed in sheep native to Morococha, Peru (4900 m) that

> …despite a reduction of approximately one-half in the oxygen pressure gradient between the maternal and fetal bloods, the sheep fetus at altitude lives in an internal environment in which the oxygen pressure and the quantity available to it are not significantly different from that of a fetus at sea level. [2]

These authors concluded that alterations in each step in the oxygen supply chain helped defend fetal oxygenation at high altitude, but that an important factor, if not the most important, was one-third greater uterine artery (UA) blood flow [3]. They speculated that if a similar increase occurred in humans, intervillous oxygen tensions would be only slightly reduced at high compared to low altitude, or within 5 mmHg if UA blood flow were increased by 50% [4]. Of course, no measurements of UA blood flow in pregnant women at high altitude were available at the time, but in light of the observations reported below, these early investigators were remarkably prescient.

No altitude-associated fetal growth restriction was present in these or later studies in sheep [2,5]. Metcalfe also noted that 'not one instance of twin pregnancy' was observed in their Peruvian studies of some 50 ewes [4]. Sheep have long been domesticated and therefore subject to extensive breeding for important animal husbandry-related attributes such as twinning. Perhaps species or breeds prone to twinning have the

---

[1] 'Uterine' is used here to refer to the maternal portion of the uteroplacental circulation, including its myometrial and decidualized regions. It is used rather than 'uteroplacental' to avoid confusion as to whether the maternal or the fetal side of the placenta is being considered.

The Placenta and Human Developmental Programming, ed. Graham J. Burton, David J. P. Barker, Ashley Moffett and Kent Thornburg. Published by Cambridge University Press. © Cambridge University Press 2011.

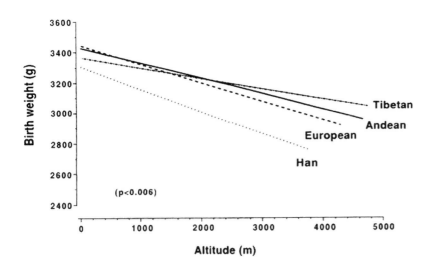

Figure 11.1 Infant birth weights at or near sea level are similar but decline with increasing altitude, with the magnitude of decline being inversely proportional to the duration (in generations) of high-altitude residence ($p < 0.006$). Shown are the best-fit regression lines for 4.6 million births as reported in 31 studies, weighted by sample size and variance and drawn for the altitude-ranges sampled. Adapted from the original [12] where reference citations may be found.

capacity to raise UA blood flow to a greater extent than those usually limited to singleton pregnancies. If so, then it might be speculated that only one member of what was initially a twin pregnancy could survive at high altitude, given finite limitations as to how much elevation of blood flow is possible. By extension, perhaps such a capacity to raise UA blood flow protects species or breeds prone to twinning (when experiencing a singleton pregnancy) from fetal growth restriction at high altitude.

High-altitude studies of human fetoplacental development also began in the 1950s. John Lichty and his team at the University of Colorado demonstrated that birth weights in Leadville (3100 m) were lower than at sea level, with intermediate values present in Denver (1600 m) [6]. Careful analysis of other factors that influence birth weight, such as gestational age, maternal nutrition, Hispanic ethnicity, and obstetrical practices (e.g. induction of labour, frequency of caesarean sections) led the authors to conclude that it was chronic hypoxia and not some other attribute of the high-altitude physical or socio-economic environment that was responsible. Importantly, this study demonstrated that the decline in birth weight was due principally to slowed fetal growth, not shortened gestation, and thus was the first to demonstrate the existence of fetal growth restriction on a population scale [7].

More recent human studies also support the notion that chronic hypoxia and not some other factor associated with high-altitude residence is responsible for this decline [8]. We [9] and Giussani [10] have shown that socio-economic status or other risk factors (e.g. primiparity, lack of prenatal care) could not account for the altitude-related reduction in birth weight. Studies across a 150–4300 m gradient in Peru demonstrated that the decline became apparent above 2500 m or the elevation at which arterial oxygen saturation or content begins to fall [11]. Given that arterial oxygen tensions fall linearly but oxygen content curvilinearly (owing to the shape of the haemoglobin – oxygen dissociation curve), this suggests that it is the reduction in oxygen content rather than partial pressure that is most important for deciding fetal growth.

Population variation in the effect of altitude on birth weight

Although an altitude-related decline in birth weight occurs in every population studied so far, its magnitude varies in relation to the duration of high-altitude residence [12]. First noted in the Andes by Haas [13] and then by our group in the Himalayas [14], the data summarized in Figure 11.1 show that the birth weight decline is greatest in Han ('Chinese'), the group with the shortest history of high-altitude residence; next greatest in Europeans, a group that has resided at high altitudes in the Americas for up to 500 years;[2] and least in Andeans and Tibetans, for which archaeological and palaeontological evidence of human residence

---

[2] There are no permanently inhabited communities in Europe above 2500 m.

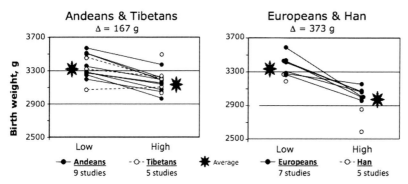

Figure 11.2 Long-resident groups, Andeans and Tibetans, have half the birth weight decline at high versus low altitude ($\Delta = 167$ g) compared to shorter-resident European and Han ('Chinese') groups ($\Delta = 373$ g). Variation among these populations shows the same pattern as seen in Figure 11.1; Tibetans have the least decline ($\Delta = 71$ g), then Andeans ($\Delta = 208$ g), followed by Europeans ($\Delta = 340$ g) and lastly the Han ($\Delta = 410$ g). Stars represent values averaged across all reports and points connected by lines are from studies conducted by the same investigator at low and high altitudes, some of which were included in Figure 11.1 [9–11,13,80–85]. More recent studies in Andeans [17,23,54,86–88], Tibetans [68,89,90], Europeans [17,23,54] and Han [68,90] are included as well.

at high altitude extends back 10 000 and 20 000 years, respectively [15]. Haas and co-workers investigated the possibility that being born and raised at high altitude exerts a protective effect, but found that one or a few generations of high-altitude residence is not sufficient [16]. Studies in larger samples and with better controls for the duration of high-altitude residence will be useful for settling this question.

The idea that Andeans and Tibetans are resistant to altitude-associated reductions in fetal growth has recently been questioned by Zamudio and co-workers, who stated that 'Andeans and Tibetans appear to have increased fetal growth, regardless of altitude' and therefore that their heavier birth weights do not represent a specific adaptation to high altitude [17]. To re-examine the protective effect of multigenerational residence at high altitude, we compared birth weights for all studies conducted by a single investigator at low and high altitudes and for newer studies appearing since our 2001 review (Figure 11.2). Most of the high-altitude studies were conducted at 3600 m, a coincidence occasioned by the fact that the two largest high-altitude cities – Lhasa, Tibet Autonomous Region, China, and La Paz, Bolivia – are at the same altitude. The low-altitude sites are more variable but all below 2500 m.

Key for the interpretation of these studies is to recognize that there are multiple influences on birth weight at any altitude, and hence such factors must be controlled as far as possible. This is particularly important at high altitude, where comprehensive vital statistics are not available and considerable variation exists among samples as the result of pronounced socio-economic inequality. Studies conducted by the same investigative team have the advantage of controlling for such factors to the extent possible, and show that altitude clearly depresses birth weight in all groups (Figure 11.2). Importantly, these data support our earlier conclusion that the magnitude of the birth weight decline is twice as great in short- than in long-resident groups, and that variation among groups in the magnitude of decline parallels the pattern shown in Figure 11.1, with Han 'Chinese' having the greatest decline in birth weight and Tibetans the least (see Figure 11.2 legend for details). Figure 11.2 also shows that although low-altitude birth weights are variable, they are not consistently greater in Andeans and Tibetans than in Europeans or Han. It is of interest that the 3367 g birth weight reported by Zamudio is the greatest Andean high-altitude value in Figure 11.2. This may be related to the fact that this important study required that all women be undergoing elective caesarean, as it was designed to collect placentas and assess neonatal oxygenation at delivery. Women electing caesarean sections in Bolivia are drawn disproportionately from higher socioeconomic groups, and so. factors associated with high socio-economic status may have contributed to the heavier birth weights observed in that study. Supporting this possibility is the secular trend for birth weights to increase over time at high altitudes in Colorado with improvements in standards of living [18], and the high value (3580 g) reported for babies born to Bolivian women of high socio-economic status residing at low altitude [10]. Thus comprehensive examination of past as well as more recent data does not support the contention that Andeans

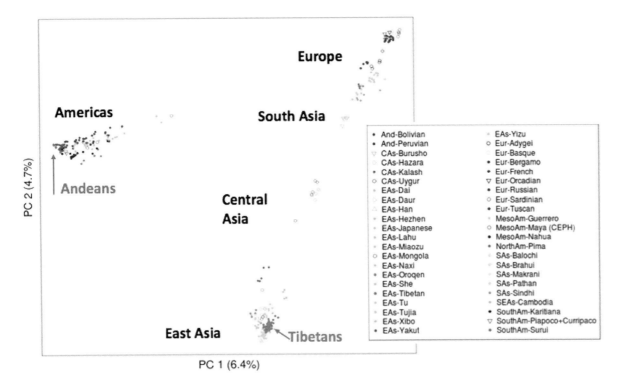

Figure 11.3 The first and second principle components (PC1, PC2) express the major relationships among 906 000 single nucleotide polymorphisms (SNPs) for the study populations shown in the legend and as described previously ([19,20], M. Bauchet, pers. comm.; A. Bigham, pers. comm.). Major geographic regions and the Andean and Tibetan samples are labelled. The clustering of data points demonstrates that Andeans and Tibetans are much more closely related to their geographic neighbours than to each other. (See colour plate section.)

or Tibetans weigh more at any altitude, but rather supports the notion that these groups are preferentially protected from altitude-associated fetal growth restriction.

The attribution of a close genetic relationship between Andeans and Tibetans in the Zamudio report [17] cited older studies of mitochondrial (mt) DNA. Such studies were limited by the relatively small amount of genetic variation able to be ascertained by the restriction-fragment length polymorphism-based methods available at the time. In addition, mtDNA comprises a small portion of the human genome. We therefore used the extensive recent studies of worldwide variation in nuclear DNA to address the genetic relationship between Tibetans and Andeans. Figure 11.3 shows high-density multilocus single nucleotide polymorphism (SNP) scores from 347 individuals in 44 populations obtained with the Affymetrix Inc (Santa Clara, CA) Genome-Wide Human SNP Array 6.0 chip, which uses 869 225 SNPs located on autosomes and 36 775 located on X chromosomes (M. Bauchet, pers. comm., A. Bigham, pers. comm.). As detailed elsewhere [19,20], principle component (PC) analysis demonstrates a striking concordance with geographic location, with PC1 reflecting a north–south axis and PC2 an east–west axis (Figure 11.3). Hence we concluded that Andeans and Tibetans are much more closely related to their geographic neighbours than to each other, and therefore that it is unlikely that their higher birth weights at high altitude are due to a close genetic relationship.

In summary, high altitude lowers birth weight by slowing fetal growth. Reduction in birth weight varies among populations but is much less marked in groups with longer – rather than shorter – residential histories. Understanding the basis of this population variation is useful not only for identifying the physiological mechanisms responsible for hypoxia-associated reductions in fetal growth, but also for testing the role of genetic factors.

Table 11.1 Determinants of calculated arterial oxygen content during pregnancy at low and high altitude.

Variable	Ref no.	Altitude, m	Europeans			Andeans		
			Non-preg	Wk 20	Wk 36–38	Non-preg	Wk 20	Wk 36–38
$P_{ET}CO_2$ or P_aCO_2, mmHg	17	400			35.2			35.7
	21, 23	3600	33.5	29.1	28.9	35.9	31.1	31.1
	17	3600			28.0			26.5
SaO_2, %	17	400			97.6			97.8
	23	400	98.8	98.2	98.5	99.0	98.1	98.6
	17	3600			91.4			91.1
	23	3600	91.6	94.4	94.2	92.3	94.5	94.0
Hb, g/dL	17	400			12.0			11.1
	23	400	11.9	10.4	10.7	11.3	10.3	10.3
	31	1600	14.0		12.6			
	31	3100	15.8		14.5			
	17	3600			15.3			14.6
	23	3600	14.5	13.3	13.5	14.3	13.3	13.2
CaO_2, mL O_2 dL min	17	400			15.7			14.6
	23	400	16.0	13.9	14.3	15.2	13.7	13.8
	31	1600	18.0		16.4			
	31	3100	19.4		18.2			
	17	3600			18.7			17.8
	23	3600	17.9	17.3	17.3	17.7	17.0	16.9

Uterine oxygen delivery in European versus. Andean high-altitude residents

Uterine oxygen delivery is the product of the concentration of oxygen in the arterial blood (CaO_2) and the volumetric blood flow to the uterine circulation. Given the effects of high altitude and the population variation described above, we hypothesized that women from longer- versus shorter-resident high-altitude groups differed in one or more of the physiological determinants of uterine oxygen delivery. We have tested this hypothesis in studies conducted in Tibet and, more recently, in Bolivia, where we employed a longitudinal study design in which the determinants of uterine oxygen delivery – CaO_2 and UA blood flow – were measured serially during pregnancy and again postpartum in high- as well as in low-altitude residents. Bolivia was chosen as it is the only country in the world where newcomer (European) and long-resident (Andean) groups live both at low and high altitudes. A total of 212 women were enrolled in these longitudinal studies, comprising approximately equal numbers of Andeans and Europeans living at high (3600–4100 m, La Paz/El Alto) or low altitude (400 m, Santa Cruz). Further details about the composition of these samples and the conduct of the studies are available elsewhere [21–23].

Arterial oxygen content

CaO_2 can be measured directly in arterial specimens but otherwise must be calculated using haemoglobin concentration (Hb) and arterial oxygen saturation (SaO_2). Table 11.1 summarizes calculated values that have been recently reported in the literature for women living at low or high altitude. Changes are evident in each of the factors determining CaO_2. Increased progesterone and oestrogen production, together with the pregnancy-related rise in metabolic rate and ventilatory sensitivity to hypoxia, raise alveolar ventilation [21,24,25]. Using end-tidal or arterial PCO_2 as an index of alveolar ventilation per unit CO_2 production, the data in Table 11.1 show that the 14% pregnancy-associated rise in alveolar ventilation is similar in the European and Andean groups, despite the Andeans' lower non-pregnant ventilation. Whereas the rise in ventilation has little effect on maternal SaO_2 at low altitude (owing to the already nearly maximal levels of SaO_2 present), SaO_2 rises appreciably at high altitude and does so similarly in the two ancestry groups

(Table 11.1). Pregnancy also lowers [Hb] ~1 g/dL as a result of greater expansion of plasma volume than of red cell mass. This effect of pregnancy on blood volume or [Hb] does not appear to differ between high and low altitude [21,23] or between ancestry groups, but values are greater at high altitude due to the well-known erythropoietic effect of chronic hypoxia (Table 11.1). As a result, CaO_2 falls slightly with pregnancy but is higher at high than at low altitude.

Uterine artery blood flow

Uterine artery (UA) blood flow can be measured directly in experimental animals but not in humans because of ethical concerns or limitations imposed by the effects of anaesthesia. Although a portion of uterine blood flow comes from a branch of the ovarian artery, the overwhelming majority comes from the two main uterine arteries (UA). We therefore concentrated on UA blood flow. In studies carried out in the mid-1980s, we were the first to demonstrate that UA blood flow could be estimated pericutaneously by Doppler ultrasound as the product of UA cross-sectional area (πr^2, where r is the vessel radius) and the average speed with which blood flows through the vessel throughout the cardiac cycle (the time-averaged mean velocity [TAM]). Because the resultant value is strongly influenced by operator-dependent decisions as to where cursors are placed for measuring diameter and the probe positioned for obtaining the sufficiently low angle of insonation required for accurate assessment of flow velocity, we have been careful to point out the importance of restricting comparisons to measurements conducted by the same operator [23,26,27]. Krampl and co-workers and Zamudio and her study team have also made extensive measurements of UA blood flow and/or resistance indices during pregnancy at low and high altitudes [17,28].

The best-controlled studies are those recently reported by Julian *et al.*, in which the same operator and instrument were used for all measurements [27]. These studies showed that the pregnancy-associated increase in UA diameter was markedly reduced at 3100 m versus 1600 m in Colorado (Table 11.2). The pregnancy rise in UA blood flow velocity was unaffected, but as blood flow increases exponentially with vessel diameter, UA blood flow at week 36 was 22% lower at high than at low altitude.

Comparing ancestry groups, pregnancy increases UA diameter and hence blood flow to a greater extent in Andean than in European high-altitude residents [17,22,23]. Given that it was not possible for the same operator to perform serial studies at two locales in Bolivia, we controlled for operator- or other location-specific effects by expressing our data as the change in UA blood flow from the non-pregnant state to week 20 or to week 36 (Figure 11.4).

To put the values in Figure 11.4 into a broader perspective, Table 11.2 includes all published data across a range of altitudes for UA diameter, TAM, and unilateral as well as estimated bilateral volumetric flows and UA oxygen delivery. The important studies of Thaler, Bernstein and others employing a transvaginal rather than a pericutaneous probe are not included, as such studies sample a different portion of the UA [29,30]. At low altitudes, UA diameter doubles with pregnancy and does so similarly in European and Andean women (Table 11.2). TAM rises appreciably as well, with the magnitude of change varying as a function of the non-pregnant starting point and method employed for calculating TAM.[3] Low-altitude Andean and European women are alike in terms of their pregnancy-associated changes in TAM; therefore, the increases in UA blood flow and UA oxygen delivery during pregnancy are similar as well (Table 11.2). Pregnancy increases UA diameter at high altitude in all groups, but within a given study the rise in Europeans is less than that seen in Europeans at low altitude or in Andeans at high altitude (Table 11.2). In other words, whereas high-altitude residence diminishes the pregnancy-associated increase in UA diameter in Europeans, Andean women are unaffected and double their UA diameters just as they do at low altitude. Somewhat greater UA diameters were seen in the Andean women we studied than in those reported by Zamudio [17,22,23], but this is probably due to machine or protocol differences, as illustrated by the variation observed within our Colorado studies conducted at the same altitude [27,31]. Likewise, methodological factors contribute to variation in the pregnancy-associated rise in TAM for Europeans, but

---

[3] Our early flow velocity studies [26,31] used a dedicated flow velocimeter with a patented algorithm for calculating the instantaneous mean of the velocity distribution and then averaging this value throughout the cardiac cycle. With advances in Doppler ultrasound technology, subsequent studies used the time-averaged mean (TAM) value derived from the region of the Doppler trace with the greatest spectral density.

Chapter 11. Uterine blood and fetoplacental development

Table 11.2 Determinants of uterine artery (UA) blood flow during pregnancy at low and high altitude.

Variable	Ref no.	Altitude, m	Europeans			Andeans		
			Non-preg	Wk 20	Wk 36–38	Non-preg	Wk 20	Wk 36–38
UA dia, cm	17	400			0.44			0.50
	23	400	0.26	0.49	0.50	0.21	0.47	0.50
	26	1600	0.14	0.28	0.34			
	27	1600	0.34	0.53	0.50			
	31	3100	0.13		0.25			
	27	3100	0.37	0.45	0.45			
	17	3600			0.38			0.46
	23	3600	0.41	0.54	0.54	0.39	0.63	0.65
UA TAM, cm s	17	400			36			35
	23	400	8	25	31	6	25	31
	26[1]	1600	8	38.5	61.4			
	27	1600	5	32	45			
	31[1]	3100	15		70			
	27	3100	9	38	46			
	17	3600			33			27
	23	3600	10	25	30	10	28	36
Unilateral UA flow, mL min	17[2]	400			312			380
	23	400	20	280	380	20	270	320
	26	1600	9	142	353			
	27	1600	31	463	613			
	31	3100	10		210			
	27	3100	63	398	477			
	17[2]	3600			225			300
	23	3600	50	380	400	50	550	730
Bilateral flow (mL min kg newborn)§		400			183, 227			214, 200
		3600			150, 269			181, 452
Unilateral UA O$_2$ delivery, mL/min	17[2]	400			50			55
	23	400	4		41 57	2	41	49
	17[2]	3600			41			49
	23	3600	12	66	70	11	92	130
Bilateral O$_2$ delivery (mL min kg newborn)§		400			32, 34			32, 31
		3600			29, 47			30, 80

Notes:
[1] As noted in the text, a dedicated flow velocimeter was employed in these studies, which used a different algorithm for calculating TAM.
[2] Values were reported as bilateral flows or oxygen delivery and therefore have been divided in half here.
§ = The first set of values are those reported in (17) and the second set are those provided in (23).

ancestry groups have similar near-term TAM within a given study.

Because blood flow rises exponentially with diameter but only linearly with velocity, Andeans had much greater pregnancy-associated increases in UA blood flow than did the European women [22,23]. Curiously, this is not the case in the Zamudio study [17], where despite the Andeans' larger diameters, UA flows were only slightly greater in the Andean and European groups at term. The source of this discrepancy between studies is unclear: we suspect that differences in the phase of gestation, operator and/or instrument effects may be responsible. In particular, the later gestational ages in the Zamudio study may be important, as fetal growth slows shortly before delivery. If UA blood flow decreases as well, then even small differences in gestational age could have an important effect on the values obtained. Because the Andean women

Chapter 11. Uterine blood and fetoplacental development

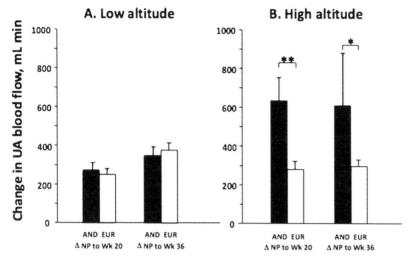

Figure 11.4 Shown is the change in uterine artery (UA) blood flow from the non-pregnant state to week 20 or week 36 of pregnancy as modified from the original [23]. Panel A: At low altitude (400 m, Santa Cruz, Bolivia), the change in UA blood flow from the non-pregnant to pregnant state is the same in women of Andean (AND) or European (EUR) ancestry. Panel B: Pregnancy at high altitude (3600 m, La Paz, Bolivia) increases UA blood flow more than twice as much in Andean as in European women. *$p < 0.05$, **$p < 0.01$.

were studied at somewhat later gestational ages, this might have also depressed Andean blood flows and contributed to the lesser ancestry-group differences. Regardless of the source of differences between studies, high-altitude Europeans as well as Andeans either maintained or had greater UA oxygen delivery near term than the values seen at low altitude. In the Zamudio report this was due to higher CaO_2, whereas in our study both greater CaO_2 and UA blood flow were responsible (Table 11.2).

Given the birth weight differences between Andeans and Europeans at high altitude, Table 11.2 also expresses UA blood flow and oxygen delivery in relation to fetal weight. The birth weight-adjusted values are similar in the European and Andean women at low altitude and in the European women residing at low versus high altitude. However, the high-altitude Andeans studied by Julian (but not by Zamudio) have markedly greater UA blood flow and oxygen delivery in relation to fetal weight (Table 11.2). These differences are entirely the result of higher UA blood flows, as Andean birth weights are similar and CaO_2 is modestly lower in the Julian than in the Zamudio report.

In summary, Andean women living at high altitude increase their UA diameter and blood flow during pregnancy to a degree that is similar to what occurs at low altitude, but European women do not. Pregnancy raises UA blood flow principally by doubling UA diameter, which translates into a fourfold increase in vessel cross-sectional area. The three- to fivefold rise in TAM also makes a significant contribution to UA blood flow, but the blood-flow differences between ancestry groups or altitudes are due entirely to UA diameter and not TAM. As a result, the Andeans' UA blood flow or oxygen delivery per kg fetal weight is approximately twice as great as that of the high-altitude Europeans and greater than that seen at low altitude as well.

Vasoactive, angiogenic and other mediators

It is interesting to consider the vasoactive, angiogenic and other mediators that may be responsible for these ancestry and altitude differences in UA blood-flow responses to pregnancy. In relation to vasoactive substances, levels of the potent vasoconstrictor endothelin-1 (ET1) generally decline with pregnancy. Whereas the production of the vasodilator nitric oxide (NO) rises [32], circulating levels of NO metabolites (NOx) are more variable [33]. We recently reported that NOx levels were lower and the ratio of ET1 to NOx higher in pregnant high- versus low-altitude women in Colorado [27]. Future studies are required to compare circulating levels of these substances in pregnant Andean and European women to see if such factors contribute to the differences in vessel diameter observed.

Very little is known about the combined effects of altitude and pregnancy on oxidative stress in the uterine circulation, although increased oxidative stress is thought to play an important role in the endothelial damage characteristic of pre-eclampsia [34]. Therefore, also needed are comparisons of markers of oxidative stress and antioxidants in Andean versus European women at low and high altitude to address their contribution to the determination of the differences in UA diameter and blood flow observed.

In isolated vessels, it is well known that pregnancy increases the UA vasodilator response to flow, acetylcholine, bradykinin and other agonists. This is thought to be due to greater endothelial NO, cyclooxygenase and/or hyperpolarizing factor production or activity (see [35] for review). UA from high-altitude-gestated animals failed to show greater pregnancy-associated vasodilator response to flow, analogous to what has been seen in myometrial vessels from pre-eclamptic women [36], and had altered distensibility as well [37,38]. UA from high-altitude-gestated guinea pigs also had less pregnancy-associated rise in NO activity and eNOS protein [39], which is the opposite of what has been reported in sheep UA [40]. Because sheep are resistant to altitude-associated reductions in fetal growth, White and Zhang speculated some years ago that species differences in NO production or activity could be playing a key role [41].

Both pregnancy and altitude influence UA growth and remodelling. Not only does UA diameter enlarge, but it and the other uterine vessels increase in length as the uterus grows to accommodate the fetus, and anastomoses develop between the UA and a branch of the ovarian artery, thus forming a bilateral dually perfused circuit. The increase in diameter reduces vascular resistance exponentially (by a power of two), whereas the increase in vessel length increases vascular resistance linearly (by a factor of eight), with the major effect on vascular resistance therefore being exerted by the increased diameter. The doubling of UA diameter is distinctive relative to that which usually occurs in other vessels in adult organisms, insofar as it is eutrophic or 'outward', with luminal diameter enlarging but the ratio of wall mass to diameter remaining constant [35]. New cellular growth is involved, as attested to by greater UA DNA synthesis *in vivo* and a greater proliferative response to serum stimulation in UA vascular smooth muscle cells *in vitro* [42,43]. Altitude influences these vascular growth response(s) to pregnancy: the rise in UA DNA synthesis and increased proliferative response of cultured UA vascular smooth muscle cells were both reduced by half in guinea pigs gestated at high. versus low altitude [44,45]. Thus it appears that chronic hypoxia interferes with the uterine vessels' normal growth and remodelling during pregnancy in guinea pigs as well as in non-native, human high-altitude populations.

Steroid hormones contribute to UA enlargement, as attested to by the increase in UA diameter seen in pseudopregnant. versus non-pregnant mice. However, they cannot fully explain the enlargement, as the rise was only half as great as that seen in normal pregnancy [46]. Steroid hormone concentrations are altered at high compared to low altitudes in Colorado, with progesterone being higher and oestradiol lower [47]. Pregnancy also greatly alters the circulating levels of members of the vascular endothelial growth factor (VEGF) family, including placenta-like growth factor (PlGF) and the soluble VEGF/PlGF receptor (sFlt-1) (see [35] for review). Serial determinations of circulating levels of steroid hormones and members of the VEGF family in Andean versus. European residents of low and high altitude are needed to determine whether they contribute to the observed differences in UA diameter.

Local factors are also important, as attested to by the incomplete nature of the vascular changes seen in pseudopregnancy and the absence of vessel enlargement in the non-pregnant horn of unilaterally pregnant animals [35]. The local factors that have received the most attention are placentation/trophoblast invasion, increased shear stress, and veno-arterial exchange [35]. Albeit important in their own right, placentation and trophoblast invasion cannot fully explain UA enlargement, given that trophoblasts never reach as far upstream in human pregnancy and the enlargement of the UA has occurred by week 6.5, which is well before trophoblast invasion is complete [48] or even placenta blood flow has been established [49]. An increase in shear stress is probably important for stimulating NO release and helping to initiate eutrophic remodelling, but the fact that the rise in flow velocity is greater than that in UA diameter argues against shear stress being the major factor [35]. Veno-arterial signalling is an intriguing idea for explaining how feto-placentally produced growth or vasoactive factors could reach maternal arteries [50], but again it is difficult to reconcile this mechanism with the early date at which UA enlargement occurs.

In summary, the factors responsible for UA enlargement remain unclear, as do the signalling pathways involved. At low altitude, a full explanation for UA enlargement requires accounting for the early date by which UA enlargement is largely complete. Whereas in most circulations conduit vessels contribute relatively little to the determination of vascular resistance, this not the case in the uterine circulation during pregnancy, where the UA is a major resistance vessel, constituting some 40% of the total vascular resistance [51]. This is because the end spiral arterioles and the arcuate or mesometrial arteries, the first vessels emanating from the UA, are both larger than the UA by mid gestation [48]. Thus the major sites of vascular resistance in the uterine vascular bed are the UA and the basal or myometrial portion of the spiral arteries. Further studies are needed to determine the contribution of alterations in these vessels' growth/remodelling, as well as of their vasoactive characteristics, to the determination of ancestry-group differences in UA diameter at high altitude.

Influence of genetic factors on UA blood flow and birth weight at high altitude

As shown in Figures 11.1 and 11.2, multigenerational high-altitude residents are protected from the depressant effects of high altitude on infant birth weight relative to shorter-resident groups. Because birth weight is subject to genetic influences [52] and 'normal' infant birth weight is that which is associated with the lowest mortality risk [53], we hypothesized that genetic variants that increased the delivery of oxygen and other nutrients to the uterine circulation have been selected for in the Andean (and Tibetan) population and therefore act to preserve normal fetal growth. We have proceeded to test this hypothesis by conducting surname and gene marker studies aimed at determining whether genetic factors were involved, genome scans to identify responsible gene regions, and examination of these regions in order to determine the physiological pathways likely to be involved.

Surname analyses

Because ancestry contains genetic components, we determined whether ancestry influenced the birth weight response to high altitude by collecting ∼4000 medical records in Bolivia, comprising 1000 consecutive deliveries per 1000 m altitude interval [54]. As women (and men) in Bolivia do not change their surnames with marriage and both parents have one paternal and one maternal surname, we used the four surnames available for each newborn to classify babies as Andean if at least three surnames were Aymara or

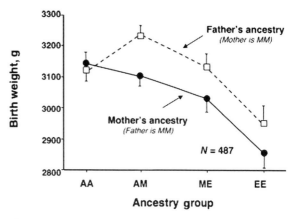

Figure 11.5 Birth weights for 1343 residents of La Paz/El Alto, Bolivia (3600–4100 m) were classified by parental surnames that had been judged to be Andean (A), European (E), or mixed ('Mestizo', M). Shown here are the 487 babies for whom one parent had only M surnames, with birth weight being classified by the 'other' (non MM) parent. Birth weights decline with increasing number of E surnames but for a given surname combination, paternal ancestry raises birth weight relative to maternal ancestry, which is consistent with the possibility that imprinting is involved. Figure adapted from [55].

Quechua (the indigenous languages), European if at least three were non-Hispanic or nationality was foreign, or Mestizo ('mixed') for everyone else. Although there were no appreciable ancestry-group differences in the frequency of small (<10 percentile) for gestational age and sex (SGA) infants at low altitude (400 m), three times as many European as Andean babies were SGA at high altitude (3600–4100 m), with intermediate increases being present for Mestizos [54].

We also used a more fine-grained system in which all possible surname combinations were considered for the 1343 babies born at high altitude [55]. Babies whose two parents had only Andean surnames weighed 252 g more than those born to parents with only European surnames. Further, the protective effect was proportional to the number of Andean surnames present (Figure 11.5) and independent of maternal parity, body size, smoking history or socio-economic status.

As babies born to Mestizo parents comprised the largest single group among these high-altitude newborns, we used women or men with two Mestizo surnames to determine whether parent of origin influenced the effect of Andean ancestry. As shown in Figure 11.5, when we classified the baby's birth weight by the ancestry of the other (non-Mestizo–Mestizo) parent, paternal ancestry raised birth weight an average of 74 g relative to that seen if maternal ancestry was employed, or 81 g higher if the effects of gestational age on birth weight were taken into account (both $p < 0.05$). These results were consistent with genetic imprinting, in which it has been shown that maternally transmitted genes restrict and paternally transmitted ones enhance fetal growth [56]. Because maternal surnames survive for only two generations, such parent-of-origin effects appear to occur over a relatively short time span, which is also consistent with recent evidence indicating that the effects of genomic imprinting endure for only a few generations [57]. The absence of any birth weight differences between Andean–Andean mothers or fathers further suggests that it is the non-Andean component of paternal parentage that raises birth weight. Clearly, further studies are required of known imprinted genes in these Andean and European groups.

Gene marker studies

To confirm genetic ancestry and test its relationship with physiological measures, we measured 100 ancestry-informative gene markers (AIMs) in our serially studied Bolivian women. Gene markers are not currently available with which Andean ancestry can be distinguished from other Amerindian groups, but gene markers have been validated for assessing Amerindian versus European or West-African ancestry (see [22,23] for details). At high altitude, the per cent maternal Amerindian AIMs exhibited a strong positive correlation with UA blood flow and oxygen delivery, a relationship that was due entirely to UA diameter and not flow velocity [23]. The per cent Amerindian AIMs also correlated positively with fetal head circumference at weeks 20 and 36, as well as with infant birth weight at high altitude. No such relationships were present between the per centAmerindian AIMs and any of the UA or fetal measures at low altitude, implying that the protective effect of Amerindian ancestry was confined to high altitude. These observations differ from those of Zamudio and co-workers, who found that the per centfetal Amerindian AIMs was also related to birth weight at low altitude [17], which as we have suggested elsewhere may be due to the influence of paternal factors on fetal growth in their study gives that fetal DNA was being sampled [23].

Based on these gene markers as well as our surname analyses, we concluded that genetic factors influenced UA blood flow and birth weight at high altitude. Because such influences were present only at high altitude, we considered that gene–environment interaction was involved or, in other words, that genetic influences were only apparent under the environmental circumstances of high altitude. As such relationships were not present at low altitude, being Andean (or Amerindian) *per se* did not emerge as a significant determinant of UA blood flow or infant birth weight. Rather, the data were consistent with the hypothesis that genetic factors stemming from Andean ancestry influenced the ways in which the maternal circulation responded to the dual challenges of pregnancy and high altitude.

Identification of genetic factors

To identify the genetic factors involved in protection from altitude-associated reduction in fetal growth and maintenance of high UA blood flow in Andeans, we conducted genome scans. We then used four statistics to identify areas of the Andean genome that differ from those of low-altitude control populations (i.e. 'departure from neutrality'). These tests – the locus-specific branch length (*LSBL*), the natural log of the ratio of heterozygosities (ln*RH*), Tajima's *D*, and the whole-genome long-range haplotype (*WGLRH*) test (see [19] for details) – are complementary and therefore best used in combination [58]. Given our overall hypothesis that protection from altitude-associated reduction in fetal growth in Andean high-altitude residents was due to natural selection having acted on oxygen-sensitive genes, we focused on the ~75 hypoxia-inducible factor (*HIF*)- regulatory or targeted genes [59]. Using the Affymetrix Inc. (Santa Clara, CA) GeneChip® Mapping 500 k array, 490 032 autosomal SNPs were evaluated in 50 persons of high-altitude ancestry (25 Aymara women from our 3600 m sample and 25 Quechua living at 4300 m in Peru) and 195 persons residing at low altitudes (105 low-altitude Amerindians and 90 East Asian Chinese or Japanese) [19]. Because the platform did not contain SNPs within the gene boundaries for several HIF-family genes, the 50 kb upstream and downstream of the start- and end-coordinates for each gene were also included.

No HIF family genes were identified by all four tests, nor was there evidence that the ~75 HIF family genes had evolved in the Andean population as a group, but there were eight HIF-targeted or HIF-regulatory gene regions that were distinctive in Andeans versus low-altitude controls [19]. All eight regions contain genes that are involved either in the

Table 11.3 Gene regions with two or more significant test statistics for signatures of natural selection (for details, see [19]).

		Test for natural selection			
Gene name	Common name	LSBL	LnRH	Tajima's D	WGLRH
EDNRA	Endothelin receptor A	X	X	X	
NOS2A	Nitric oxide synthase 2, inducible	X	X	X	
PRKAA1	Protein kinase, AMP-activated, alpha 1 catalytic subunit	X	X	X	
ELF2	E74-like factor 2 or ets domain transcription factor	X	X		
CDH1	Cadherin 1, type 1 or E-cadherin	X	X		
PIK3CA	Phosphoinositide-3-kinase, catalytic, alpha polypeptide	X		X	
TNC	Tenascin C	X	X		
VEGF	Vascular endothelial growth factor	X		X	

regulation of the HIF-pathway itself or in maternal and/or fetal vascular adjustments to pregnancy (Table 11.3). Of interest is that the HIF-targeted insulin growth factor (*IGF2, IGFBP1, IGFBP2, IGFBP3*) and inflammatory (*IL1A, IL1B, IL-6*) genes were not among those identified.

The gene regions implicated by three of the four tests were *EDNRA, NOS2A* and *PRKAA1*. *EDNRA* is the 'A' type receptor for endothelin-1 (ET1), which is expressed on vascular smooth muscle cells and is responsible for ET1's potent vasoconstrictor effects. *NOS2A*, also called iNOS, is the inducible form of NO synthase and hence an important source of NO. Both ET1 and NO have important roles in the regulation of uterine blood flow during normal and abnormal pregnancy via their vasoconstrictor and vasodilator actions, as reviewed above. *PRKAA1* or protein kinase AMP-activated α_1 catalytic subunit (also called *AMPK* or *AMPKa1*) plays a key role in the regulation of protein synthesis by acting via inhibitory effects on the mTOR (mammalian target of rapamycin) pathway. *AMPK* functions as a cellular energy sensor, being activated by the accumulation of AMP and hence detecting energy depletion, and has recently been implicated in the aetiology of intrauterine growth restriction (IUGR) and pre-eclampsia. As recently proposed by Yung [60] and Burton [49], deficient spiral artery conversion would be expected to lead to spontaneous constriction, repetitive ischaemia/reperfusion (I/R) injury, increased production of reactive oxygen species and endoplasmic reticulum (ER) stress. ER stress is thought, in turn, to provoke the unfolded protein response, which attempts to restore ER homoeostasis by one several pathways, many of which involve eukaryotic initial factor (elF2, which is not the same as ELF2 in Table 11.3). A four- to seven-fold increase in the phosphorylated or inactivated form of elF2 has been found in IUGR or pre-eclamptic compared to normal placentas, and suggests that a feed-forward mechanism may be operating that leads to reduced *AKT/mTOR* activity and suppression of protein synthesis [60]. Future studies should be directed at detecting whether this pathway is involved in altitude-associated IUGR and/or the altitude-associated protection from IUGR seen in the Andeans.

ELF2, CDH1, PIK3CA, TNC and *VEGF* are the five additional gene regions identified by two of the tests for natural selection (Table 11.3). *ELF2* or ets domain transcription factor (also known as NERF) has not been well studied during pregnancy but is thought to be important for cell survival and proliferation under cytokine stress. *CDH1*, also known as cadherin-1 or e-cadherin, is a calcium-dependent cell–cell adhesion glycoprotein involved in trophoblast invasion, and therefore important for vascular growth and remodelling. *PIK3CA* is implicated in various kinds of cancer, including cervical and breast, but has not to our knowledge been studied during pregnancy. *TNC* or tenascin-3 is an extracellular matrix glycoprotein that has anti-adhesive properties and is important in cell growth. *VEGF* or vascular endothelial growth factor, as reviewed above, plays a key role in pregnancy-related vascular growth and remodelling.

In summary, the population differences in altitude-associated reductions in fetal growth (Figures 11.1 and 11.2), together with the results of our surname, gene-marker and genome-scan studies, clearly implicate genetic factors in the ancestry-group differences in UA blood flow and birth weight at high altitude. We have recently identified eight HIF-targeted gene regions that are distinctive in Andean compared to low-altitude control populations. Of note is that these gene regions play key roles in the vasoactive and vascular growth/remodelling processes central to maternal vascular responses to pregnancy and in the regulation of protein synthesis. Further studies are warranted using higher-density genome screens and gene-sequencing technologies to identify the precise genetic changes involved in Andean adaptation to high altitude.

Does greater UA blood flow improve fetal oxygenation?

The final issue to be addressed concerns the relationship between maternal uterine oxygen delivery and fetoplacental demand. Central to our hypothesis is that the Andean women's high UA blood flow during pregnancy helps maintain normal fetal growth.

UA blood flow and fetal growth

It is clear from the data summarized in Table 11.2 and Figures 11.1, 11.2 and 11.4 that Andean women have higher levels of UA blood flow and give birth to babies that are relatively protected from altitude-associated reductions in fetal growth. Supporting our hypothesis are our observations that Andean ancestry is positively related to UA diameter, UA blood flow, UA oxygen delivery, fetal biometry and birth weight at high altitude. Although UA blood flow was not significantly associated with infant birth weight in our healthy, prospectively studied subjects, there was a strong such relationship when the full range of UA blood flow and birth weights was considered by including women with pre-eclampsia or gestational hypertension at high altitude [61]. The women studied by Zamudio *et al.* also showed a strong positive relationship between UA blood flow or oxygen delivery and birth weight, especially in Europeans or the group in whom birth weight was most strongly depressed. However, these authors concluded that 'decreased maternal oxygen delivery does not cause the progressive reduction in fetal growth observed after 24 weeks of human pregnancy at high altitude' [17]. What accounts for this apparent discrepancy as to whether the Andean women's greater UA blood flow or oxygen delivery does or does not help protect their infants from altitude-associated reductions in fetal growth? We think there are two issues involved; the first concerns study design and data interpretation, and the second pertains to the fundamental nature of the relationship between uterine oxygen delivery and fetoplacental oxygen consumption.

In relation to study design, the important project conducted by Zamudio and co-workers was designed for collecting placentas and not for addressing the contribution of increases in UA blood flow during pregnancy to the determination of fetal growth at high altitude. Given the slowing of fetal growth shortly before delivery, the differences in gestational age between the Andean and European groups, and the possibility that operator effects influenced the comparison of absolute values between locations, their measurements of UA blood flow at term are unable to address the role of UA blood flow or oxygen delivery in the determination of third trimester fetal growth. Longitudinal studies are required, certainly beginning in the second trimester but ideally even earlier, given our observation that European – Andean differences in fetal biometry and changes in UA blood flow are already present by week 20 [23].

In terms of data interpretation, we agree that the fetus at high altitude is well oxygenated, but note that fetal umbilical or scalp vein pO_2 values at delivery are slightly lower than at low altitude [62–64]. It is important to point out that birth is a stimulatory event and only a snapshot of longer-term developmental processes. The intrauterine environment almost certainly influences the neonatal transition during which SaO_2 declines over the first week of life in a manner that varies by altitude and ancestry, with the benefit accruing at high altitude to multigenerational residents [65]. Whereas pO_2 was only slightly lower in European than Andean babies at delivery, the Andeans' higher UA blood flow and oxygen delivery appeared to improve fetal oxygenation insofar as Andean newborns had higher umbilical vein pH, higher umbilical vein blood flow, and lower erythropoietin concentrations than the Europeans in the Zamudio study. Additionally, high altitude increased fetal oxygen extraction in the Europeans but not the Andeans, and birth weight at high altitude was positively related to umbilical vein blood flow, umbilical oxygen delivery and UA oxygen delivery [17,64]. Measurements of fetal oxygen consumption (VO_2) are difficult and we credit these investigators for acquiring these important data. However, we consider that the equivalence in fetal VO_2 per kg weight between Andean and European babies at delivery is to be expected, given that VO_2 is matched to

Table 11.4 Fetal oxygen supply (UA oxygen delivery) in relation to fetal and fetoplacental oxygen demand (O_2 consumption, VO_2).

Variable	Ref no.	Europeans		Andeans	
		Low altitude	High altitude	Low altitude	High altitude
Bilateral UA oxygen delivery (absolute, mL min)	17	100	82	110	98
	23	113	140	98	260
per kg fetal weight (mL min kg)	17	29.4	27.4	31.2	29.6
	23	34.5	44.5	30.7	85.4
Fetal VO_2 (mL min)	64	20	18	21	20
Estimated placental VO_2 (mL min)[1]	64	13	12	14	13
Total fetoplacental VO_2 (mL min)	64	33	30	35	33

Note: [1] Placental VO_2 was estimated by assuming it to be 40% of total fetoplacental VO_2 [66].

body size. Absolute fetal VO_2 was marginally but probably not significantly higher in the Andeans than in the Europeans at high altitude (Table 11.4), consistent with the Andean women's greater UA oxygen delivery (Table 11.2). In short, we interpret the data reported by Zamudio and previous investigators as being consistent with the likelihood that the Andeans' greater UA blood flow and oxygen delivery are important contributors to their improved fetal oxygenation and preservation of fetal growth at high altitude.

Relationship between uterine oxygen delivery and fetoplacental oxygen consumption

Comparing near-term uterine oxygen delivery at low and high altitudes reveals that week 36–38 uterine oxygen delivery is well maintained or even greater in the high- than in the low-altitude women, whether expressed as absolute values or normalized by fetal weight (Table 11.4). UA oxygen delivery is also greater than fetal VO_2 or total fetoplacental VO_2, as estimated by assuming that placental VO_2 constitutes 40% of the total [66]. At low altitude, maternal oxygen delivery is five times greater than fetal VO_2 or three times greater than the estimated fetoplacental VO_2. At high altitude, maternal oxygen delivery is five to 13 times greater than fetal VO_2 or three to eight times greater than the fetoplacental VO_2, depending on whether the values reported by Zamudio or our group are used (Table 11.4). Because UA blood flow is the major factor determining UA oxygen delivery, the question becomes why is UA blood flow so high, especially in high-altitude Andeans? And if uterine oxygen delivery is maintained, why then is fetal growth reduced, albeit by differing amounts in the Andean and European groups? Answering these questions requires consideration of the fundamental nature of the relationship between uterine oxygen delivery and fetoplacental VO_2. We suggest that one or more of five factors are likely involved in resolving this apparent dilemma.

One possibility is that UA blood flow is being overestimated owing to inaccuracies introduced by non-laminar flow, or the tendency of pulsed-wave techniques to overestimate mean velocity due to overweighting of the higher velocity components at the centre of the vessel lumen [67]. However, the major factor elevating UA blood flow is the pregnancy-associated increase in vessel diameter, which would not be affected by these considerations pertaining to the measurement of velocity. Other sources of error could be involved, such as operator or instrumentation effects, postural variation or other characteristics of the women being studied, but it is not apparent why such sources of error would be selectively greater in the Andean women studied by our group at high altitude.

A second possibility is that Andean placental VO_2 at high altitude is more than 40% of total fetoplacental VO_2. There are no measurements of human placental VO_2 at high altitude to our knowledge, but consistent with the possibility that values are greater is a recent report showing a 71% increase in oxidative phosphorylation in placentas from native Tibetans compared with immigrant Han residents at 3600 m [68]. Further study of placental mitochondria, in terms of both numbers and metabolic activity, will be important for deciding the appropriate values to use for estimating placental VO_2 at high altitude.

A third possibility has to do with the mechanical consequences of the very high levels of UA blood flow achieved during pregnancy. High levels of UA blood flow occur as the result of the fall in systemic vascular resistance and expansion of blood volume that together raise cardiac output, and the even greater fall in uterine vascular resistance that serves to direct a large fraction of the cardiac output rise to the uterine vascular bed. The fall in uterine vascular resistance is due to vessel enlargement as reviewed above, but also to the formation of arteriovenous shunts. The placenta itself has long been likened to an arteriovenous fistula. More recently Schaaps et al. pointed to another kind of shunt, namely, anastomoses that can be observed by ultrasound between arterial and venous vessels in the myometrial segment of the uterine circulation [69]. Although such shunts lower vascular resistance and hence increase blood flow, Burton and co-workers have suggested that they also act to slow velocity and thereby protect fetal villi from mechanical damage due to high velocity and turbulence, or to collapse from high intervillous pressure [70]. Several studies of placental morphometry have been conducted at high altitude [71–75]. In particular, the Mayhew studies showed a 40% increase in placental volume that was due primarily to greater maternal intervillous space, which in turn raised morphometric conductance. Whereas the altitude-associated increase in conductance tended to be greater in the Andean. than in the European or Mestizo women, group differences were not significant [72]. Future studies with appropriate controls for genetic ancestry are required, as the inclusion of Mestizo subjects in the non-native group may have influenced the comparisons in the Mayhew reports. Evaluation of myometrial specimens will also be helpful for deciding whether myometrial shunts contribute to the very high UA blood flows achieved and, in turn, offer protection to fetal villi from mechanical damage or collapse. Given that collapse would lead to reduced umbilical vein blood flow, it is of interest that umbilical vein blood flow was lower in the European than the Andean women at high altitude [64].

The fourth possibility is that high UA blood flow and hence oxygen delivery provides a reserve that protects against intermittent hypoxia and ischaemia–reperfusion (I/R) injury. Burton and co-workers have pointed out that trophoblast invasion extends well upstream of the end-spiral arterioles, serving not only to enlarge lumenal diameter but also to reduce contractile capacity [70]. Blood flow is likely to be highly variable during pregnancy, as attested to by the considerable reductions in UA diameter and blood flow seen, for example, with change from the supine to the left-lateral posture [76]. Intermittent hypoxia and I/R injury stimulate the production of reactive oxygen species (ROS), which in turn can damage the endothelium, reduce vasodilator production or activity and thereby decrease uterine vessel diameter. Although existing data point to reduced oxidative stress in high-altitude placentas at delivery [77,78], this may be the cumulative result of hypoxic preconditioning or greater antioxidant production. It will be particularly important to have information concerning markers of oxidative stress and antioxidant capacity during pregnancy in Andean and European women to determine whether Andeans may be protected from I/R injury.

The fifth possibility is not independent of the above factors but involves a consideration of the nature of human placental exchange. Many consider the human placenta to operate as a venous equilibration exchanger for which a large uterine–umbilical venous pO_2 gradient is required to draw oxygen across the placental barrier [79]. At 3600 m, maternal arterial pO_2 during pregnancy is likely to be 55–60 mmHg, or at least 40 mmHg lower than at sea level. If uterine venous pO_2 were proportionately reduced, it would be below the threshold required to maintain umbilical venous pO_2. Thus, as concluded long ago [4], high UA blood flow serves to raise uterine venous pO_2 and thereby sustain fetal oxygenation. The presence of shunts, as noted above, could also be another factor serving to raise uterine venous pO_2. However, as the higher uterine venous pO_2 would be 'downstream' of the maternofetal exchange site and such shunts would also serve to lower uterine arterial pO_2, increased UA blood flow is more likely to be playing a major role in uterine–umbilical venous oxygen exchange, rather than shunts *per se*.

In short, the relationships between UA blood flow, fetoplacental oxygenation and fetoplacental development are likely to be complex, with no single factor being narrowly deterministic. We consider that the Andeans' higher UA blood flows are beneficial and contribute to their infants' protection from altitude-associated reductions in fetal growth. Existing data are, however, insufficient to determine whether the importance of higher blood flows relates to increased placental VO$_2$, operates via mechanical considerations by serving to slow velocity and prevent damage to fragile fetal villi from high velocity, constitutes a reserve that protects against I/R injury, or helps maintain high uterine venous pO$_2$ in order to facilitate placental exchange. Of course it is possible that all of these factors and others, such as increasing the delivery of other nutrients such as glucose or amino acids, are involved. Future studies using higher-density genome scans, gene sequencing of implicated regions, and functional studies of the variants detected will prove informative for deciding among these intriguing possibilities.

Summary and conclusion

High altitude regions continue to provide an instructive natural laboratory for understanding fetoplacental development. The early studies begun some 50 years ago pointed not only to the important effects of chronic hypoxia on fetal growth but also to the role of high UA blood flow in maintaining the oxygen pressure gradient from maternal to fetal blood. Over the intervening 50 years, studies at high altitude have been instructive in clarifying the factors that decide uterine oxygen delivery to the fetoplacental circulation. An important discovery has been the recognition that the magnitude of altitude-associated reduction in fetal growth varies among human populations in relation to their duration (in generations) of high-altitude residence. Such population variation is accompanied by marked differences in UA diameter and blood flow which, in turn, serve to raise uterine oxygen delivery in long-relative to shorter-resident groups, and reinforces earlier observations as to the importance of high UA blood flow for deciding fetal growth.

Future studies are required to determine the factors responsible for raising UA blood flow during pregnancy and for explaining the UA blood flow differences between native and newcomer groups. Important will be population comparisons of vasodilator relative to vasoconstrictor substances, markers of oxidative stress, and angiogenic factors. Genome and genetic studies offer new means of identifying the physiological pathways involved in explaining such differences. Truly high-altitude populations continue to offer a unique model system for improving our understanding and approaches for the treatment and ultimately the prevention of disorders adversely affecting maternal wellbeing and fetoplacental development.

References

1. **Eliot T**. *Four Quartets*, (*No. 4, Little Giddings*) London: Harcourt Books, 1971.
2. **Barron D, Metcalfe J, Meschia G** *et al.* Adaptations of pregnant ewes and their fetuses to high altitude. In: Weihe WH, ed. *The Physiological Effects of High Altitude*. New York: Macmillan, 1964.
3. **Huckabee W, Metcalf J, Prystowsky H** *et al.* Uterine blood flow and metabolism in pregnant sheep at high altitude. *Fed Proc* 1959; **18**: 72.
4. **Metcalfe J, Novy MJ, Peterson EN**. Reproduction at high altitude. In: Benirschke K, ed. *Comparative Aspects of Reproductive Failure*. New York: Springer-Verlag, 1967.
5. **Kamitomo M, Longo LD, Gilbert RD**. Right and left ventricular function in fetal sheep exposed to long-term high-altitude hypoxemia. *Am J Physiol* 1992; **262**: H399–H405.
6. **Lichty JA, Ting RY, Bruns PD, Dyar E**. Studies of babies born at high altitude. I. Relation of altitude to birth weight. *Am J Di Child* 19 1957; **93**: 666–9.
7. **Moore LG**. Small babies and big mountains: John Lichty solves a Colorado mystery in Leadville. In: Reeves JT, Grover FT, ed. *Attitudes on Altitude*. Boulder, CO: University Colorado Press, 2001; 137–59.
8. **Moore LG**. Fetal growth restriction and maternal oxygen transport during high-altitude pregnancy. *High Altitude Medicine & Biology* 2003; **4**(3): 141–56.
9. **Jensen GM, Moore LG**. The effect of high altitude and other risk factors on birth weight: independent or interactive effects? *Am J Pub Health* 1997; **87**: 1003–7.
10. **Giussani DA, Phillips S, Anstee S, Barker DJP**. Effects of altitude versus economic status on birth weight and body shape at birth. *Pediatr Res* 2001; **49**(4): 490–4.
11. **Mortola JP, Frappell PB, Aguero L, Armstrong K**. Birth weight and altitude: a study in Peruvian communities. *J Pediatr* 2000; **136**: 324–9.
12. **Moore LG**. Human genetic adaptation to high altitude. *High Altitude Medicine & Biology* 2001; **2**(2): 257–79.
13. **Haas JD, Frongillo EJ, Stepcik C, Beard J, Hurtado L**. Altitude, ethnic and sex differences in birth weight and length in Bolivia. *Hum Biol* 1980; **52**: 459–77.
14. **Moore LG**. Maternal O$_2$ transport and fetal growth in Colorado, Peru and Tibet high-altitude residents. *Am J Hum Biol* 1990; **2**: 627–37.
15. **Niermeyer S, Zamudio S, Moore LG**. The people. In: Hornbein T, Schoene RB, eds. *Adaptations to Hypoxia*. New York: Marcel Dekker and Co., 2001; 43–100.

16. **Weinstein RS**, **Haas JD**. Early stress and later reproductive performance under conditions of malnutrition and high altitude hypoxia. *Med Anthropol* 1977; **1**: 25–54.

17. **Zamudio S**, **Postigo L**, **Illsley NP** et al. Maternal oxygen delivery is not related to altitude – and ancestry-associated differences in human fetal growth. *Journal of Physiology* 2007; **582**(2): 12.

18. **Unger C**, **Weiser JK**, **McCullough RE**, **Keefer S**, **Moore LG**. Altitude, low birth weight, and infant mortality in Colorado. *J Am Med Assoc* 1988; **259**: 3427–32.

19. **Bigham AW**, **Mao X**, **Mei R** et al. Identifying positive selection candidate loci for high-altitude adaptation in Andean populations. *Hum Genomics* 2009 Dec; **4**(2): 79–90.

20. **Lopez Herraez D**, **Bauchet M**, **Tang K** et al. Genetic variation and recent positive selection in worldwide human populations: evidence from nearly 1 million SNPs. *PLoS One* 2009; **4**(11): e7888.

21. **Vargas M**, **Vargas E**, **Julian CG** et al. Determinants of blood oxygenation during pregnancy in Andean and European residents of high altitude. *Am J Physiol Regul Integr Comp Physiol* 2007; **293**(3): R1303–12.

22. **Wilson MJ**, **Lopez M**, **Vargas M** et al. Greater uterine artery blood flow during pregnancy in multigenerational (Andean) than shorter-term (European) high-altitude residents. *Am J Physiol Regul Integr Comp Physiol* 2007; **293**(3): R1313–24.

23. **Julian CG**, **Wilson MJ**, **Lopez M** et al. Augmented uterine artery blood flow and oxygen delivery protect Andeans from altitude-associated reductions in fetal growth. *Am J Physiol Regul Integr Comp Physiol* 2009; **296**(5): R1564–75.

24. **Moore LG**, **McCullough RE**, **Weil JV**. Increased HVR in pregnancy: relationship to hormonal and metabolic changes. *J Appl Physiol* 1987; **62**: 158–63.

25. **Hannhart B**, **Pickett CK**, **Weil JV**, **Moore LG**. Influence of pregnancy on ventilatory and carotid body neural output responsiveness to hypoxia in cats. *J Appl Physiol* 1989; **67**: 797–803.

26. **Palmer SK**, **Zamudio S**, **Coffin C** et al. Quantitative estimation of human uterine artery blood flow and pelvic blood flow redistribution in pregnancy. *Obstet Gynecol* 1992; **80**: 1000–6.

27. **Julian CG**, **Galan HL**, **Wilson MJ** et al. Lower uterine artery blood flow and higher endothelin relative to nitric oxide metabolite levels are associated with reductions in birth weight at high altitude. *Am J Physiol Regul Integr Comp Physio* 2008; **295**(3): R906–15.

28. **Krampl E**, **Espinoza-Dorado J**, **Lees CC**, **Moscoso G**, **Bland JM**. Maternal uterine artery Doppler studies at high altitude and sea level. *Ultrasound Obstet Gynecol* 2001; **18**: 578–82.

29. **Thaler I**, **Manor D**, **Itskovitz J** et al. Changes in uterine blood flow during human pregnancy. *Am J Obstet Gynecol* 1990; **162**(1): 121–5.

30. **Bernstein I**, **Ziegler W**, **Leavitt T**, **Badger G**. Uterine artery hemodynamic adaptations through the menstrual cycle into early pregnancy. *Obstet Gynecol* 2002; **99**(4): 620–4.

31. **Zamudio S**, **Palmer SK**, **Droma T** et al. Effect of altitude on uterine artery blood flow during normal pregnancy. *J Appl Physiol* 1995; **79**: 7–14.

32. **Tuttle JL**, **Nachreiner RD**, **Bhuller AS** et al. Shear level influences resistance artery remodeling: wall dimensions, cell density, and eNOS expression. *Am J Physiol Heart Circ Physiol* 2001; **281**(3): H1380–9.

33. **Davidge ST**, **Stranko CP**, **Roberts JM**. Urine but not plasma nitric oxide metabolites are decreased in women with preeclampsia. *Am J Obstet Gynecol* 1996; **174**: 1008–13.

34. **Hubel CA**, **Kagan VE**, **Kisin ER**, **McLaughlin MK**, **Roberts JM**. Increased ascorbate radical formation and ascorbate depletion in plasma from women with preeclampsia: implications for oxidative stress. *Free Radic Biol Med* 1997; **23**(4): 597–609.

35. **Osol G**, **Mandala M**. Maternal uterine vascular remodeling during pregnancy. *Physiology (Bethesda)* 2009; **24**: 58–71.

36. **Cockell AP**, **Poston L**. Flow-mediated vasodilation is enhanced in normal pregnancy but reduced in preeclampsia. *Hypertension* 1997; **30**(Part 1): 247–51.

37. **Mateev S**, **Sillau AH**, **Mouser R** et al. Chronic hypoxia opposes pregnancy-induced increase in uterine artery vasodilator response to flow. *Am J Physiol Heart Circul Physiol* 2003; **284**(3): H820–H9.

38. **Mateev S**, **Mouser R**, **Young D**, **Mecham RP**, **Moore LG**. Chronic hypoxia augments uterine artery distensibility and alters the circumferential wall stress-strain relationship during pregnancy. *J Appl Physiol* 2006; **100**(6): 1842–50.

39. **White MM**, **Horan MP**, **LeCras TD** et al. Pregnancy increases and chronic hypoxia decreases endothelial nitric oxide synthase (eNOS) protein expression in guinea pig uterine arteries (UA). *Fed Am Soc Exp Biol J* 1997; **11**: A479.

40. **Xiao D**, **Bird IM**, **Magness RR**, **Longo LD**, **Zhang L**. Upregulation of eNOS in pregnant ovine uterine arteries by chronic hypoxia. *Am J Physiol Heart Circ Physiol* 2001; **280**(2): H812–20.

41. **White M**, **Zhang L**. Effects of chronic hypoxia on maternal vasodilation and vascular reactivity in guinea

pig and ovine pregnancy. *High Alt Med Biol* 2003; **4**(2): 12.

42. **Cipolla M**, **Osol G**. Hypertrophic and hyperplastic effects of pregnancy on the rat uterine arterial wall. *Am J Obstet Gynecol* 1994; **171**: 805–11.

43. **Keyes LE**, **Majack R**, **Dempsey EC**, **Moore LG**. Pregnancy stimulation of DNA synthesis and uterine blood flow in the guinea pig. *Pediatr Res* 1997; **41**: 708–15.

44. **Rockwell LC**, **Keyes LE**, **Moore LG**. Chronic hypoxia diminishes pregnancy-associated DNA synthesis in guinea pig uteroplacental arteries. *Placenta* 2000; **21**(4): 313–19.

45. **Rockwell LC**, **Dempsey EC**, **Moore LG**. Chronic hypoxia diminishes the proliferative response of guinea pig uterine artery vascular smooth muscle cells in vitro. *High Alt Med Biol* 2006; **6**: 237–44.

46. **Van Der Heijden OW**, **Essers YP**, **Spaanderman ME** *et al*. Uterine artery remodeling in pseudopregnancy is comparable to that in early pregnancy. *Biol Reprod* 2005; **73**(6): 1289–93.

47. **Zamudio S**, **Leslie K**, **White M**, **Hagerman D**, **Moore LG**. Low serum estradiol and high serum progesterone concentrations characterize hypertensive pregnancies at high altitude. *J Soc Gynecol Invest* 1994; **1**: 197–205.

48. **Burchell RC**. Arterial blood flow in the human intervillous space. *Am J Obstet Gynecol* 1967; **98**: 303–11.

49. **Burton G**, **Jauniaux E**, **Charnock-Jones DS**. The influence of the intrauterine environment on human placental development. *Int J Dev Biol* 2010; **54**(2–3): 303–12.

50. **Celia G**, **Osol G**. Venoarterial communication as a mechanism for localized signaling in the rat uterine circulation. *Am J Obstet Gynecol* 2002; **187**(6): 1653–9.

51. **Moll W**, **Kunzel W**. The blood pressure in arteries entering the placentae of guinea pigs, rats, rabbits, and sheep. *Pflugers Archiv Eur J Physiol* 1973; **338**: 125–31.

52. **Magnus P**. Causes of variation in birth weight: a study of offspring of twins. *Clin Genet* 1984; **25**: 15–24.

53. **Karn MN**, **Penrose LS**. Birth weight and gestational time in relation to maternal age, parity and infant survival. *Eugenics* 1957; **16**: 147–64.

54. **Julian CG**, **Vargas E**, **Armaza JF** *et al*. High-altitude ancestry protects against hypoxia-associated reductions in fetal growth. *Arch Dis Child Fetal Neonatal Ed* 2007; **92**(5): F372–7.

55. **Bennett A**, **Sain SR**, **Vargas E**, **Moore LG**. Evidence that parent-of-origin affects birth-weight reductions at high altitude. *Am J Hum Biol* 2008; **20**(5): 592–7.

56. **Moore T**, **Haig D**. Genomic imprinting in mammalian development: a parental tug-of-war. *Trends Genet* 1991; **7**: 2361–6.

57. **Hitchins MP**, **Wong JJ**, **Suthers G** *et al*. Inheritance of a cancer-associated MLH1 germ-line epimutation. *N Engl J Med* 2007; **356**(7): 697–705.

58. **Storz JF**, **Payseur BA**, **Nachman MW**. Genome scans of DNA variability in humans reveal evidence for selective sweeps outside of Africa. *Mol Biol Evol* 2004; **21**(9): 1800–11.

59. **Moore LG**, **Shriver M**, **Bemis L**, **Vargas E**. An evolutionary model for identifying genetic adaptation to high altitude. *Adv Exp Med Biol* 2006; **588**: 101–18.

60. **Yung HW**, **Calabrese S**, **Hynx D** *et al*. Evidence of placental translation inhibition and endoplasmic reticulum stress in the etiology of human intrauterine growth restriction. *Am J Pathol* 2008; **173**(2): 451–62.

61. **Browne V**, **Toledo-Jaldin L**, **Davila R** *et al*. High end-arteriolar resistance limits uterine artery blood flow and restricts fetal growth in preeclampsia and gestational hypertension at high altitude. *Am J Physiol Regul Integr Comp Physiol* 2010; in press.

62. **Howard RC**, **Bruns PD**, **A**. Studies of babies born at high altitude. III. Arterial oxygen saturation and hematocrit values at birth. *Am Med Assoc J Dis Child* 1957; **93**: 674–8.

63. **Sobrevilla LA**, **Cassinelli MT**, **Carcelen A**, **Malaga JM**. Human fetal and maternal oxygen tension and acid-base status during delivery at high altitude. *Am J Obstet Gynecol* 1971; **111**: 1111–18.

64. **Postigo L**, **Heredia G**, **Illsley NP** *et al*. Where the O2 goes to: preservation of human fetal oxygen delivery and consumption at high altitude. *J Physiol* 2009; **587**(3): 15.

65. **Niermeyer S**, **Yang P**, **Shanmina**, **Drolkar**, **Zhuang J**, **Moore LG**. Arterial oxygen saturation in Tibetan and Han infants born in Lhasa, Tibet. *N Engl J Med* 1995 1995; **333**: 1248–52.

66. **Carter AM**. Placental oxygen consumption. Part I: in vivo studies – a review. *Placenta* 2000; 21 Suppl A: S31–7.

67. **Abi-Nader KN**, **Mehta V**, **Wigley V** *et al*. Doppler ultrasonography for the noninvasive measurement of uterine artery volume blood flow through gestation in the pregnant sheep. *Reprod Sci* 2010; **17**(1): 13–9.

68. **Zhao X**. Placental mitochondrial respiratory function of native Tibetan at high altitude. *Zhonghua Yi Xue Za Zhi* 2007; **87**(13): 3.

69. **Schaaps JP**, **Tsatsaris V**, **Goffin F** *et al*. Shunting the intervillous space: new concepts in human uteroplacental vascularization. *Am J Obstet Gynecol* 2005; **192**(1): 323–32.

70. **Burton GJ**, **Woods AW**, **Jauniaux E**, **Kingdom JC**. Rheological and physiological consequences of conversion of the maternal spiral arteries for uteroplacental blood flow during human pregnancy. *Placenta* 2009; **30**(6): 473–82.

71. **Jackson MR**, **Mayhew TM**, **Haas JD**. Morphometric studies on villi in human term placentae and the effects of altitude, ethnic grouping and sex of newborn. *Placenta* 1987; **8**: 487–95.

72. **Mayhew TM**, **Jackson MR**, **Haas JD**. Oxygen diffusive conductances of human placentae from term pregnancies at low and high altitudes. *Placenta* 1990; **11**: 493–503.

73. **Reshetnikova OS**, **Burton GJ**, **Milovanov AP**. Effects of hypobaric hypoxia on the fetoplacental unit: the morphometric diffusing capacity of the villous membrane at high altitude. *Am J Obstet Gynecol* 1994; **171**: 1560–5.

74. **Zamudio S**. The placenta at high altitude. *High Alt Med Biol* 2003; **4**(2): 20.

75. **Tissot van Patot MC**, **Valdez M**, **Becky V** et al. Impact of pregnancy at high altitude on placental morphology in non-native women with and without preeclampsia. *Placenta* 2009; **30**(6): 523–8.

76. **Jeffreys RM**, **Stepanchak W**, **Lopez B**, **Hardis J**, **Clapp JF**, III. Uterine blood flow during supine rest and exercise after 28 weeks of gestation. *Br J Obstet Gynaecol* 2006; **113**(11): 1239–47.

77. **Zamudio S**, **Kovalenko O**, **Vanderlelie J** et al. Chronic hypoxia in vivo reduces placental oxidative stress. *Placenta* 2007; **28**(8–9): 846–53.

78. **Tissot van Patot MC**, **Murray AJ**, **Beckey V** et al. Human placental metabolic adaptation to chronic hypoxia, high altitude: hypoxic preconditioning. *Am J Physiol Regul Integr Comp Physiol* 2010; **298**(1): R166–72.

79. **Wilkening RB**, **Meschia G**. Current topic: comparative physiology of placental oxygen transport. *Placenta* 1992; **13**(1): 1–15.

80. **Yip R**. Altitude and birth weight. *J Pediatr* 1987; **111**: 869–76.

81. **Yip R**, **Li Z**, **Chong WH**. Race and birth weight: the Chinese example. *Pediatrics* 1991; **87**: 688–93.

82. **Zamudio S**, **Droma T**, **Norkyel KY** et al. Protection from intrauterine growth retardation in Tibetans at high altitude. *Am J Phys Anthropol* 1993; **91**: 215–24.

83. **Wen S**, **Kramer M**, **Usher R**. Comparison of birth weight distributions between Chinese and Caucasian infants. *Am J Epidemiol* 1995; **141**(12): 1177–87.

84. **Smith CF**. The effect of maternal nutritional variables on birth weight outcomes of infants born to Sherpa women at low and high altitudes in Nepal. *Am J Hum Biol* 1997; **9**: 751–61.

85. **Moore LG**, **Young D**, **McCullough RE**, **Droma T**, **Zamudio S**. Tibetan protection from intrauterine growth restriction (IUGR) and reproductive loss at high altitude. *Am J Hum Biol* 2001; **13**(5): 635–44.

86. **Hartinger S**, **Tapia V**, **Carrillo C**, **Bejarano L**, **Gonzales GF**. Birth weight at high altitudes in Peru. *Int J Gynaecol Obstet* 2006; **93**(3): 275–81.

87. **Gonzales GF**, **Tapia V**, **Carrillo CE**. Stillbirth rates in Peruvian populations at high altitude. *Int J Gynaecol Obstet* 2008; **100**(3): 221–7.

88. **Gonzales GF**, **Tapia V**. Birth weight charts for gestational age in 63 620 healthy infants born in Peruvian public hospitals at low and at high altitude. *Acta Paediatr* 2009; **98**(3): 454–8.

89. **Tripathy V**, **Gupta R**. Birth weight among Tibetans at different altitudes in India: are Tibetans better protected from IUGR? *Am J Hum Biol* 2005; **17**(4): 442–50.

90. **Yangzom Y**, **Qian L**, **Shan M** et al. Outcome of hospital deliveries of women living at high altitude: a study from Lhasa in Tibet. *Acta Paediatr* 2008; **97**(3): 317–21.

Discussion

SMITH: It seems a bit paradoxical that in hypoxia, there are lower ratios of s-Flt to PLGF and there are higher rates of pre-eclampsia. How would you tie this in with observations about s-Flt elsewhere?

MOORE: All these were normal women. We have measured s-Flt in pre-eclamptic women but they are not included in the data presented here. If there was greater placental oxygen delivery, there might be less hypoxic stress in the high-altitude Andeans and that, in turn, might lower their s-Flt levels relative to Europeans.

GIUSSANI: I have a comment relating to yesterday's discussion which was centred on nutrient delivery to the baby and its relationship to fetal growth and programming. When we were talking about nutrients we weren't considering oxygen, which is also a nutrient. Work like this highlights the contribution of fetal underoxygenation to fetal growth restriction. The next step is to try to relate the contribution of fetal underoxygenation in these populations to programming of cardiovascular and metabolic disease in adulthood. It would be interesting to see whether this protection against fetal growth restriction in individuals with a prolonged high-altitude residence ancestry (i.e.

Andeans versus Europeans) extends to protection against developmental programming of cardiometabolic disease in adulthood.

JANSSON: Is there any evidence that there is a deeper trophoblast invasion into spiral arteries in the high-altitude women?

MOORE: We haven't looked at the placenta, but my colleague Stacy Zamudio has, but I am not sure whether she has published these data yet. I would suspect that there is. Some years ago, Elizabeth Krampl did some work in Peru showing lower uterine artery vascular resistance indices in cross-sectional samples of high- versus low-altitude Andeans. We have not seen this in our women, but her sample sizes were somewhat greater. My colleague, Dr Vaughn Browne, has observed that Andean women with pre-eclampsia or gestational hypertension have aberrant wave-forms. So not all Andean women are protected. Interestingly, those women in whom the aberrant wave-forms are present have fully enlarged uterine arteries, so different sectors of the vascular tree appear to be affected by Andean ancestry versus pre-eclampsia. The upstream regions of the uterine vasculature are comparatively understudied: I don't know anyone other than us and a few other groups who are measuring volumetric uterine artery blood flow. We chose that level of the vascular tree because we couldn't assure ourselves that we were measuring blood flow at the same anatomic location if more downstream (e.g. arcuate artery) locations were used. Years ago, in our guinea pig work, we showed that the uterine artery itself was affected by chronic hypoxia. So we think hypoxia is altering vascular responses to pregnancy in the uterine artery as well as, potentially, more downstream locations of the vascular tree.

CETIN: You presented data on uterine blood flows in absolute terms (ml per min). Were they also increased if the differences in uterine, placental and fetal mass were also considered?

MOORE: In my oral presentation here[4] I did not consider the placenta, because we did not collect placentas at delivery. The babies were delivered in a great many different hospitals so this wasn't practical. Nor have we normalized by fetal size. The differences in uterine artery blood flow are so large that they are going to be increased relative to fetal mass. Placental sizes aren't particularly different in these groups.

CETIN: What is the placental: fetal ratio like?

MOORE: We don't have data, but from the work of Martha Tissot van Patot and others, the ratio is fairly closely maintained and not increased at high versus low altitude within a given ancestry group.

THORNBURG: What happens to these people who are born small (in the European population especially) in terms of their longevity and disease in later life?

MOORE: Susan Niermeyer has studied European as well as Andean babies over the first several months of life, but I don't believe she has published these data. She did a study in Tibet some years ago in which Tibetan and Han Chinese babies were compared over the first four months of life. At birth Tibetan and Han babies maintain their arterial oxygen saturation in the low 90s. Values remain at this level in the Tibetans while awake or asleep, but Han values progressively fall so that by four months of age their average saturation during sleep was 76%. Han in Tibet more frequently suffer from a condition called subacute infantile mountain sickness at about 6–8 months, which is essentially right heart failure. It is not clear the extent to which this occurs in Andean or European babies, but Dr Colleen Julian and colleagues at the Bolivian High-Altitude Biology Institute are studying whether size at birth influences susceptibility to chronic mountain sickness, a condition that occurs with surprisingly high frequency: 10% of the male population aged 40 and over. Chronic mountain sickness is also marked by pulmonary hypertension and right heart failure.

THORNBURG: Lubo Zhang at Loma Linda showed that in rats, a short exposure of hypoxia in the womb leads to cardiac cell enlargement in later life, and also to propensity to heart failure and damage with ischemia. I thought you might find heart failure in these people who had not adjusted well to altitude in later life.

MOORE: I think he was seeing left heart failure, and these conditions result mostly in right heart failure. Both could be potentially affected.

GIUSSANI: While there is work occurring in the postnatal consequences of fetal growth restriction, there have not been studies conducted of cardiovascular disease or metabolic disease in adulthood in these people. Together with David Barker and funded by the MRC, we now have a cohort of birth records of people who are now 30 or 40 years old, but the work hasn't been done in terms of trying to match their cardiovascular or metabolic status to their birth weight and European or Andean ancestry. This is a key question.

---

[4] In the chapter submitted for this volume, uterine artery flows as well as oxygen delivery have been normalized by newborn weight and considered in relation to estimated fetoplacental oxygen consumption.

MAGNUS: What about people born in Denver, Colorado? Are there any data on the incidence of cardiovascular disease in them?

MOORE: Denver is only 5000 feet (1600 m). Babies are a little bit smaller and there is intermediate placental morphology. Perhaps of relevance, chronic lung disease is a surprisingly high contributor to mortality in Colorado; it rises about 50% as you go from the lower to higher altitudes in the state of Colorado (Moore LG, Rohr AL, Maisenbach JK, Reeves JT. Emphysema mortality is increased in Colorado residents of high altitude. *Am Rev Resp Dis* 1982; **126**(2): 225–228).

COOPER: You were focusing on fixed genetic variation and teleology over a long period. Of course, one of the plastic phenomena might be alterations in expression to insults over relatively shorter periods. We have data showing that *eNOS* and *SOD1* promoter region methylation status relate to the bone mass of the child, but this is work done with cord samples. In your work, has there been any research on methylation status of those oxidative systems on the placental/maternal side that links to the uterine environment?

MOORE: That's a fascinating question and one certainly worth looking into. I don't think any work has been done on this.

HUNT: Looking at your gene scans, you have two that are highly expressed during inflammation, *iNOS* and *IL1β*, both of which are primarily products of the macrophages. Is there any evidence for inflammation being more prevalent in European than Andean women?

MOORE: My colleague Dr. R. Daniela Davila has looked at the cytokines. There are lower IL1β levels in the Andeans, but there is not much difference in anything else.

Chapter 12

Placental amino acid transporters
The critical link between maternal nutrition and fetal programming?

Thomas Jansson and Theresa L. Powell

Introduction

An adverse intrauterine environment may program the fetus for metabolic and cardiovascular disease in childhood and later in life. In particular, the ability of the maternal supply line to provide oxygen and nutrients to the growing placenta and fetus represents a key factor determining fetal outcome. The placenta constitutes the interface between the maternal and fetal circulations, serving as an immune barrier as well as the organ for exchange and metabolism of nutrients between mother and fetus. The placenta also controls maternal metabolism during pregnancy by secreting a variety of hormones into the maternal circulation. A large number of studies has now shown that variations in maternal diet (amount and composition) can alter fetal growth and/or development and program the fetus for later disease. In order to affect the fetus, perturbations in maternal nutrient availability must be transmitted across the placenta and there is an increasing awareness that the placenta responds to and modulates changes in the ability of the mother to provide nutrients [1–5]. Therefore, the placenta is in a powerful position to influence fetal development by changes in expression and activity of, for example, specific nutrient transporters that affect fetal growth. Altered fetal growth (intrauterine growth restriction [IUGR] and fetal overgrowth) in human pregnancy is associated with specific changes in placental amino acid transporters [2,3,6–8], which may cause changes in fetal growth rate and mediate programming of the fetus. In this chapter we will review the evidence suggesting that regulation of placental amino acid transporters constitutes a critical link between maternal nutrient availability and fetal growth and programming.

Amino acid availabilty, fetal growth and programming

Amino acid availability influences fetal growth in a multitude of ways. Amino acids constitute the building blocks for fetal protein synthesis, and in addition are potent stimulators of the secretion of fetal insulin [9], the primary growth-stimulating hormone during fetal life. Adequate amino acid supply is also critical for normal cellular growth, mediated by nutrient-sensing pathways such as mammalian target of rapamycin (mTOR) signalling [10]. Furthermore, the availability of specific individual amino acids influences fetal growth. For example, animal experiments have demonstrated that decreased availability of taurine, an essential amino acid in fetal life, results in IUGR [11]. Epigenetic mechanisms are believed to be important in mediating the developmental programming of adult disease [12], and gene methylation constitutes one of the principal mechanisms for epigenetic regulation. All biological methylation reactions are dependent on the availability of amino acids, such as methionine. Therefore, a restricted supply of methionine may program the fetus, even without effects on overall growth. Because fetal amino acid availability is critically dependent on the transfer of amino acids across the placenta, changes in placental amino acid transport influence fetal growth and may contribute to fetal programming.

Principles of placental amino acid transport

In the human placenta, transfer of nutrients from the maternal to the fetal circulation requires movement

---

The Placenta and Human Developmental Programming, ed. Graham J. Burton, David J. P. Barker, Ashley Moffett and Kent Thornburg. Published by Cambridge University Press. © Cambridge University Press 2011.

across two cell layers, the syncytiotrophoblast and the endothelium of the fetal capillaries (Figure 12.1). The syncytiotrophoblast constitutes the placental transporting epithelium and has two polarized plasma membranes: the microvillous plasma membrane (MVM) directed towards the maternal blood in the intervillous space, and the basal plasma membrane (BM) facing the fetal capillaries. It is primarily the syncytiotrophoblast, and in particular its two plasma membranes, that restricts the movement of amino acids across the placental barrier. Placental amino acid transport, mediated by transporter proteins in the MVM and BM, is an active process resulting in fetal amino acid concentrations higher – or much higher – than maternal ones. The uptake of amino acids from the maternal blood into the syncytiotrophoblast across the MVM constitutes the active step in placental amino acid transfer. Subsequently, amino acids diffuse across the BM down their concentration gradient, facilitated by an array of transporter proteins expressed in the BM of the syncytiotrophoblast. Placental amino acid transport is complex because there are more than 20 different amino acid transporters present in the placenta [13–15], each of which mediates the transfer of several amino acids with similar structure or properties. Furthermore, amino acid transporters have overlapping specificities, and so each individual amino acid is typically transported by several different transporter systems.

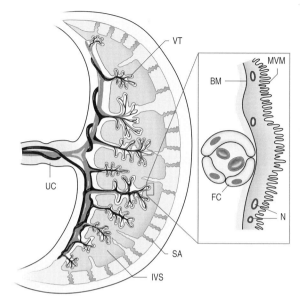

Figure 12.1 The placental barrier. The illustration represents a cross-section of the human placenta at term. The insert to the right shows a simplified representation of the placental barrier, which primarily consists of the syncytiotrophoblast cell and the fetal capillary (FC) endothelial cell. Of these structures it is primarily the two polarized plasma membranes, the microvillous (MVM) and the basal plasma membrane (BM) of the syncytiotrophoblast that restrict the transfer of molecules such as glucose and amino acids. N, nuclei of the syncytiotrophoblast cell; IVS, intervillous space; SA, spiral artery; VT, villous tree; UC, umbilical cord.

Two key amino acid transporters in the placenta

Only system A and system L amino acid transporters have been studied in any detail in the placenta and found to be present in multiple isoforms, highly regulated and altered in abnormal fetal growth. Consequently, this chapter will focus on these two transporters.

System A is a sodium-dependent transporter mediating the uptake of non-essential neutral amino acids with short, unbranched side chains, such as alanine, glycine and serine, into the cell [16]. All three known isoforms of system A, SNAT 1 (*SLC38A1*), SNAT 2 (*SLC38A2*) and SNAT 4 (*SLC38A4*), are expressed in the placenta [17]. System A has a unique ability to transport MeAIB, a non-metabolizable amino acid analogue which has been used extensively to study system A transporter activity in a wide range of cells and tissues, including the placenta. System A activity is particularly important, because system A transporter substrates can be transported by the obligatory exchanger system L. System L transporters can only exchange one amino acid molecule from outside the cell for one from the inside of the cell. As a result of system A activity, a steep outwardly directed concentration gradient of some non-essential amino acids, such as glycine, is created. Glycine molecules can bind to system L transporters on the inside of the cell and be exchanged for extracellular essential amino acids, thus driving the uptake of essential amino acids against their concentration gradients. This exchange is facilitated by a higher affinity for essential amino acids to bind to the extracellular portion of the transporter rather than binding to the intracellular portion.

System L is a sodium-independent amino acid transporter mediating cellular uptake of many branched-chain and aromatic amino acids such as leucine, phenylalanine and tyrosine [18]. System L is a heterodimer consisting of a light chain LAT 1 (*SLC7A5*) or LAT 2 (*SLC7A8*) and a heavy chain 4F2hc (*SLC3A2*) [18]. Another LAT isoform, LAT 4

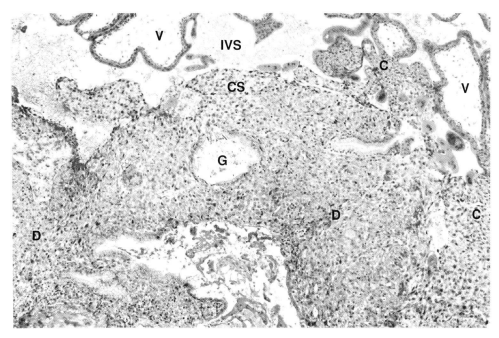

Figure 5.1 Photomicrograph of the maternofetal interface at 6 weeks of pregnancy, showing early villi (V), cytotrophoblast cell columns (C) and the cytotrophoblastic shell (CS), and the underlying decidua (D). Endometrial glands (G) can be seen discharging secretions, staining blue, into the intervillous space (IVS).

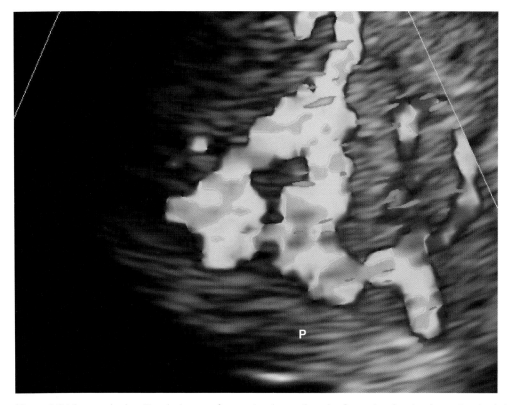

Figure 5.2 Ultrasound colour Doppler images of a gestational sac at 6 weeks of gestation showing the uterine circulation under the placental bed. Note that the maternal circulation stops before entering the centre of the primitive placenta (P).

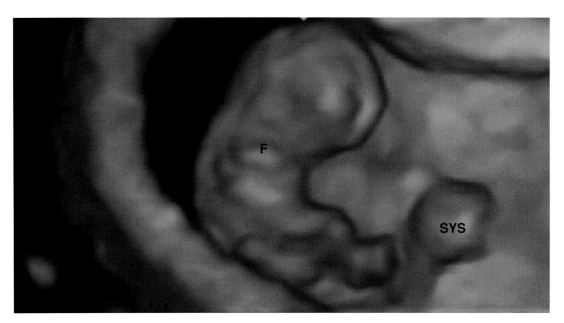

Figure 5.3 Three-dimensional ultrasound view of a gestational sac at 8 weeks of gestation showing the fetus (F) and the secondary yolk sac (SYS).

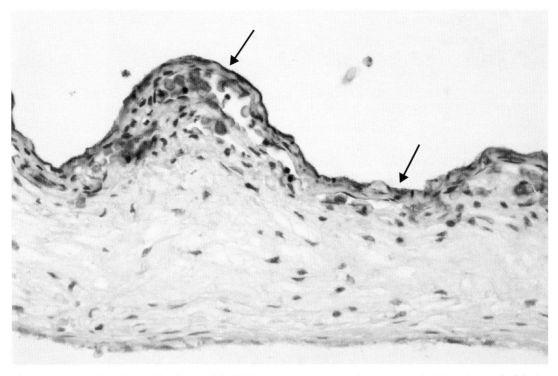

Figure 5.4 Immunohistochemical localization of the GLUT1 transporter protein to the outer mesothelial layer (arrowed) of the human yolk sac at 11 weeks of gestational age.

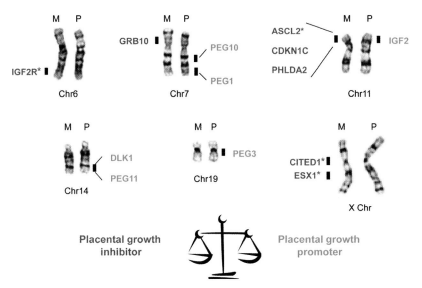

Figure 6.1 Chromosomal locations and affects on placental growth of selected imprinted genes. Maternally expressed genes are shown next to the maternal chromosome homologue (M), and paternally expressed genes next to the paternal homologue (P). Genes promoting placental growth (placental stunting in knockout mice) are in green; genes inhibiting placental growth (placentomegaly in knockout mice) are in red. ASCL2 is the human homologue of *Mash2/Ascl2*, which is essential for placental development in mice. The severe disruption of placental development in *Ascl2* KO mice makes this gene difficult to assign with respect to its effect on growth of the intact organ, although future experiments with transgenic lines might be informative. Several of these genes, including *ASCL2*, *IGF2R*, *CITED1* and *ESX1* are imprinted in mice, but are never or only rarely functionally imprinted in humans (imprinted genes on the mouse X-chromosome may in part be explained by selective paternal X-inactivation in that species). This fact raises the interesting hypothesis that imprinting is slowly being lost in humans relative to rodents, possibly due to weaker intergenomic conflict in our species.

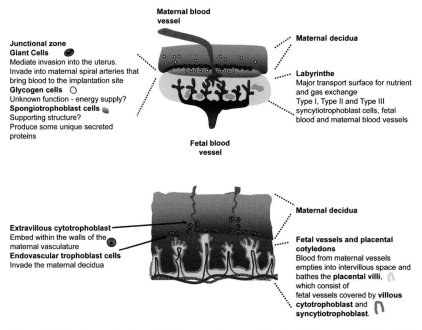

Figure 6.2 Structure of the placenta in human and mouse. The placenta is composed of trophoblast and fetal mesenchyme and blood vessels, which are the embryonic component, and deciduas and maternal vessels, which are the maternal component. In humans, the fetal blood vessels are covered by a layer of cytotrophoblast cells and an adjacent layer of multinucleated syncytiotrophoblast. These structures, called placental villi, are bathed in maternal blood. In mouse, the functionally equivalent region is termed the labyrinth layer. Here fetal blood vessels are encased in a bilayer of syncytiotrophoblast and a layer of mononuclear trophoblast cells which are in contact with the maternal blood. Both humans and mice possess invasive cell types. In humans, extravillous cytotrophoblast cells, which appear to arise from anchoring villi, invade the maternal decidua. Some of these cells embed within the walls of the maternal vasculature and are called endovascular trophoblast cells. Mice possess giant trophoblast cells, which are aligned between the junctional zone and the decidua and are behaviourally somewhat equivalent to the invasive extravillous cytotrophoblast, both being polyploid cell types. In the mouse there is a specialized region termed the junctional zone, which is composed primarily of spongiotrophoblast cells for which there is no clear human parallel. Glycogen cells, which arise within the junctional zone, also possess invasive properties migrating into the maternal decidua at E16.5 where they may provide an easily exploitable energy source to support late gestational growth. Adopted from Rossant J, Cross JC. Placenta development: lessons from mouse. mutants. *Nat Rev Genet* 2001; **2**(7): 538–48 [92].

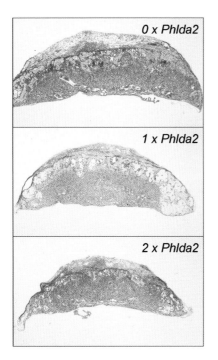

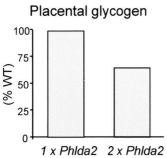

Figure 6.3 *Phlda2*: an example of an imprinted gene acting as a rheostat for placental growth. Shown are haemotoxylin/eosin midline sections of E14.5 placenta from *Phlda$^{-/+}$* (0 X *Phlda2*), wild type and 5D3 transgenic (2 X *Phlda2*) embryos and biochemically determined placental glycogen shown as a percentage of wild type. Deletion of the maternal allele of *Phlda2* (0 X *Phlda2*) produces placentomegaly, with disproportionate overgrowth of the junctional zone whereas overexpression of *Phlda2* in the transgenic model with a double dose of *Phlda2* (2 X *Phlda2*) causes placental stunting, with a reduced volume of the junctional zone. In addition, there is a 35% reduction in stored glycogen at E14.5. Although no definitive data are available concerning the function of placental glycogen stores, one possibility is that they provide an easily exploitable source of energy to support very late embryonic growth or parturition.

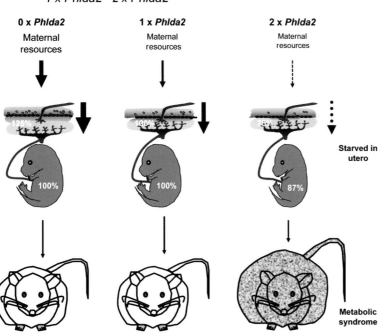

Figure 6.4 Phlda2 KO and Tg mice as a possible example of a genetic system for modeling DOHAD. Embryonic growth and placental support are carefully balanced so that the genetic growth potential of the fetus is fully supported by the nutrient capacity of the placenta. *Phlda2*-deficiency (0 X *Phlda2*) increases placental weight and glycogen stores but does not benefit embryonic growth – an example whereby excessive growth of the placenta depletes maternal nutritional resources and potentially limits her current and future reproductive success, i.e. reduced maternal fitness. In contrast, excess *Phlda2* (2 X *Phlda2*) restricts placental growth, which in turn restricts late embryonic growth. The consequences later in life can include a phenotype similar to human metabolic syndrome.

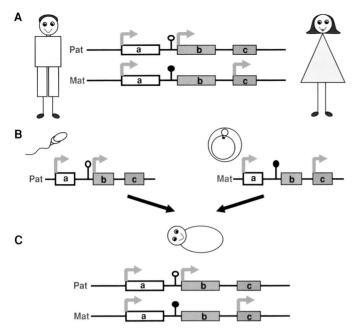

Figure 7.1 Genomic imprinting. (A) The majority of genes in the genome are like gene a, they are expressed or repressed biallelically from both parentally inherited chromosomes. A small group of genes are imprinted, they are expressed monoallelically according to their parental origin. In this case gene b is expressed only from the paternally inherited chromosome and gene c from the maternally inherited chromosome. This process is controlled by epigenetic marks which distinguish the two chromosomes, represented here by unfilled and filled lollipops. (B) In the mammalian gametes – sperm and egg – these epigenetic marks must be erased and re-established according to the sex of the individual. This ensures that all potential offspring (C) will inherit the appropriate complement of epigenetic marks. After fertilization, these marks are stable and heritable in all the somatic cells and maintain the imprinted state.

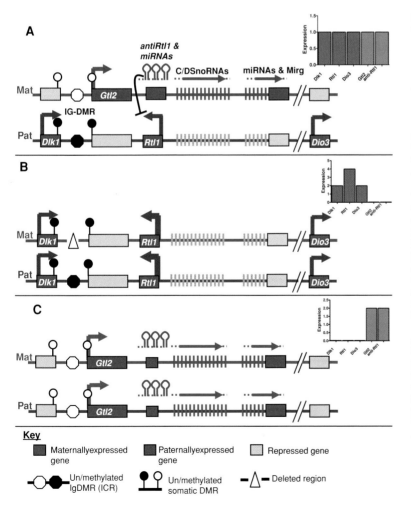

Figure 7.2 Imprinting control at the chromosome 12 *Dlk1-Dio3* domain (A) The imprinting control region of the *Dlk1-Dio3* locus is an intergenic germline differentially methylated region (IG-DMR) which is methylated on the paternally inherited chromosome and unmethylated on the maternally inherited chromosome. This results in the expression of the non-coding RNAs *Gtl2*, *anti-Rtl1*, *Mirg* and several clusters of microRNAs and SnoRNAs from the maternally inherited chromosome, whereas the protein coding genes *Dlk1*, *Rtl1* and *Dio3* are expressed from the paternally inherited chromosome. (B) The deletion of the maternally inherited unmethylated copy of the IG-DMR results in the paternalization of the maternal chromosome, the gain of methylation at post-fertilization somatic DMRs, de-repression of the normally paternally expressed protein-coding genes and silencing of non-coding RNAs on the maternally inherited chromosome. The loss of maternally expressed microRNA which normally reduce levels of *Rtl1* in *trans* results in approximately a fourfold upregulation of *Rtl1*, whereas *Dlk1* and *Dio3* are expressed at a double dose, reflecting their loss of imprinting. (C) Loss of methylation on the paternally inherited copy causes maternalization of the paternal chromosome, the loss of methylation at somatic DMRs, repression of the protein-coding genes and activation of the normally repressed non-coding RNAs [9,10]. Adapted from Lin et al. 2003 [9].

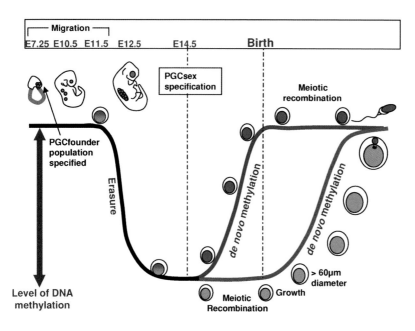

Figure 7.3 Reprogramming of DNA methylation at imprinting control regions during primordial germ cell development Genome-wide epigenetic reprogramming, including at imprinting control regions during primordial germ cell (PGC) development is required for the viability of future offspring. This begins once the primordial germ cells (PGCs) have been specified at E7.25 and continues throughout the migration of the PGCs to the genital ridge. As the PGCs enter the ridge around E11.5 extensive demethylation occurs, including the dramatic and rapid erasure of methylation at imprinted loci. Following sex determination *de novo* methylation reinstates imprints according to PGC sex. In males this occurs between E14.5 and birth, concomitantly with *de novo* methylation of retrotransposon sequences in the G1 arrested PGCs. In the female germline methylation of maternally derived imprints occurs postnatally, concurrent with oocyte growth. Adapted from Bourc'his and Proudhon [76].

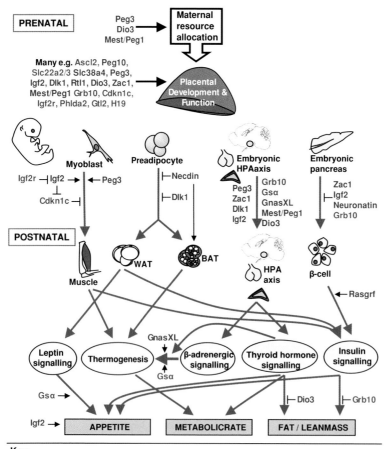

Figure 7.4 Imprinted genes act at multiple levels during prenatal and postnatal life to control energy homoeostasis The correct dosage of imprinted gene expression is vital for energy homoeostasis during both prenatal and postnatal life. Imprinted genes play critical roles in the control of nutrient supply to the fetus through effects on maternal metabolism and energy partitioning, placentation and placental function. There is also evidence that interactions between imprinted genes act together in the fetus to regulate growth, thus altering fetal demand for maternal resources. Imprinted genes play key roles in the growth and differentiation of metabolic organs, with lasting implications for adult metabolism. Furthermore, there is increasing evidence that imprinted genes modulate the function of key adult metabolic pathways. These varied roles have largely been revealed by the careful analysis and comparison of mouse genetic models *in vivo* where the dosage of a particular gene or cluster has been altered. The terms maternally and paternally expressed refer to imprinted gene expression from maternally inherited and paternally inherited chromosomes. Adapted from Charalambous et al. 2007 [40].

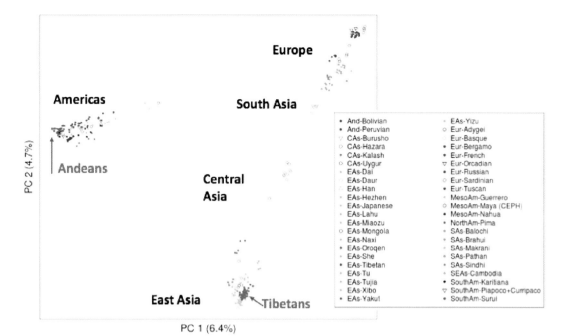

Figure 11.3 The first and second principle components (PC1, PC2) express the major relationships among 906 000 single nucleotide polymorphisms (SNPs) for the study populations shown in the legend and as described previously ([19,20], M. Bauchet, pers. comm.; A. Bigham, pers. comm.). Major geographic regions and the Andean and Tibetan samples are labelled. The clustering of data points demonstrates that Andeans and Tibetans are much more closely related to their geographic neighbours than to each other.

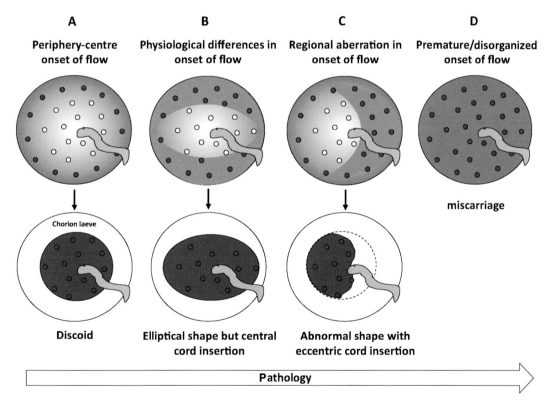

Figure 13.1 Diagrammatic *en face* view of villus regression towards the end of the first trimester. A) In normal pregnancies onset of the maternal circulation starts in spiral arteries (small red circles) at the periphery of the placenta, whereas arteries in the central region (white circles) remain plugged. Locally high levels of oxidative stress (diffuse orange) cause villus regression and formation of the chorion laeve. B) Regional differences in trophoblast invasion or density of the spiral arteries may cause more asymmetrical regression and an elliptical placental shape, which may have consequences for developmental programming. C) Excessive villus regression results in more abnormal placental shapes with eccentric attachment of the umbilical cord. D) In miscarriage, precocious and disorganized onset of the maternal circulation leads to overwhelming placental oxidative stress. Adapted from [70] with permission.

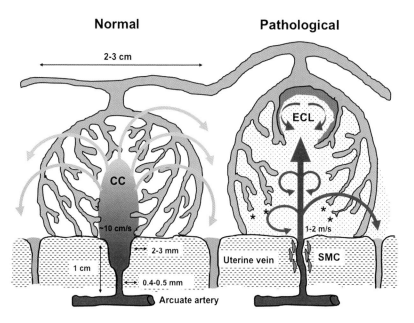

Figure 13.3 Diagrammatic representation of the haemodynamic consequences of deficient spiral artery conversion. In the absence of conversion, maternal arterial blood enters the intervillous space at a high velocity and in a turbulent manner. The momentum of the incoming blood may rupture anchoring villi (asterisked), allowing the chorionic and basal plates to move apart, increasing placental thickness. Besides this mechanical damage, and the formation of echoic cystic lesions (ECL), a mismatch between the maternal and fetal rates of flow may result in reduced placental efficiency. Adapted from [18] with permission.

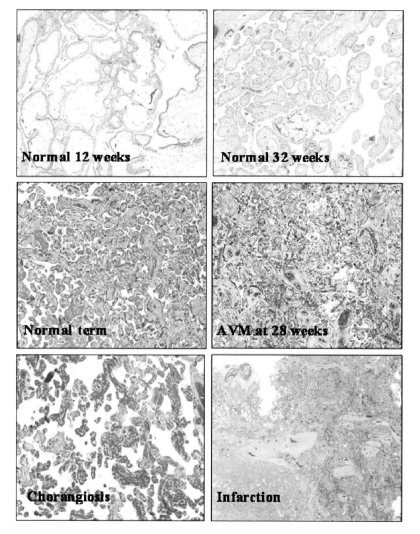

Figure 16.3 Histologic features of uteroplacental insufficiency. As the placenta ages it arborizes like a tree with increasing architectural complexity (i.e. tertiary villi are like the leaves). Accelerated villous maturation (AVM) is premature hypermaturation of the chorionic villi with syncytial knotting, and perivillous fibrosis. Chorangiosis is excessive angiogenesis in the villi. Infarction is the consequence of ischaemia and failed maternofetal compensation.

(*SLC43A2*), as well as TAT1 (*SLC16A10*), has been implicated in the net efflux of amino acids such as leucine across the BM into the fetal compartment [13].

Placental amino acid transport in altered fetal growth

The MVM activity of the system A transporter is reduced in IUGR [19–22], a finding which has been confirmed in primary villous fragments [23]. In fetal overgrowth associated with diabetes, MVM activity of system A was found to be upregulated in one study [24] and decreased in another [25]. The reason for these discrepancies is unclear, but may be related to the different study populations. This possibility is supported by the fact that placental weight was increased in one study [24] and unaffected in the other [25]. The activity of system L is decreased in both plasma membranes of the syncytiotrophoblast in IUGR [26]. These *in vitro* findings are compatible with a study in pregnant women in which Paolini *et al.* [27], using stable isotope techniques, demonstrated that placental transfer of the essential amino acids leucine and phenylalanine is reduced in IUGR. System L activity is increased in MVM vesicles isolated from placentas from mothers with gestational diabetes giving birth to large for gestational age (LGA) infants [24].

Collectively these data from the human clearly show that reduced fetal growth is associated with a downregulation of placental amino acid transporters, whereas fetal overgrowth appears to be associated with an increased placental amino acid transporter capacity. Whether these associations reflect a cause-and-effect relationship is difficult to address in the pregnant woman. However, in pregnant rats, placental system A amino acid is reduced several days before IUGR is observed in response to a low-protein diet [28], which is compatible with the hypothesis that changes in placental amino acid transport contribute directly to the development of IUGR.

Regulation of placental amino acid transport

Classic regulators

Several hormones and growth factors have been shown to stimulate placental system A transporter activity in cultured trophoblasts and placental explants, including insulin, IGF-I, EGF and leptin [29–34]. It is well established that the activity of system A is upregulated in response to amino acid deprivation, a phenomenon referred to as adaptive regulation. This has been demonstrated in many cell types, including trophoblast cells [35]. Low oxygen levels reduce system A activity through downregulation of both SNAT1 and SNAT2 mRNAs [36]. In the BeWo cell model, 24-hour cortisol treatment increased system A activity and SNAT2 mRNA expression [37]. Furthermore, incubating primary villous fragments with SIN-1, which releases NO and O_2^-, has been shown to inhibit system A uptake. This inhibition was proposed to be a direct effect of free radicals on the transporter [38].

The regulation of the placental system L amino acid transporter has not been extensively studied. It has been reported that lowering the extracellular pH, treating cells with the PKC activator PMA, and calmodulin antagonists all stimulate system L transport in the human placental choriocarcinoma cell line JAR [39,40]. Combined treatment of BeWo cells with PMA and a calcium ionophore stimulated system L activity by increasing mRNA and protein expression of both 4F2hc and LAT1 [41].

Emerging concepts in placental amino acid transporter regulation

We recently reported that physiological concentrations of the proinflammatory cytokines interleukin-6 (IL-6) and tumour necrosis factor-α (TNF-α) stimulate the activity of amino acid transporter system A, but not system L, in cultured human primary trophoblast cells [42]. Both cytokines increased the gene and protein expression of the SNAT2 isoform and upregulated SNAT1 protein expression. IL-6 increased Tyr705 phosphorylation of signal transducer and activator of transcription 3 (STAT3). In STAT3 silenced cells, the RNA and protein expression of SNAT2, but not SNAT1, was reduced and the stimulating effect of IL-6 on system A activity was abolished. In contrast, TNF-α effects on system A activity were not mediated through the JAK/STAT pathway [42].

Emerging evidence suggests that full-length adiponectin (fAd) has marked effects on placental insulin signalling and amino acid transport [43]; fAd alone had no effect on System A activity or

SNAT expression. Insulin increased Akt and IRS-1 phosphorylation, system A activity and SNAT2 expression. When combined with insulin, fAd abolished insulin-stimulated Akt(T308) and IRS-1(Y612) phosphorylation, system A activity and SNAT2 expression. Furthermore, fAd increased PPARα expression and PPARα (S21) phosphorylation [43]. These *in vitro* data are supported by preliminary studies from our laboratory in which fAd was administered as a 3-day infusion by means of mini-osmotic pumps to pregnant mice in late gestation. In this series of experiments maternal adiponectin infusion resulted in a marked inhibition of placental insulin signalling and reduced fetal growth. Thus, in contrast to the insulin-sensitizing actions of adiponectin in liver and muscle, fAd attenuates insulin signalling in primary human trophoblast cells. As a result, fAd inhibits insulin-stimulated amino acid transport, which may have important implications for placental nutrient transport and fetal growth in pregnancy complications, such as obesity, associated with altered maternal adiponectin levels.

mTOR is an ubiquitously expressed serine/threonine kinase that controls cell growth. Although it is well established that the mTOR signalling pathway affects the transcription of genes involved in amino acid, lipid and nucleotide metabolism, protein synthesis and immune modulation [44], the primary mechanism by which mTOR regulates cell growth is via effects on protein translation [45–48]. mTOR exists in two complexes, mTORC1 and 2, and the primary difference between these is that mTOR is associated with the protein raptor in mTORC1 and with rictor in mTORC2. mTORC1 phosphorylates S6K1 (p70 S6 kinase) and 4E-BP1 (eukaryotic initiation factor 4E binding protein 1), which releases eIF-4E. We have recently shown that mTORC1 is a positive regulator of trophoblast system A and L amino acid transporters [49,50], an effect that may not be mediated through altered transcription/translation [50]. Furthermore we have provided evidence that glucose and growth factors are upstream regulators of trophoblast mTORC1 [51]. The role of mTORC2 in the regulation of cell function remains to be established; however, mTORC2 is believed to influence the actin skeleton. Our preliminary data indicate that both mTORC 1 and 2 regulate trophoblast amino acid transporters, as silencing of either raptor or rictor results in marked inhibition of both system A and system L activity.

Amino acid deprivation has two opposing effects on trophoblast system A activity: an increase due to adaptive regulation (see above) and a reduction due to mTOR inhibition. System L amino acid transport activity, which is not subjected to adaptive regulation, is inhibited in response to amino acid deprivation. Similarly, hypoglycemia inhibits system L in an mTOR-dependent manner [51]. Interestingly, preliminary data from our laboratory suggest that lipids, in the form of oleic acid, stimulate trophoblast system A amino acid transporter activity mediated by Toll-like receptor-4 signalling.

Regulation of placental amino acid transport: who is in charge?

Placental functional capacity, including amino acid transporter activity, is likely to be subjected to regulation by fetal, placental and maternal factors (Figure 12.2).

Fetal signals

Evidence from the human indicates the presence of fetal signals that, in certain circumstances, adapt placental nutrient transport capacity to fetal demand. The activity of the BM calcium pump is increased in the IUGR placenta [52], possibly owing to elevated fetal levels of a mid-molecule fragment of PTHrp [53]. Additional evidence for the adaptability of placental transport functions comes from a study by Godfrey and co-workers showing that MVM system A amino acid transporter activity is inversely correlated to fetal size within the normal range of birth weights [54]. These findings suggest that a smaller fetus has the ability to upregulate placental amino acid transport, thereby maintaining a growth trajectory along the lower limit of the normal range. This is in sharp contrast to true IUGR, which is associated with decreased MVM system A activity [19–22]. Furthermore, there is support for the existence of a fetal demand signal from studies of mice with placenta-specific knockout of the insulin-like growth factor 2 gene (*igf*-2). In this model, which is associated with placental growth restriction in mid-gestation, there is a temporary upregulation of placental SNAT 4 expression and system A activity, which maintains fetal growth in the normal range until late pregnancy when compensatory mechanisms fail and IUGR develops [55,56].

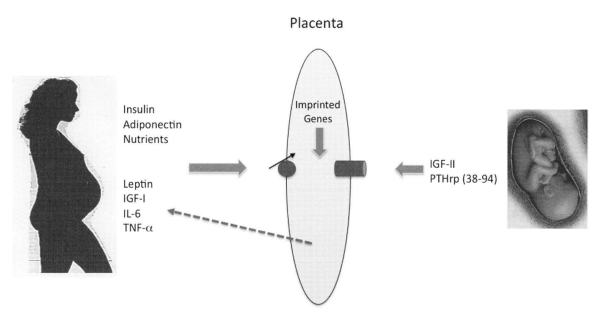

Figure 12.2 Maternal, placental and fetal regulation of placental nutrient transport. Placental functional capacity, including amino acid transporter activity, is subjected to regulation by fetal, placental and maternal factors. Fetal factors may include IGF-II and PTHrp (38–94). Some genes, including *igf-2* and the system A amino acid transporter isoform SNAT 4, are imprinted only in the placenta and are involved in the regulation of placental nutrient transport and growth. Furthermore, the placenta produces several growth factors and hormones, including IGF-I and leptin, which are secreted into the maternal circulation where they can interact with the receptors on the microvillous plasma membrane, constituting an autocrine/paracrine regulation of placental transport of amino acids. The placenta responds to nutritional cues in the maternal compartment causing changes in placental transport function and therefore fetal growth. These maternal signals include hormones such as insulin, adiponectin, IGF-I, leptin and cytokines, as well as nutrients.

Placental imprinted genes and placental growth factor production

Imprinted genes are expressed predominantly from one of two parental alleles and are subjected to epigenetic regulation. More than 70 imprinted genes have been discovered in the mouse, and a subgroup of these are imprinted only in the placenta and are involved in the regulation of fetal and placental growth [57]. An example of a paternally expressed/maternally repressed placental genes is *igf-2* [55], which regulates the growth of the placenta and therefore indirectly its transport capacity. Interestingly, the system A amino acid transporter isoform SNAT 4 (*SLC38A4*) is expressed from the paternal allele in the placenta. Thus, the paternal genome may exert influence on placental amino acid transport capacity and fetal growth by means of paternally imprinted placental genes. Furthermore, the placenta produces several growth factors and hormones, including IGF-I and leptin [58,59], that have been shown to stimulate placental amino acid transport *in vitro*. These factors are secreted into the intervillous space, where they can interact with the receptors on the microvillous plasma membrane, constituting an autocrine/paracrine regulation of placental transport of amino acids.

Maternal signals

As discussed previously, studies from *in vitro* preparations of the human placenta have demonstrated that placental amino acid transporters are under the regulation of hormones and growth factors, including IGF-I, insulin and leptin. The receptors for these factors in the placenta are localized primarily in the microvillous plasma membrane of the syncytiotrophoblast [60–62], and therefore directly exposed to maternal blood. Thus it appears that placental amino acid transporters may be regulated by hormones and growth/factors circulating in the maternal circulation. It is interesting to note that in IUGR and fetal overgrowth the maternal serum concentrations are changed in a direction compatible with the possibility that these factors play a critical role in regulating placental transport and fetal growth in clinically important pregnancy complications. In IUGR, for example, maternal serum

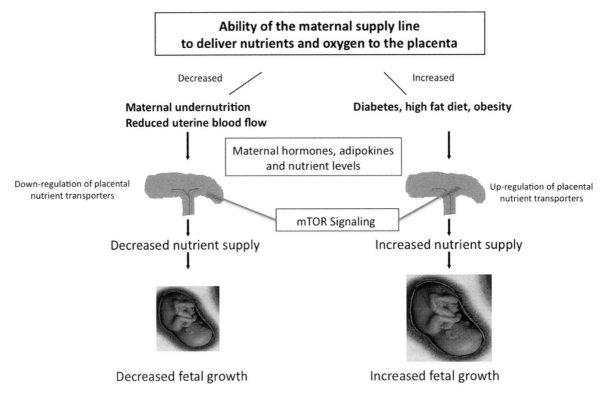

Figure 12.3 The placenta as a nutrient sensor – a hypothesis. It is suggested that placental nutrient transporters are regulated in response to a primary event, such as a lack of increase in placental blood flow, maternal malnutrition or hyperglycaemia. Alterations in placental transport activity result in changes in nutrient delivery to the fetus, which, in turn, affects fetal growth. Hormones and cytokines produced by the placenta or the mother, nutrient levels, hypoxia and nutrient-sensing mechanisms 'intrinsic' to the placenta (such as mammalian target of rapamycin-mTOR) may constitute mechanisms mediating placental nutrient sensing.

concentrations of IGF-I are decreased [63], and some studies indicate that maternal serum leptin concentrations are reduced [64]. In contrast, circulating maternal IGF-I is increased in pregnancies with diabetes and accelerated fetal growth [65]. Furthermore, in maternal obesity, a condition associated with a high risk for fetal overgrowth, maternal serum insulin, leptin and IGF-I are typically increased [66].

The placenta as a nutrient sensor – a model

In order to model the mechanisms linking maternal nutrient availability (in a broad sense) to fetal growth and programming, we have proposed that the placenta functions as a nutrient sensor, regulating placental amino acid transporters in response to changes in the ability of the maternal supply line to deliver nutrients to the placenta (Figure 12.3) [1,2]. According to this theoretical model, the ability of the mother to supply nutrients to the placenta and growing fetus is the primary determinant of fetal growth. The ability of the maternal supply line to deliver nutrients could be reduced by, for example, maternal malnutrition or reduced uterine blood flow (Figure 12.3). The central premise in our proposal is that these alterations are 'sensed' by the placenta, leading to alterations in placental growth and transport capacity. We specifically propose that, in response to undernutrition or reduced placental blood flow, placental amino acid transporters are downregulated, which reduces fetal nutrient supply and growth. The maternal supply line could also have an increased ability to deliver nutrients, such as in maternal diabetes, a high-fat diet and obesity. In this situation, we propose that placental nutrient transporters are upregulated, which increases fetal nutrient supply and fetal growth. We further postulate that the mechanisms responsible for placental nutrient sensing include maternal hormones and adipokines as well as trophoblast mTOR signalling, which have been shown

Chapter 12. Placental amino acid transporters

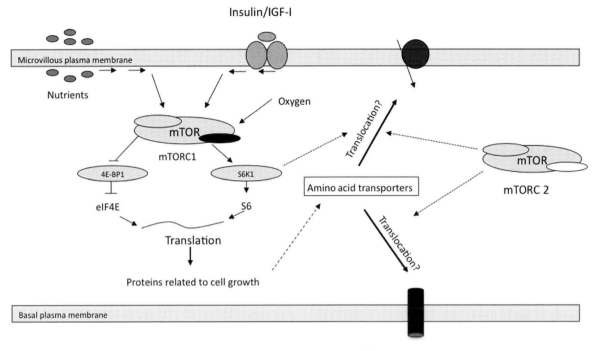

Figure 12.4 A model of mTOR signaling as a placental nutrient-sensing pathway. mTOR integrates nutrient and growth factor signalling (and possibly a number of other upstream signals) to control nutrient transport from the mother to the fetus. mTOR forms two complexes in the cell, mTORC1 and mTORC2. The exact mechanism for mTOR regulation of nutrient transport in the placenta remains to be established, but it appears to involve both mTOR complexes. As placental amino acid transporters are important regulators of fetal growth, we propose that placental mTOR constitutes a mechanistic link between maternal nutrient availability and fetal growth. mTORC1 and mTORC2, mammalian target of rapamycin complexes 1 and 2; 4E-BP1, eukaryotic initiation factor binding protein; eIF-4E, eukaryotic initiation factor; S6K1, p70 S6 kinase; S6, ribosomal protein s6.

to be important regulators of placental amino acid transporters (Figure 12.4). To test the validity of this model, three different situations in which maternal nutrient availability is perturbed will be discussed: maternal low-protein diet in the rat, maternal high-fat diet in the mouse and obesity in pregnant women.

Maternal low-protein diet in the rat

Maternal protein restriction in the rat is a well-established model for the study of developmental programming. Feeding the dam a low-protein diet throughout pregnancy typically results in the development of IUGR in late gestation. However, the mechanisms causing the reduced fetal growth in this model have been elusive, as maternal circulating amino acid concentrations are relatively unaffected [28,67], most likely due to maternal protein catabolism. In a study by Malandro and co-workers, protein restriction in the pregnant rat was shown to downregulate the *in vitro* activity of specific placental amino acid transporters at term (system A and transporters for cationic amino acids) [67]. However, it is unclear from these studies whether placental transport changes are a cause, or a secondary consequence, of IUGR.

We investigated the time course of changes in placental and fetal growth, placental nutrient transport *in vivo* and the expression of placental nutrient transporters in pregnant rats subjected to protein malnutrition [28]. Pregnant rats were given either a low-protein (LP) diet or an isocaloric control diet throughout pregnancy. Fetal and placental weights in the LP group were unaltered compared to those on the control diet at gestational days (GD) 15, 18 and 19, but significantly reduced at GD 21 (term = GD 23). Placental system A transport activity was reduced at GD 19 and 21 in response to a low-protein diet. Placental protein expression of SNAT2 was decreased at GD 21 [28]. Thus, placental amino acid transport was down regulated prior to the development of IUGR, suggesting that these placental transport changes are a cause, rather than a consequence, of IUGR. In this model maternal insulin, leptin and IGF-I levels decreased in late pregnancy, whereas maternal amino acid concentrations increased moderately in response to the LP diet. Thus, reduced maternal serum concentrations of insulin, leptin and IGF-I may link maternal protein restriction to reduced fetal growth by downregulation of key placental amino acid transporters. In preliminary studies, we have explored placental mTOR (by measuring expression of phosphorylated S6K1 and 4E-BP1), insulin (phosphorylated Akt) and leptin signalling (phosphorylated STAT3) in response to a LP diet in the rat. The results indicate that the activity of placental insulin, leptin and mTOR signalling was decreased in response to maternal protein restriction. In particular, the activity of the placental mTOR signalling pathway was markedly decreased at GD 19, coinciding with the downregulation of placental amino acid transport and preceding fetal growth restriction by 2 days. Collectively, these data are in agreement with the placental nutrient-sensing model, where maternal hormones and placental mTOR signalling regulate placental amino acid transporters in response to changes in maternal nutrient availability, resulting in altered fetal growth. It is also notable that we found no evidence, at any gestational age under study, for a compensatory upregulation of placental amino acid transport in response to maternal protein restriction. This suggests that there are no strong fetal demand signals operating in this model.

Maternal high-fat diet in the mouse

We recently developed and published [68] an experimental model in which feeding female mice a diet with a moderately high fat content is associated with increased adiposity, elevated maternal circulating levels of leptin and triglycerides and decreased serum adiponectin concentrations as well as fetal overgrowth. Thus, this model has some of the key features of obesity in pregnant women. Interestingly, both placental glucose and system A mediated transport *in vivo* were markedly increased in high-fat-fed animals [68]. Furthermore, using microvillous plasma membranes isolated from the mouse placenta, we found that the placental protein expression of glucose transporter 1, GLUT 1 (but not GLUT 3) and SNAT 2 (but not SNAT 4) was markedly upregulated in response to a high-fat diet [68]. Preliminary studies show that placental mTOR signalling is activated in high-fat-fed animals. Thus, in line with our placental nutrient-sensing model, increased maternal nutrient availability (high-fat diet) in this animal model results in up regulation of placental amino acid transport and fetal overgrowth, possibly mediated by high maternal triglycerides, leptin and low adiponectin.

Obesity in pregnant women

More than 50% of all pregnant women in the USA today are overweight or obese [69]. In obese women pregnancy complications are much more common than in lean women [70,71]. Furthermore, the baby of the obese woman is often large at birth, which is associated with traumatic birth injuries [70–72] and an increased risk of developing obesity, diabetes and hypertension in childhood [73]. However, the mechanisms underlying fetal overgrowth in pregnancies of obese women remain to be fully established. In preliminary studies, we found that in placentas of large babies born to overweight and obese women, the activity of the system A amino acid transporter and the protein expression of SNAT 2 (but not SNAT 1 or 4) was upregulated compared to placentas from normal-sized babies in overweight/obese women as well as lean women. Furthermore, preliminary data show that placental mTOR signalling may be upregulated in overweight/obese women giving birth to large babies, compatible with our overall model. It is possible that the upregulation of placental mTOR and amino acid transport is due to the hormonal and metabolic changes in obese women, which include increased circulating levels of insulin, leptin, IGF-I, IL-6, TNF-α, triglycerides and amino acids and decreased maternal serum concentrations of adiponectin [66,74,75]. These observations in obese pregnant women, albeit preliminary, are in general agreement with the predictions of the placental nutrient-sensing model (Figure 12.3).

Conclusion

Notwithstanding that associations between fetal growth and programming of adult disease often extend across the normal birth weight range, it is the IUGR and large-for-gestational age babies that are at particular risk of developing diabetes and cardiovascular disease later in life. Thus, placental amino acid transport plays a central role in fetal programming by directly regulating fetal amino acid supply and fetal growth. There are a large number of signals impinging on the regulatory pathways that determine the expression and activity of placental amino acid transporters. These signals are likely to originate in the fetus, placenta and mother. In this chapter we have reviewed the evidence demonstrating that maternal factors have a particularly important influence on placental amino acid transport. We conclude that there is significant experimental support for the overall model of placental nutrient sensing, postulating that regulation of placental amino acid transporters, mediated by maternal hormones, adipokines and nutrient levels as well as placental mTOR signalling, constitutes an important link between maternal nutrient availability and fetal growth. A better understanding of the molecular mechanisms linking maternal nutrient availability, placental transport functions and fetal growth will allow for the design of novel intervention strategies to alleviate abnormal fetal growth and, as a result, prevent cardiovascular and metabolic disease in future generations.

The conflict theory of genomic imprinting proposes that maternally imprinted placental genes tend to limit fetal growth, whereas paternally imprinted genes promote fetal growth. We believe that maternally imprinted placental genes and placental nutrient sensing are complementary, where placental nutrient sensing limits fetal growth in response to conditions in the index pregnancy, whereas maternally imprinted genes can carry an epigenetic memory across generations. The proposed placental nutrient sensing model may be counterintuitive from a classic homoeostatic point of view, which predicts an upregulation of placental nutrient transporters in situations of restricted fetal nutrient availability. However, we speculate that the strong maternal influence on placental nutrient transport associated with the placental nutrient-sensing model constitutes an evolutionary strategy to ensure successful maternal reproduction at times of less than optimal nutrient availability. For example, when food is very scarce placental nutrient sensing will result in a smaller fetus that requires fewer total resources, or alternatively, abortion or fetal demise, allowing the mother to reproduce successfully at a later time, when food availability is ample. In contrast, strong homoeostatic regulation of placental transport capacity, resulting in up regulation of placental transport capacity, in a situation with severely restricted maternal food intake, would potentially jeopardize maternal (and therefore also fetal) survival.

Acknowledgements

Supported by grants NICHD/NIH HD058032 (Jansson), HD058030 (Powell), and NHLBI/NIH HL093532 (Powell). Figure 12.1 is reproduced by permission from Elsevier Ltd. This figure was published in the chapter 'Placental function and maternofetal exchange' in *Fetal Medicine: Basic Science and Clinical Practice*, 2nd edn, 2008, ISBN 978-0-443-10408-4.

References

1. **Jansson T, Powell TL.** Placental nutrient transfer and fetal growth. *Nutrition* 2000; **16**: 500–2.
2. **Jansson T, Powell TL.** Human placental transport in altered fetal growth: does the placenta function as a nutrient sensor? – a review. *Placenta* 2006; **27** Suppl: 91–7.
3. **Jansson T, Powell TL.** Role of the placenta in fetal programming: underlying mechanisms and potential interventional approaches. *Clin Sci* 2007; **113**: 1–13.
4. **Myatt L.** Placental adaptive responses and fetal programming. *J Physiol* 2006; **572**: 25–30.
5. **Fowden AL, Forhead AJ, Coan PM, Burton GJ.** The placenta and intrauterine programming. *J Neuroendocrinol* 2008; **20**: 439–50.
6. **Jones HN, Powell TL, Jansson T.** Regulation of placental nutrient transport. A review. *Placenta* 2007; **28**: 763–74.
7. **Sibley CP, Turner MA, Cetin I** *et al.* Placental Phenotypes of Intrauterine Growth. *Pediatr Res* 2005; **58**: 827–32.
8. **Marconi AM, Paolini CL.** Nutrient transport across the intrauterine growth-restricted placenta. *Semin Perinatol* 2008; **32**: 178–81.
9. **Hoffman L, Mandel TE, Carter WM, Koulmanda M, Martin FIR.** Insulin secretion by fetal human human pancreas in organ culture. *Diabetologia* 1982; **23**: 426–30.
10. **Gulati P, Gaspers LD, Dann SG** *et al.* Amino acids activate mTOR Complex 1 via Ca2+/CaM signaling to hVps34. *Cell Metab* 2008; **7**: 456–65.
11. **Sturman J.** Taurine in development. *J Nutr* 1988; **118**: 1169–76.
12. **Gluckman PD, Hanson MA, Cooper C, Thornburg KL.** Effect of in utero and early-life conditions on adult health and disease. *N Engl J Med* 2008; **359**: 61–73.
13. **Cleal JK, Lewis RM.** The mechanisms and regulation of placental amino acid transport to the human fetus. *J Neuroendocrinol* 2008; **20**: 419–26.

14. **Moe AJ**. Placental amino acid transport. *Am J Physiol* 1995; **268**: C1321-C31.

15. **Jansson T**. Amino acid transporters in the human placenta. *Pediatr Res* 2001; **49**: 141–7.

16. **Mackenzie B, Erickson JD**. Sodium-coupled neutral amino acid (System N/A) transporters of the SLC38 gene family. *Pflugers Arch* 2004; **447**: 784–95.

17. **Desforges M, Lacey HA, Glazier JD** et al. The SNAT4 isoform of the system A amino acid transporter is expressed in human placenta. *Am J Physiol Cell Physiol* 2006; **290**: C305–C12.

18. **Verrey F, Closs EI, Wagner CA, Palacin M, Endou H, Kanai Y**. CATs and HATs: the SLC7 family of amino acid transporters. *Pflugers Arch* 2003.

19. **Dicke JM, Henderson GI**. Placental amino acid uptake in normal and complicated pregnancies. *Am J Med Sci* 1988; **295**(3): 223–7.

20. **Mahendran D, Donnai P, Glazier JD** et al. Amino acid (System A) transporter activity in microvillous membrane vesicles from the placentas of appropriate and small for gestational age babies. *Pediatr Res* 1993; **34**: 661–5.

21. **Glazier JD, Cetin I, Perugino G** et al. Association between the activity of the system A amino acid transporter in the microvillous plasma membrane of the human placenta and severity of fetal compromise in intrauterine growth restriction. *Pediatr Res* 1997; **42**: 514–19.

22. **Jansson T, Ylvén K, Wennergren M, Powell TL**. Glucose transport and system A activity in syncytiotrophoblast microvillous and basal membranes in intrauterine growth restriction. *Placenta* 2002; **23**: 386–91.

23. **Shibata E, Hubel CA, Powers RW** et al. Placental System A amino acid transport is reduced in pregnancies complicated with small for gestational age (SGA) infants but not in preecalmpsia with SGA infants. *Placenta* 2008; **29**: 879–82.

24. **Jansson T, Ekstrand Y, Björn C, Wennergren M, Powell TL**. Alterations in the activity of placental amino acid transporters in pregnancies complicated by diabetes. *Diabetes* 2002; **51**: 2214–19.

25. **Kuruvilla AG, D'Souza SW, Glazier JD** et al. Altered activity of the system A amino acid transporter in microvillous membrane vesicles from placentas of macrosomic babies born to diabetic women. *J Clin Invest* 1994; **94**: 689–95.

26. **Jansson T, Scholtbach V, Powell TL**. Placental transport of leucine and lysine is reduced in intrauterine growth restriction. *Pediatr Res* 1998; **44**: 532–7.

27. **Paolini CL, Marconi AM, Ronzoni S** et al. Placental transport of leucine, phenylalanine, glycine, and proline in intrauterine growth-restricted pregnancies. *J Clin Endocrinol Metab* 2001; **86**: 5427–32.

28. **Jansson N, Pettersson J, Haafiz A** et al. Downregulation of placental transport of amino acids precedes the development of intrauterine growth restriction in rats fed a low protein diet. *J Physiol (Lond)* 2006; **576**: 935–46.

29. **Karl PI, Alpy KL, Fischer SE**. Amino acid transport by the cultured human placental trophoblast: effect of insulin on AIB transport. *Am J Physiol* 1992; **262**: C834–C9.

30. **Bloxam DL, Bax BE, Bax CMR**. Epidermal growth factor and insulin-like growth factor I differentially influence the directional accumulation and transfer of 2-aminoisobutyrate (AIB) by human placental trophoblast in two-sided culture. *Biochim Biophys Res Commun* 1994; **199**: 922–9.

31. **Kniss DA, Shubert PJ, Zimmerman PD, Landon MB, Gabbe SG**. Insulin growth factors: their regulation of glucose and amino acid transport in placental trophoblasts isolated from first-trimester chorionic villi. *J Reprod Med* 1994; **39**: 249–56.

32. **Karl PI**. Insulin-like growth factor-1 stimulates amino acid uptake by the cultured human placental trophoblast. *J Cell Physiol* 1995; **165**: 83–8.

33. **Nandakumaran M, Makhseed M, Al-Rayyes S, Akanji AO, Sugathan TN**. Effect on transport kinetics of alpha-aminoisobutyric acid in the perfused human placental lobule in vitro. *Pediatr Int* 2001; **43**: 581–6.

34. **Jansson N, Greenwood S, Johansson BR, Powell TL, Jansson T**. Leptin stimulates the activity of the system A amino acid transporter in human placental villous fragments. *J Clin Endocrinol Metab* 2003; **88**: 1205–11.

35. **Jones HN, Ashworth CJ, Page KR, McArdle HJ**. Expression and adaptive regulation of amino acid transport system A in a placental cell line under amino acid restriction. *Reproduction* 2006; **131**: 951–60.

36. **Nelson DM, Smith SD, Furesz TC** et al. Hypoxia reduces expression and function of system A amino acid transporters in cultured term human trophoblasts. *Am J Physiol* 2003; **284**: C310–C5.

37. **Jones HN, Ashworth CM, Page KR, McArdle HJ**. Cortisol stimulates System A amino acid transport and SNAT2 expression in a human placental cell line (BeWo). *Am J Endocrinol Metab* 2006; **291**: E596–E603.

38. **Khullar S, Greenwood SL, McCord N, Glazier JD, Ayuk PT**. Nitric oxide and superoxide impair human placental amino acid uptake and increase Na^+ permeability: implications for fetal growth. *Free Radical Biol Med* 2004; **36**: 271–7.

39. **Ramamoorthy S, Leibach FH, Mahesh VB, Ganapathy V**. Modulation of the activity of amino acid transport system L by phorbol esters and calmodulin antagonists in a human placental choriocarcinoma cell line. *Biochim Biophys Acta* 1992; **1136**: 181–8.

40. **Brandsch M, Leibach FH, Mahesh VB, Ganapathy V**. Calmodulin-dependent modulation of pH sensitivity of the amino acid transport system L in human placental choriocarcinoma cells. *Biochim Biophys Acta* 1994; **1192**: 177–84.

41. **Okamoto Y, Sakata M, Ogura K** et al. Expression and regulation of 4Fhc and hLAT1 in human trophoblasts. *Am J Physiol Cell Physiol* 2002; **282**: C196–C204.

42. **Jones HN, Jansson T, Powell TL**. IL-6 stimulates System A amino acid transporter activity in trophoblast cells through STAT3 and increased expression of SNAT2. *Am J Physiol Cell* 2009; **297**: C1228–35.

43. **Jones HN, Jansson T, Powell TL**. Full length adiponectin attenuates insulin signaling and inhibits insulin-stimulated amino acid transport in human primary trophoblast cells. *Diabetes* 2010; **59**: 1161–70.

44. **Peng T, Golub TR, Sabatini DM**. The immunosuppressant rapamycin mimics a starvation-like signal distinct from amino acid and glucose deprivation. *Mol Cell Biol* 2002; **22**(15): 5575–84.

45. **Tee AR, Blenis J**. mTor, translational control and human disease. *Semin Cell Dev Biol* 2005; **16**: 29–37.

46. **Martin DE, Hall MN**. The expanding TOR network. *Curr Opin Cell Biol* 2005; **17**: 158–66.

47. **Hay N, Soneneberg N**. Upstream and downstream of mTOR. *Genes Dev* 2004; **18**: 1926–45.

48. **Jacinto E, Hall MN**. TOR signalling in bugs, brain and brawn. *Nature Rev Mol Cell Biol* 2003; **4**: 117–26.

49. **Roos S, Jansson N, Palmberg I, Säljö K, Powell TL, Jansson T**. Mammalian target of rapamycin in the human placenta regulates leucine transport and is downregulated in restricted foetal growth. *J Physiol* 2007; **582**: 449–59.

50. **Roos S, Kanai Y, Prasad PD, Powell TL, Jansson T**. Regulation of placental amino acid transporter activity by mammalian target of rapamycin. *Am J Physiol Cell* 2009; **296**: C142–C50.

51. **Roos S, Lagerlöf O, Wennergren M, Powell TL, Jansson T**. Regulation of amino acid transporters by glucose and growth factors in cultured primary human trophoblast cells is mediated by mTOR signaling. *Am J Physiol Cell* 2009; **297**: C723–C31.

52. **Strid H, Bucht E, Jansson T, Wennergren M, Powell T**. ATP-dependent Ca^{2+} transport across basal membrane of human syncytiotrophoblast in pregnancies complicated by diabetes or intrauterine growth restriction. *Placenta* 2003; **24**: 445–52.

53. **Strid H, Care AD, Jansson T, Powell TL**. PTHrp midmolecule stimulates Ca^{2+} ATPase in human syncytiotrophoblast basal membrane. *J Endocrinol* 2002; **175**: 517–24.

54. **Godfrey KM, Matthews N, Glazier J** et al. Neutral amino acid uptake by microvillous plasma membrane of the human placenta is inversely related to fetal size at birth in normal pregnancy. *J Clin Endocrinol Metab* 1998; **83**: 3320–6.

55. **Constancia M, Hemberger M, Hughes J** et al. Placental-specific IGF-II is a major modulator of placental and fetal growth. *Nature* 2002; **417**: 945–8.

56. **Constancia M, Angiolini E, Sandovici I** et al. Adaptation of nutrient supply to fetal demand in the mouse involves interaction between the Igf2 gene and placental transporter systems. *Proc Natl Acad Sci USA* 2005; **102**: 19219–24.

57. **Coan PM, Burton GJ, Ferguson-Smith AC**. Imprinted genes in the placenta: a review. *Placenta* 2006; **26** (Suppl A): S10–S20.

58. **Masuzaki H, Ogawa Y, Sagawa N** et al. Nonadipose tissue production of leptin: leptin as a novel placenta-derived hormone in humans. *Nature Med* 1997; **3**: 1029–33.

59. **Reis FM, Florio P, Cobellis L** et al. Human placenta as a source of neuroendocrine factors. *Biol Neonate* 2001; **79**: 150–6.

60. **Desoye G, Hartmann M, Blaschitz A** et al. Insulin receptors in syncytiotrophoblast and fetal endothelium of human placenta. Immunohistochemical evidence for developmental changes in distribution pattern. *Histochemistry* 1994; **101**: 277–85.

61. **Ebenbichler CF, Kasser S, Laimer M** et al. Polar expression and phosphorylation of human leptin receptor isoforms in paired syncytial, microvillous and basal membranes from human term placenta. *Placenta* 2002; **23**: 516–21.

62. **Fang J, Furesz TC, Lurent RS, Smith CH, Fant M**. Spatial polarization of insulin-like growth factor receptors on the human syncytiotrophoblast. *Pediatr Res* 1997; **41**: 258–65.

63. **Holmes R, Montemagno R, Jones J** et al. Fetal and maternal plasma insulin-like growth factors in pregnancies with appropriate or retarded fetal growth. *Early Hum Dev* 1997; **49**: 7–17.

64. **Yildiz L, Avci B, Ingec M**. Umbilical cord and maternal blood leptin concentrations in intrauterine growth retardation. *Clin Chem Lab Med* 2002; **40**: 1114–17.

65. **Lauszus FF**, **Klebe JG**, **Flyvbjerg A**. Macrosomia associated with maternal serum insulin-like growth factor-I and -II in diabeteic pregnancy. *Obstet Gynecol* 2001; **97**: 734–41.

66. **Jansson N**, **Nilsfelt A**, **Gellerstedt M** et al. Maternal hormones linking maternal body mass index and dietary intake to birth weight. *Am J Clin Nutr* 2008; **87**: 1743–9.

67. **Malandro MS**, **Beveridge MJ**, **Kihlberg MS**, **Novak DA**. Effect of low-protein diet-induced intrauterine growth retardation on rat placental amino acid transport. *Am J Physiol* 1996; **271**: C295–C303.

68. **Jones HN**, **Woollett LP**, **Barbour N** et al. High fat diet before and during pregnancy causes marked upregulation of placental nutrient transport and fetal overgrowth in C57/Bl6 mice. *FASEB J* 2009; **23**: 271–8.

69. **Ogden CL**, **Caroll MD**, **Curtin LR** et al. Prevalence of overweight and obesity in the United States, 1999–2004. *JAMA* 2006; **295**: 1549–55.

70. **Baeten JM**, **Bukusi EA**, **Lambe M**. Pregnancy complications and outcomes among overweight and obese nulliparous women. *Am J Publ Health* 2001; **91**: 436–40.

71. **Sebire NJ**, **Jolly M**, **Harris JP** et al. Maternal obesity and pregnancy outcome: a study of 287,213 pregnancies in London. *Int J Obes Relat Metab Disord* 2001; **25**: 1175–82.

72. **Cnattingius S**, **Bergstrom R**, **Lipworth L**, **Kramer MS**. Prepregnancy weight and the risk of adverse pregnancy outcomes. *N Engl J Med* 1998; **338**(3): 147–52.

73. **Boney CM**, **Verma A**, **Tucker R**, **Vohr BS**. Metabolic syndrome in childhood: Association with birth weight, maternal obesity, and gestational diabetes. *Pediatrics* 2005; **115**: e290–e6.

74. **Ramsay JE**, **Greer I**, **Sattar N**. Obesity and reproduction. *Br Med J* 2006; **333**: 1159–62.

75. **Hendler I**, **Blackwell SC**, **Metha SH** et al. The levels of leptin, adiponectin, and resistin in normal weight, overweight, and obese pregnant women with and without preeclampsia. *Am J Obstet Gynecol* 2005; **193**: 979–83.

Discussion

JACKSON: The low-protein diet you used is 4% protein, which is generally considered to be insufficient. What happened to maternal body weight in those studies?

JANSSON: They lost weight. Starting at gestation day 18 there was a difference.

SFERRUZZI-PERRI: With your knockdown studies involving *Raptor* and *Rictor*, and the downregulation of the trophoblast substrate uptake, have you tried knocking down both *Rictor* and *Raptor*?

JANSSON: Yes, this is something we are currently doing.

SFERRUZZI-PERRI: With the liver-specific *igf-1* knockout mouse, other people have published that in the non-pregnant state there are metabolic disturbances. How are these mums doing during pregnancy? Are they losing fat? Are they insulin resistant?

JANSSON: That is a good point. We didn't develop the model; it was developed by our colleagues in Gothenburg for non-pregnant animals, and hasn't been used in pregnancy before. In non-pregnant animals there are changes in insulin concentration, leptin concentrations, and they are slightly hyperglycaemic. In the pregnant state we haven't completely phenotyped them, but in order to interpret the changes in IGF-I we have to know what happens with hormone levels and nutrient partitioning within the mother.

SFERRUZZI-PERRI: What are the effects on placental *igf-2* expression?

JANSSON: We haven't looked at that either.

POSTON: You are suggesting that there is an increase in nutrient transport. How much do you think there is a need for a change in this transport in the relationship between maternal hormonal status and fetal size. The HAPO study published last year showed that even very small increments of plasma glucose led to increased fetal weight. You are looking at an extreme model with the high-fat diet. Is it just a matter of diffusion through the known number of transporters that are there?

JANSSON: In terms of glucose, it is a facilitated transport mechanism across the placenta, which means exactly as you are alluding to: that any change in gradient across the placenta will actually affect net transport. In contrast, amino acids are actively transported and small changes in concentration don't do much to transport. It is the transporter activity and expression that is limiting. In the high-fat diet animals there is no change in glucose, so we don't believe it is glucose driving this. At first we were a little disappointed with this result, because we had expected that our model would look a little more like obesity in that at least we would see some sort of evidence of insulin resistance, but they have normal glucose tolerance and are normoglycaemic. But then we realized that this is perhaps isolating the effect of high-fat diet *per se*, so we changed our mind and thought it was a good result. When we saw a direct effect of fatty acids on placental function *in vitro* this fitted the picture.

POSTON: There is a lot of interest in raised triglycerides and birth weight in human obese pregnancies. In relation to your high fat model did you look at the fatty acid transporters?

JANSSON: No. The changes in this model that are similar to pregnant obese women is that they have low adiponectin, high leptin and high triglycerides. One of the major contributions of Irene Cetin is that she is looking at fatty acid and lipid transport. So little of this work has been done.

POSTON: It might be a good model of triglyceride-induced fetal growth.

FLEMING: With the siRNA *Raptor* and *Rictor* studies, have you tried them on the rat low-protein model?

JANSSON: No. Right now we don't have a means to do trophoblast-specific knockdowns, although we have an idea how to do it. It is difficult to mimic this low-protein model *in vitro*. It would be very interesting.

SIBLEY: My view is that the placenta is integrating signals from both mum and fetus and then trying to integrate the two. What could be doing the integrating in the placenta? There is no reason why mTOR couldn't be receiving signals from the fetal side. Is mTOR a central integrator between maternal and fetal signals? How would we investigate this in terms of fetal-side signals on mTOR?

JANSSON: I focused on the maternal signals, but clearly the fetal signals are also important. The effect on calcium transport which is upregulated in IUGR is probably evidence of a fetal signal. I think we agree on this. The only thing that speaks against this idea is that, in terms of the growth factor signalling, the growth factor receptors (insulin, IGF-I, leptin) are almost exclusively localized in the microvillous membrane. In terms of feeding in from the growth factor signalling to mTOR, they are definitely more predominant on the microvillous membrane. If mTOR is downregulated in response to low maternal nutrient availability, the fetal signal would be to counteract this. This would feed through the mTOR system not on the level of mTOR but maybe further down. Perhaps we'd have to look at larger animal models where we can get to both circulations and both surfaces.

SFERRUZZI-PERRI: There's some work by Jane Harding in which she infused IGF-I into fetal sheep, showing that placental glucose transfer and even structure were changed in response to just a short-term infusion.

CETIN: The placenta is definitely integrating signals. I am convinced that the mother is conditioning placenta mTOR and transport (I am thinking of human data) in the case of obesity, gestational diabetes, high-fat diet and so on. In the IUGR model, what we know from the maternal phenotype in IUGR is that it is very tricky because the mother has high levels of fatty acids (twice normal), pretty high triglycerides (this depends on whether she is beginning to go towards pre-eclampsia or not) and she has high amino acid levels (1.5 times higher than normal). They also have high lactate levels. It is a very particular maternal phenotype not reflected in any of these animal studies. I wonder whether oxygen is more relevant?

JANSSON: You illustrate a good point here: in a particular condition we can always find several factors that could counteract the main effect. We feel that this is a system upon which a large number of factors are impinging. Some are more important than others. In the IUGR mother, it is true that there are high amino acid levels. But uterine bloodflow is low, and we feel that this overrides any changes in hormones. Exactly how uterine bloodflow downregulates amino acid transport is unknown, but oxygen is a good candidate in the case where there is hypoxia. Many IUGR mothers have low IGF-I, and this is a powerful regulator of placental function.

LOKE: From this diagram I take away the rather sad conclusion that the placenta is not a very efficient organ in compensating for adverse conditions. I would expect that, being a nutrient sensor, with maternal malnutrition instead of downregulating transporters it should upregulate transporters and thereby protect the fetus. But this is not the case, which means that if something goes wrong, the placenta suffers and the fetus suffers. So my regard for the placenta has been diminished!

JANSSON: You bring up an important point. This is one reason that people haven't bought into this model. We have been talking about this for a few years: the assumption that the fetus has to be able to upregulate things. In the human especially, what we see is the end of a process expressed in pathology; we don't know what has happened throughout pregnancy. It is possible that there are fetal/placental compensation mechanisms that we will not be able to see at the end as a pathophysiology. In the rat model we didn't see much compensation going on. The other thing is that it makes evolutionary sense that the mother has not only a mechanism through the imprinted genes in the placenta, but actually a physiological mechanism in the pregnancy to decrease or influence the growth in the face of severe undernutrition. She can build a small fetus or in extreme cases abort the fetus and go for reproduction a year or two later when the nutrition is good.

COOPER: My interest would be in disproportionate composition of the offspring at birth, more than just symmetrical retardation of growth. Particularly for mTOR, is there any evidence in the human data that there is a selective effect on lean mass and muscle, compared with the adipose composition? It would be possible to do DEXA assessments in human neonates.

JANSSON: In general, these fetuses are asymmetrically growth restricted. They are lean and relatively tall, with low deposits of fat. Weight is restricted more than length.

CETIN: It has been shown that the fat mass is reduced in IUGR babies, but I don't think there are any papers showing a direct relationship with mTOR.

JANSSON: There are some sheep experiments showing not unexpectedly that mTOR signalling in skeletal muscle of IUGR fetuses is markedly downregulated. I don't recall any studies on mTOR in fat in IUGR.

COOPER: For mTOR I would expect the effect to be more on muscle.

RICHARD BOYD: Colin Sibley raised the issue of integration, and I think it is important we don't lose focus on this. A lot of work is being done on amino acid transporters using methyl-AIB as a substrate, for good reasons, as you pointed out. You also pointed out that it is a sodium-dependent transporter, in contrast, for example, to the glucose or leucine story. My reason for emphasizing this sodium dependence is that we are in danger of losing sight of another major integrative phenomenon, which is membrane potential. For sodium-coupled transporters, we look just at blots at our peril.

JANSSON: I think you are right. If you look at the sodium-dependent transport *in vitro* like we do, we don't take account of any change *in vivo* of membrane potential. In the case of sodium gradients, we have shown that the Na/K ATPase activity in the membranes is decreased in IUGR, which could decrease the Na gradient, for example. This could also affect membrane potential.

ROBERT BOYD: I'd like to finish the discussion with a general comment. Thinking in terms of placenta, we have talked all the time in terms of transport from the mother to the fetus. I have not heard a single word so far about transport in the other direction. It's something we need to think about more.

Chapter 13

The maternal circulation and placental shape

Villus remodelling induced through haemodynamics and oxidative and endoplasmic reticulum stress

Graham J. Burton and Eric Jauniaux

Placental shape varies greatly between species, ranging from the cotyledonary placenta of the ruminants, through the zonary placenta of some carnivores to the discoid placenta of rodents and primates [1]. In the vast majority of species the shape is determined by either pre-existing specializations in the non-pregnant uterus, or differences in the receptivity of the uterine wall to invasion by the trophoblast. Thus, in the ruminant, where there is no invasion and the conceptus remains within the uterine lumen throughout pregnancy, villi only develop opposite the non-glandular caruncles, the number and size of which are species specific. In the rhesus monkey, where there is superficial implantation, one placental disc develops at the animal pole of the blastocyst that first attaches to the uterine wall, and a second slightly later, when the chorionic sac enlarges and impinges on the opposite wall of the uterus.

The situation is very different in the human, however, and presumably in the other great apes, although little is known about early pregnancy in the majority of these species (see Chapter 8). The human displays the most highly invasive form of implantation – interstitial implantation – during which the conceptus becomes completely embedded in the endometrium during the second week post-fertilization. Initially, villi form over the entire surface of the chorionic sac, giving rise to the chorion frondosum [2], but towards the end of the first trimester the villi over the superficial pole begin to regress. Those that persist at the deep pole form the definitive placenta, whereas the remainder of the chorionic sac becomes the chorion laeve, or the smooth chorion. This remodelling is essential to allow the opening of the amniotic sac and delivery of the fetus, without tearing through the highly vascular substance of the placenta [3].

Because the human definitive placenta forms at the pole of a sphere, it is usually circular or oval in shape, with a smooth outline. Other shapes occasionally occur, such as a bilobed placenta, which is thought to reflect an implantation that is particularly lateral in the uterus, with attachment to both the anterior and posterior walls [4]. In the past, the shape of the placenta has not been considered of great importance, as long as overall growth is adequate [4]. Normally, it is 18–20 cm in diameter, and 1.5–3.0 cm thick, with the umbilical cord attaching centrally to the disc [4,5]. However, it is notable that abnormal shapes, in particular those with irregular outlines and eccentric attachment of the cord, are frequently associated with poor obstetric outcome, such as low birth weight [6,7]. The recent finding that more subtle variations in the ellipticity of the placenta are associated with alterations in birth weight [8,9] and developmental programming of the fetus [10], with the more elliptical placenta being associated with a poorer outcome, has renewed interest in the factors determining placental shape.

Impaired development of the early placenta has long been implicated in aberrations of placental shape at term, although the mechanisms were unknown. In the last few years it has become possible to construct a hypothesis linking placental shape and complications of pregnancy based on oxidative and endoplasmic reticulum stress. The hypothesis centres on trophoblast invasion and conversion of the uterine spiral arteries that ultimately supply the placenta with maternal blood, and defects in this process. We propose that a spectrum of defects may occur, ranging from

---

The Placenta and Human Developmental Programming, ed. Graham J. Burton, David J. P. Barker, Ashley Moffett and Kent Thornburg. Published by Cambridge University Press. © Cambridge University Press 2011.

variations at the physiological end in clinically 'normal' placentas that may relate to programming, to frank pathological changes associated with severe complications of pregnancy.

Spiral artery conversion

The spiral arteries arise from the arcuate arteries that run circumferentially around the uterus, between the outer and middle thirds of the myometrium. In the non-pregnant state each spiral artery supplies an area of endometrium equivalent to 4–9 mm^2 at the uterine luminal surface. Ultimately, around 40–60 arteries supply the term placenta, and there is generally a one-to-one relationship between the opening of a spiral artery and the formation of a fetal villous lobule [11]. Each lobule therefore functions as an independent maternofetal exchange unit.

The wall of a spiral artery contains large quantities of smooth muscle equipped with a rich autonomic innervation, and so these vessels are highly responsive to both exogenous and endogenous adrenergic stimuli [12]. Conversion of the arteries is of fundamental importance to the development of the placenta, and to a successful pregnancy. Many of the common complications of pregnancy, such as miscarriage, intrauterine growth restriction (IUGR), pre-eclampsia and preterm rupture of the membranes, have been associated with deficient spiral arterial conversion [13–16]. Conversion involves the loss of the smooth muscle and the elastic lamina from the wall of the distal segment of an artery, from the point where it opens into the intervillous space as far as the inner third of the myometrium. The process is associated with dilation of the artery, but there are very few data on which to base the magnitude of this effect. Harris and Ramsey performed the most complete study when they generated 3D reconstructions from serial sections of pregnant hysterectomy specimens at different stages of gestation [17]. Their specimen at term demonstrated an approximately fivefold dilation, from 0.4–0.5 mm diameter at the myometrial/endometrial boundary to 2.4 mm at the artery's mouth, creating an antechamber before its opening into the placenta. The creation of this antechamber has a profound impact on the rheological properties with which the maternal blood enters the intervillous space.

Haemochorial placentation poses unique challenges, for the conceptus taps into an inherently high-pressure, high-velocity maternal arterial system. Somehow, in the absence of resistance arteries, the velocity and the pressure of the inflowing arterial blood must be reduced so that the fetal villi are not damaged by the momentum of the inflowing blood, and the fetal capillaries contained within are not compressed. At the same time, there must be a high volume of flow to supply sufficient nutrients and oxygen, particularly in late gestation. Dilation of the distal segment of the artery is one way of meeting these requirements. Mathematical modelling based on the dimensions quoted above reveals that the velocity of the inflowing blood is reduced by an order of magnitude, from around 2–3 m/s in the non-dilated part of an artery to around 10 cm/s at the mouth, depending on the precise dimensions and the viscosity of the maternal blood [18]. This reduction will ensure that the maternal blood enters the intervillous space with sufficient momentum to displace that already present, but with insufficient momentum to damage the fetal villous trees, many of which are floating and only attached at their proximal ends. It is notable that despite the physiological conversion, a villus-free cavity develops in the centre of each lobule when the maternal arterial circulation is first established at the start of the second trimester [19]. Equally important is the pressure within the intervillous space, which is thought to be in the region of 10 mmHg [20]. A pressure differential across the villous membrane is essential to keep the fetal capillaries distended, and to maintain the thin barrier necessary for effective diffusional exchange [21].

Whereas the formation of an antechamber at the mouth of the artery slows the maternal inflow, conversion of the deeper parts of the artery most likely has a different function. The section of a spiral artery just below the endometrial/myometrial boundary is particularly vasoreactive, and highly sensitive to steroid hormones [22]. It is thought that this section constricts at the time of menstruation, limiting blood loss [23]. Equivalent constriction during gestation would have a severe deleterious impact on placental blood flow. Physiological conversion appears to be a way of resolving the conflicting demands of the non-pregnant cycle and pregnancy, and most likely explains why trophoblast invasion extends as far as the inner third of the myometrium but no further. The more proximal parts of the uterine vasculature do undergo dilation to accommodate the overall increase in uteroplacental blood flow, but this is mediated by the effects of hormones, growth factors and flow-mediated effects, as in species displaying non-invasive placentation [18]. Invasion is not as deep in the rhesus monkey as in the human [24], and spontaneous vasoconstriction of this segment is believed to account for the intermittent perfusion of the intervillous space observed, independent of uterine contractions [25]. Whether placental perfusion is similarly intermittent in the human is not known, although narrowings, almost to the point of occlusion, have been outlined radiographically in the inner myometrial segments of the spiral arteries at mid-pregnancy [26].

Deficient conversion will therefore affect both the velocity and the constancy of the inflow of maternal blood into the placenta. It is notable that conversion of the spiral arteries is not an 'all or nothing' phenomenon [24,27]. Thus, absence of physiological changes has been reported to varying degrees in normal pregnancies [16,28,29]. Because trophoblast invasion occurs from the endometrial surface progressively towards the myometrium, it is likely that conversion of the myometrial segment will be the most vulnerable in cases of compromise. There are some data indicating that this is the case in normal pregnancies [16], and in pregnancies associated with growth restriction or pre-eclampsia [30], although accurate quantification is problematic owing to the small size of the biopsies generally available. The consequences of deficient spiral artery conversion for placental pathology will be considered later, but first we must consider the onset of the maternal arterial circulation to the placenta.

Onset of the maternal arterial circulation to the human placenta

Conversion of the spiral arteries is associated with invasion of the endometrium and the arterial walls by extravillous trophoblast. The extravillous lineage comprises two cell types, the interstitial and the endovascular trophoblast. The interstitial cells migrate from the maternal surface of the cytotrophoblastic shell into the uterine wall, reaching as far as the inner third of the myometrium, at which point they normally fuse to form multinucleated giant cells. In contrast, the endovascular trophoblast arises either from the cell columns direct, or from the shell and migrates down the lumen of the arteries. Although the molecular mechanisms underlying loss of the smooth muscle cells are currently unknown, the presence of the extravillous trophoblast cells appears to be essential for this process [24]. It is possible that interactions with maternal immune cells, in particular the uterine Natural Killer (NK) cell population, causes the latter to release angiogenic factors [31]. Such a mechanism could explain the relationship between certain pairings of NK cell receptors, trophoblast antigens and obstetric outcome [32], as discussed more fully in Chapter 9.

An important aspect of spiral artery conversion only recently realized is that the magnitude of the endovascular invasion is such that the tips of the arteries are effectively plugged during early pregnancy [33,34]. As a result, there is little, if any, maternal arterial inflow into the placenta during the first trimester [35], the conceptus being supported by the alternative means of histiotrophic nutrition from the endometrial glands [36]. The principal advantage of this arrangement is that the oxygen tension within the fetoplacental unit is relatively low during the critical phase of organogenesis, around 20 mmHg (5–10%), which may protect the embryo from oxygen free radical mediated teratogenesis [37]. At some point, however, there must be a switch from histiotrophic to haemotrophic nutrition to meet the increasing demands of the fetus. We have proposed that the onset of the maternal circulation is interlinked with remodelling of the chorion frondosum into the definitive placenta, and that oxidative stress plays a key role in this process [38].

The chorion frondosum begins to regress over the superficial pole of the chorionic sac between the sixth and eighth weeks of pregnancy, contemporaneously with the earliest ultrasonographic evidence of maternal blood flow into the placenta. In normal pregnancies, moving echoes indicative of significant blood flow can first be detected most commonly in the peripheral regions of the placenta at 8–10 weeks [38]. Differences in the degree of trophoblast invasion across the placental bed may account for these regional differences, for invasion is greatest in the centre of the implantation site and less extensive towards the margins [39]. It is to be expected, therefore, that plugging of the arteries will be less extensive in the periphery of the placental bed than in the central region.

Developmental villus regression

Onset of the circulation is associated with a threefold increase in the oxygen concentration within the placenta [40]. Adapting to this increase represents a major challenge for the placental tissues, especially for the syncytiotrophoblast as this tissue contains low levels

Chapter 13. Maternal circulation and placental shape

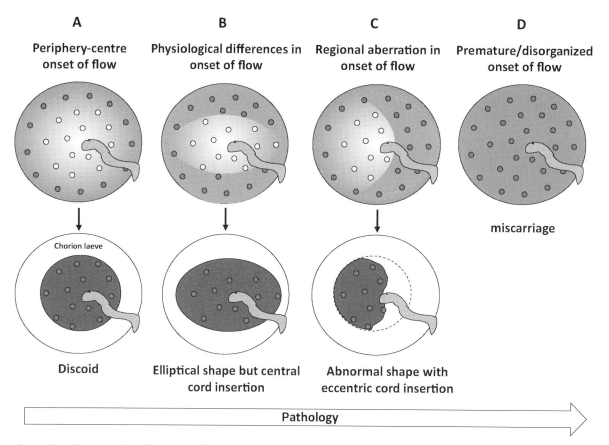

Figure 13.1 Diagrammatic *en face* view of villus regression towards the end of the first trimester. A) In normal pregnancies onset of the maternal circulation starts in spiral arteries (small red circles) at the periphery of the placenta, whereas arteries in the central region (white circles) remain plugged. Locally high levels of oxidative stress (diffuse orange) cause villus regression and formation of the chorion laeve. B) Regional differences in trophoblast invasion or density of the spiral arteries may cause more asymmetrical regression and an elliptical placental shape, which may have consequences for developmental programming. C) Excessive villus regression results in more abnormal placental shapes with eccentric attachment of the umbilical cord. D) In miscarriage, precocious and disorganized onset of the maternal circulation leads to overwhelming placental oxidative stress. Adapted from [70] with permission. (See colour plate section.)

of the principal antioxidant enzymes early in gestation [41]. The syncytiotrophoblast is thus highly vulnerable to oxidative stress, and rapidly degenerates *in vitro* when exposed to ambient concentrations of oxygen [42]. Consistent with the pattern of blood flow, placental villi sampled under ultrasound guidance from the periphery of the placenta show higher levels of immunoreactivity for nitrosylated proteins and lipid peroxidation, indicative of excessive oxygen free radical production, than do counterparts sampled from the centre [38].

Oxidative stress is well known to regulate activation of the apoptotic cascade, and so it is not surprising that levels of cleaved caspases 3 and 9 are higher in the peripheral tissues. Morphologically, trophoblast degeneration is more extensive in the periphery, and there is also regression of the fetal capillaries. Consequently, villi in the periphery become avascular, with hypocellular collagenous cores covered by a thin layer of trophoblast [38]. Regression of the villi over the superficial pole gives rise to the chorion laeve, or the smooth chorion, and as the chorionic sac expands the villous remnants become separated spatially and embedded in the membranes as ghost villi (Figure 13.1). The remainder of the original chorion frondosum persists as the definitive placenta, and as this is centred on the deep pole of the spherical chorionic sac it will usually be discoid in shape.

At the time of implantation, the blastocyst is orientated such that the trophoblast at the animal pole overlying the inner cell mass makes contact with the uterine epithelium. Thus, when the connecting stalk – the forerunner of the umbilical cord – forms, it does so opposite the deep pole of the chorionic sac, where plugging of the spiral arteries is most extensive. Consequently, the umbilical cord is usually attached to the central part of the placental disc.

Abnormal villus regression

There is therefore considerable circumstantial evidence that villus regression is mediated by locally high levels of what we have termed 'physiological oxidative stress' secondary to regional variations in the timing of onset of the maternal circulation. Further support for the hypothesis comes from the finding that qualitatively similar, but quantitatively greater, changes occur within the whole placenta in cases of missed miscarriage [43]. In this situation, onset of the maternal circulation is both premature and disorganized, occurring throughout the placenta [40] (Figure 13.1). This aberrant pattern is thought to be due to the shallow trophoblast invasion that typifies approximately 70% of these cases [44], leading to incomplete plugging of the spiral arteries. Retention of the products of conception for several days after onset of the blood flow results in thinning of the placenta, assessed ultrasonographically, as villus regression is induced.

Clearly, the situation in missed miscarriage is towards the extreme end of the pathological spectrum of deficient trophoblast invasion that encompasses pre-eclampsia and IUGR [45]. As mentioned earlier, spiral artery conversion is not an all-or-nothing phenomenon, and gradations in the extent of conversion have been reported in placental bed biopsies between cases of hypertension, with and without small for gestational age, and normal controls [30], and between cases of preterm rupture of the membranes and normal controls [16]. Because trophoblastic plugging of the spiral arteries during the first trimester is intimately linked with physiological conversion of the arteries, it is not unreasonable to hypothesize that onset of the maternal circulation will be abnormal in pregnancies associated with deficient conversion. Excessive early onset of maternal blood flow could lead to more extensive villus regression than normal, whereas the reverse may be true if onset is delayed.

Many factors are likely to influence the extent of trophoblast invasion into the uterine wall, and hence arterial plugging. Some may have a global impact, for example intrinsic differences in the invading trophoblast cells due to genetic or other factors, and others a more regional effect, for example variations in the thickness or quality of the decidua, reflecting proximity to anatomical features such as the uterine cornuae or sites of inflammation or previous implantations. If it is regional there may be excessive villus regression in one or more quadrants of the placenta, resulting in an abnormally shaped placenta with an eccentrically placed umbilical cord (Figure 13.1).

Although developmental villus regression induced by oxidative stress is an attractive hypothesis, it remains to be tested experimentally. Nonetheless, placental volume as assessed by 3D ultrasonography is less in pregnancies that go on to be small for gestational age, with or without pre-eclampsia, than in normal controls [46], indicating that whatever mechanism is operating does so from the first trimester onwards.

Abnormal maternal blood flow during the second and third trimesters and placental development

Abnormal onset of the maternal circulation to the placenta is highly likely to be associated with malperfusion of the organ in later pregnancy, as plugging of the spiral arteries and their physiological conversion are two facets of the same process of endovascular invasion. Deficient spiral artery conversion can potentially affect both the velocity of the incoming maternal arterial blood to the placenta and the constancy of that flow, as discussed earlier, depending on the severity of the compromise. Mild cases involving just the myometrial segment may be associated with greater intermittency of perfusion than normal, whereas the more severe cases in which the decidual segment is also compromised will have jet-like delivery superimposed on that background [47]. We speculate that the pathological changes induced by the former will revolve around oxidative and endoplasmic reticulum stress, whereas the latter will also involve mechanical damage.

Oxidative and endoplasmic reticulum stress

Intermittent perfusion will lead to fluctuations in placental oxygenation. These will not be as great as in classic ischaemia–reperfusion episodes, such as myocardial infarction and cerebrovascular incidents, and will not cause the same extensive tissue damage. However, fluctuations in placental oxygenation are a powerful stimulus for the generation of oxidative stress, and the related endoplasmic reticulum (ER) stress.

Thus, the normal process of labour, in which the uterine contractions cause profound intermittent reductions in uterine blood flow, is associated with increased markers of oxidative stress in the placental tissues, and activation of proinflammatory pathways [48]. These changes can be mimicked *in vitro* by subjecting explants from non-laboured caesarean deliveries to hypoxia–reoxygenation [49]. Immunohistochemistry localizes the principal markers of oxidative stress to the syncytiotrophoblast and the fetal endothelial cells in both situations.

Placental oxidative stress has long been recognized as playing a key role in the pathophysiology of pre-eclampsia [50,51]. Recently, we presented data confirming endoplasmic reticulum stress in placentas from cases of IUGR resulting from abnormal placental perfusion, with higher levels in cases where pre-eclampsia was also present [52]. Endoplasmic reticulum stress activates the evolutionarily conserved unfolded protein response, which aims to restore homoeostasis within the organelle through a number of integrated mechanisms. One of the first pathways to be activated in response to incremental doses of tunicamycin is phosphorylation of eukaryotic initiation factor α, eIF2α, which blocks the initiation of protein translation on the ribosomes, thereby preventing further nascent peptide chains from entering the endoplasmic reticulum [52]. Translation inhibition results in a reduction in the level of many kinases and proteins involved in the regulation of the cell cycle, particularly those with a short half-life. In other systems, a reduction in cyclin D1 has been identified following endoplasmic reticulum stress [53], and we observed the same change in our pathological placentas. This loss of cyclin D1 would be expected to be associated with a reduction in the rate of cell proliferation, and indeed the percentage of cytotrophoblast cells engaged in DNA synthesis is reduced in growth-restricted placentas [54]. This concept is consistent with the slower increase in placental volume observed ultrasonographically *in vivo* in cases of small for gestational age associated with pre-eclampsia [46]. It is further reinforced by the fact that placental choriocarcinoma cell lines proliferate at a slower rate following the induction of endoplasmic reticulum stress by either the addition of tunicamycin or exposure to hypoxia–reoxygenation [52].

In addition to the effect on cyclin D1, we also observed a dramatic decrease in AKT (protein kinase B) at the protein, but not at the mRNA, level [52]. AKT-mTOR signalling is a central regulator of cell proliferation [55], integrating the actions of growth factors and acting as an ATP and amino acid sensor to fine-tune protein synthesis to nutrient availability [56]. During placental development, binding of placentally derived IGF2, a key autocrine/paracrine regulator of fetoplacental growth [57], to the IGF receptor type I activates downstream PI3K signalling, which in turn increases AKT activity. AKT signalling regulates protein synthesis, and hence cell growth, by a number of pathways. One of the downstream targets of AKT is glycogen synthase 3 (GSK-3), which exerts an inhibitory effect on protein translation through phosphorylation of eIF2B. Activation of AKT inhibits GSK-3, thereby relieving this block. Another target of AKT is mTOR, both directly through phosphorylation of serine 2448 and indirectly through the regulatory tuberous sclerosis protein complex TSC2 [58]. mTOR in turn regulates 4E-BP1, a member of the family of 4E binding proteins that control cap-dependent translation initiation.

We observed multiple blocks to protein translation in our pathological placentas, consistent with a reduction in AKT/mTOR signalling [52]. These blocks will exacerbate the problem in a feed-forward way by lowering further levels of AKT and other kinases, and also autocrine growth factors such as IGF2 and its receptors (Figure 13.2). The net result will be reduced cell proliferation, and hence the small placental phenotype. In the mouse, homozygous disruption of the *Akt1* gene is sufficient to cause IUGR [59], and we found a strong positive correlation between the phosphorylation status of Akt and placental weight across mutants and wild types [52]. AKT/mTOR signalling is also known to regulate amino acid transporters in the human placenta [60,61], and so it is likely that it acts as a final common pathway controlling placental growth in response to a variety of stressors.

As all membrane and secreted proteins pass through the organelle, ER stress may have additional effects on other aspects of placental function. Thus, the number and activity of transporter proteins in the syncytiotrophoblast membranes may be compromised, reducing exchange capacity. Equally, placental secretion of hormones that influence maternal metabolism and enhance the supply of nutrients to the fetus, such as human placental lactogen, may be adversely

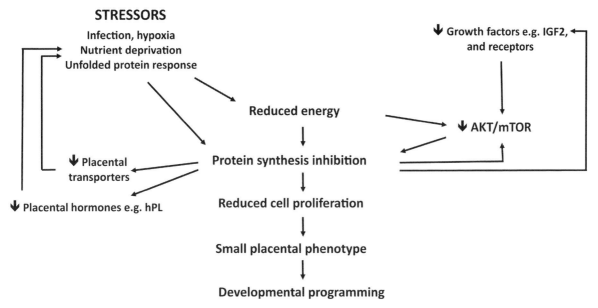

Figure 13.2 Flow diagram illustrating how protein synthesis inhibition due to various causes may lead to a feed-forward effect through reduced AKT/mTOR signalling, resulting in a small placental phenotype and developmental programming of the fetus.

affected. Thus, size alone may not provide an adequate measure of placental efficiency.

Abnormal velocity of maternal arterial inflow

Deficient conversion of the decidual portion of the spiral artery will result in maternal arterial blood entering the placental intervillous space at abnormally high velocities, as previously discussed [18]. This prediction is supported by ultrasonographic observations in pregnancies associated with poor trophoblast invasion, in which the blood enters in jet-like spurts associated with turbulent flow – the so-called 'hose-pipe' effect [62]. The momentum of these spurts is sufficient to create large villus-free cavities within the placenta, termed 'placental lakes' [62] or 'echoic cystic lesions' [63], giving the placenta a 'jelly-like' appearance on ultrasound [47]. The presence of these spaces, and the high rate at which the maternal blood passes through them, will severely affect the normal balance and matching between maternal and fetal circulations in the placenta, and hence the efficacy of exchange (Figure 13.3). The combination of an abnormal uterine circulation and placental dysmorphology on ultrasound examination identifies a subset of women who are at increased risk of adverse outcome [7,47,64].

The most severe forms of these placentas have a highly characteristic shape, in that they are much thicker than normal and have a relatively small area of attachment to the uterine wall [63]. Consequently, they tend to be globular in appearance, and to 'wobble' when the abdomen is tapped. We have speculated that the increased thickness is due to rupture of a proportion of the anchoring villi that link the chorionic and basal plates [18], although at present there is no direct evidence to support this hypothesis. In early pregnancy, the distal part of an anchoring villus attaching to the developing basal plate is represented by a cytotrophoblastic cell column. There is little extracellular matrix in a column, and although their tensile strength has yet to be measured it is probably low. Besides allowing the chorionic and basal plates to drift apart, rupture of the anchoring villi would have severe feedforward consequences for spiral artery remodelling, as it would cut off the supply of the essential extravillous trophoblast cells. The concept that the increase in thickness is driven by the dynamics of the maternal circulation is reinforced by the finding that only two out of 27 placentas identified as being excessively thick by ultrasound examination *in vivo* had a thickness <4 cm following delivery and fixation [63]. Drainage of blood from the intervillous space *ex vivo* is thought to account for this collapse.

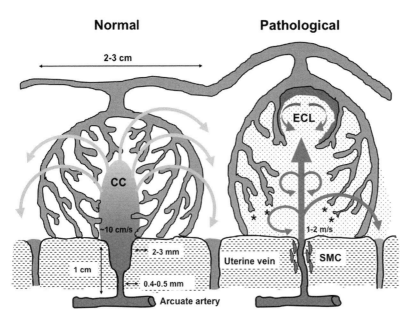

Figure 13.3 Diagrammatic representation of the haemodynamic consequences of deficient spiral artery conversion. In the absence of conversion, maternal arterial blood enters the intervillous space at a high velocity and in a turbulent manner. The momentum of the incoming blood may rupture anchoring villi (asterisked), allowing the chorionic and basal plates to move apart, increasing placental thickness. Besides this mechanical damage, and the formation of echoic cystic lesions (ECL), a mismatch between the maternal and fetal rates of flow may result in reduced placental efficiency. Adapted from [18] with permission. (See colour plate section.)

Implications for developmental programming

Establishing an appropriate maternal arterial blood supply to the placenta is fundamental to a successful pregnancy. Malperfusion will not only lead to a reduction in nutrient and oxygen delivery to the fetoplacental unit, but may also induce ER and oxidative stress, which can have major effects on placental function. Furthermore, the associated abnormal onset of the maternal circulation may result in excessive villus regression during the first trimester. The combination of these effects may alter the growth trajectory of the placenta from that stage onwards. The implications for developmental programming of the fetus are broadly twofold: the impact of altered placental development and function on nutrient and oxygen supply to the fetus, and the impact of altered resistance in the umbilical circulation for development of the fetal heart.

Pathologically small or abnormally shaped placentas clearly have reduced exchange capacity through a reduced villous surface area, potentially confounded by lower transporter activity secondary to ER stress. The impoverished supply to the fetus will undoubtedly result in growth restriction and altered organ development. Less obvious is the impact of ellipticity on placental functional capacity. It is intriguing that in a study of 209 normal-term (39–42 weeks of gestation) placentas collected at over 12 000 feet and at sea level, Chabes *et al.* found the elliptical shape to be the most stable [65]. Thus, they found 42% at sea level to be oval and 47% at high altitude, whereas for round placentas the figures were 50% and 27%, respectively. The reduction in round placentas at altitude was associated with an increase in other shapes, such as that of a triangle, rectangle or heart, and this was particularly so in association with male fetuses. The relative proportion of round placentas also fell with advancing maternal age, particularly at high altitude. Pregnancy at high altitude may be considered an experiment of nature, in which the placenta is subjected to a chronic hypobaric hypoxic stress. Whether the increase in ellipticity under these conditions reflects altered patterns of trophoblast invasion, or adaptive changes in the number and disposition of spiral arteries in the non-pregnant uterus, is not known (Figure 13.1). However, it is clear that the elliptical shape is the most stable form, and the molecular mechanisms underlying this deserve further investigation.

Normally, the umbilical placental circulation is characterized by high flow at a low resistance. However, placentas displaying an eccentric attachment of the umbilical cord are associated with a high resistance and often absent or reversed end-diastolic flow [66]. The resistance offered by the network depends in part on the branching pattern of the chorionic arteries as they ramify over the chorionic plate [67]. This pattern is altered in abnormally shaped placentas, especially when the insertion of the umbilical cord is eccentric.

Such placentas support a smaller fetal weight per unit weight of placenta than those with central attachment of the cord, indicative of a lower efficiency [9]. The resistance may also be modulated during pregnancy by oxidative stress secondary to maternal malperfusion, as fetal placental endothelial cells are highly susceptible to stress induced by fluctuations in oxygenation [68]. Excessive production of superoxide anions during oxidative stress may reduce the bioavailability of nitric oxide, leading to vasoconstriction. In more severe cases, activation of the apoptotic cascade may lead to regression of villous vascular networks, resulting in an impoverished fetal placental circulation. Either way, the raised umbilical resistance will have a profound effect on the development of the fetal heart, as discussed in Chapter 16. Furthermore, fetuses with absent end-diastolic flow are chronically hypoxic [69], and consequently both mechanical and biochemical factors may interact in the regulation of fetal cardiovascular development.

The relationships between the uterine vasculature, the establishment of the maternal circulation to the placenta, the subsequent remodelling and development of the definitive placenta, and programming of the fetus are clearly complex. However, it is likely that as far as the placenta is concerned the key events take place during the early stages following implantation. This is a challenging time for systematic research, but it should be our focus if the burden of chronic disease due to developmental programming is to be reduced.

Acknowledgements

The authors wish to thanks all their colleagues for discussions that have helped to formulate the concepts expressed, in particular Stephen Charnock-Jones, John Kingdom, Andrew Woods, Tereza Cindrova-Davies and Hong-wa Yung. They are grateful to the MRC, Wellcome Trust and Tommy's the Baby Charity for past and present funding for their research.

References

1. **Wooding FP, Burton GJ**. *Comparative Placentation. Structures, Functions and Evolution.* Berlin: Springer; 2008.
2. **Boyd JD, Hamilton WJ**. *The Human Placenta.* Cambridge: Heffer and Sons; 1970.
3. **Jauniaux E, Poston L, Burton GJ**. Placental-related diseases of pregnancy: involvement of oxidative stress and implications in human evolution. *Hum Reprod Update* 2006; **12**: 747–55.
4. **Kaplan CG**. Gross pathology of the placenta: weight, shape, size, colour. *J Clin Pathol* 2008; **61**: 1285–95.
5. **Salafia CM, Maas E, Thorp JM** et al. Measures of placental growth in relation to birth weight and gestational age. *Am J Epidemiol* 2005; **162**: 991–8.
6. **Nordenvall M, Sandstedt B, Ulmsten U**. Relationship between placental shape, cord insertion, lobes and gestational outcome. *Acta Obstet Gynecol Scand* 1988; **67**: 611–16.
7. **Toal M, Chan C, Fallah S** et al. Usefulness of a placental profile in high-risk pregnancies. *Am J Obstet Gynecol* 2007; **196**: 363 e1–7.
8. **Salafia CM, Zhang J, Miller RK** et al. Placental growth patterns affect birth weight for given placental weight. *Birth Defects Res A Clin Mol Teratol* 2007; **79**: 281–8.
9. **Yampolsky M, Salafia CM, Shlakhter O** et al. Centrality of the umbilical cord insertion in a human placenta influences the placental efficiency. *Placenta* 2009; **30**: 1058–64.
10. **Barker DJ, Thornburg KL, Osmond C** et al. The surface area of the placenta and hypertension in the offspring in later life. *Int J Dev Biol* 2009 Oct 14.
11. **Wigglesworth JS**. Vascular anatomy of the human placenta and its significance for placental pathology. *J Obstet Gynaecol Br Commonw* 1969; **76**: 979–89.
12. **Ramsey EM, Donner MW**. *Placental Vasculature and Circulation. Anatomy, Physiology, Radiology, Clinical Aspects, Atlas and Textbook.* Stuttgart: Georg Thieme; 1980.
13. **Brosens I, Dixon HG, Robertson WB**. Fetal growth retardation and the arteries of the placental bed. *Br J Obstet Gynaecol* 1977; **84**: 656–63.
14. **Ball E, Bulmer JN, Ayis S** et al. Late sporadic miscarriage is associated with abnormalities in spiral artery transformation and trophoblast invasion. *J Pathol* 2006; **208**: 535–42.
15. **Khong TY, Liddell HS, Robertson WB**. Defective haemochorial placentation as a cause of miscarriage. A preliminary study. *Br J Obstet Gynaecol* 1987; **94**: 649–55.
16. **Kim YM, Chaiworapongsa T, Gomez R** et al. Failure of the physiologic transformation of the spiral arteries in the placental bed in preterm premature rupture of membranes. *Am J Obstet Gynecol* 2002; **187**: 1137–42.
17. **Harris JWS, Ramsey EM**. The morphology of human uteroplacental vasculature. *Contrib Embryol* 1966; **38**: 43–58.
18. **Burton GJ, Woods AW, Jauniaux E** et al. Rheological and physiological consequences of conversion of the maternal spiral arteries for uteroplacental blood flow during human pregnancy. *Placenta* 2009; **30**: 473–82.

19. **Reynolds SRM, Freese UE, Bieniarz J** et al. Multiple simultaneous intervillous space pressures recorded in several regions of the hemochorial placenta in relation to functional anatomy of the fetal cotyledon. *Am J Obstet Gynecol* 1968; **102**: 1128–34.

20. **Moll W, Künzel W, Herberger J.** Hemodynamic implications of hemochorial placentation. *Eur J Obstet Gynecol Reprod Biol* 1975; **5**: 67–74.

21. **Karimu AL, Burton GJ.** The effects of maternal vascular pressure on the dimensions of the placental capillaries. *Br J Obstet Gynaecol* 1994; **101**: 57–63.

22. **Brosens JJ, Pijnenborg R, Brosens IA.** The myometrial junctional zone spiral arteries in normal and abnormal pregnancies. *Am J Obstet Gynecol* 2002; **187**: 1416–23.

23. **Markee JE.** Menstruation in intraocular endometrial transplants in the rhesus monkey. *Contrib Embyol* 1940; **28**: 219–308.

24. **Pijnenborg R, Vercruysse L, Hanssens M.** The uterine spiral arteries in human pregnancy: facts and controversies. *Placenta* 2006; **27**: 939–58.

25. **Martin CB, McGaughey HS, Kaiser IH** et al. Intermittent functioning of the uteroplacental arteries. *Am J Obstet Gynecol* 1964; **90**: 819–23.

26. **Borell U, Fernström I, Ohlson L** et al. Influence of uterine contractions on the uteroplacental blood flow at term. *Am J Obstet Gynecol* 1965; **93**: 44–57.

27. **Espinoza J, Romero R, Mee Kim Y** et al. Normal and abnormal transformation of the spiral arteries during pregnancy. *J Perinat Med* 2006; **34**: 447–58.

28. **Meekins JW, Pijnenborg R, Hanssens M** et al. A study of placental bed spiral arteries and trophoblast invasion in normal and severe pre-eclamptic pregnancies. *Br J Obstet Gynaecol* 1994; **101**: 669–74.

29. **Aardema MW, Oosterhof H, Timmer A** et al. Uterine artery Doppler flow and uteroplacental vascular pathology in normal pregnancies and pregnancies complicated by pre-eclampsia and small for gestational age fetuses. *Placenta* 2001; **22**: 405–11.

30. **Khong TY, De Wolf F, Robertson WB** et al. Inadequate maternal vascular response to placentation in pregnancies complicated by pre-eclampsia and by small-for-gestational age infants. *Br J Obstet Gynaecol* 1986; **93**: 1049–59.

31. **Hanna J, Goldman-Wohl D, Hamani Y** et al. Decidual NK cells regulate key developmental processes at the human fetal–maternal interface. *Nature Med* 2006; **12**: 1065–74.

32. **Hiby SE, Walker JJ, O'Shaughnessy KM** et al. Combinations of maternal KIR and fetal HLA-C genes influence the risk of preeclampsia and reproductive success. *J Exp Med* 2004; **200**: 957–65.

33. **Hustin J, Schaaps JP.** Echographic and anatomic studies of the maternotrophoblastic border during the first trimester of pregnancy. *Am J Obstet Gynecol* 1987; **157**: 162–8.

34. **Burton GJ, Jauniaux E, Watson AL.** Maternal arterial connections to the placental intervillous space during the first trimester of human pregnancy: the Boyd Collection revisited. *Am J Obstet Gynecol* 1999; **181**: 718–24.

35. **Jauniaux E, Gulbis B, Burton GJ.** The human first trimester gestational sac limits rather than facilitates oxygen transfer to the fetus – a review. *Placenta* 2003; **24**, Suppl. A: S86–93.

36. **Burton GJ, Watson AL, Hempstock J** et al. Uterine glands provide histiotrophic nutrition for the human fetus during the first trimester of pregnancy. *J Clin Endocrinol Metab* 2002; **87**: 2954–9.

37. **Burton GJ, Hempstock J, Jauniaux E.** Oxygen, early embryonic metabolism and free radical-mediated embryopathies. *Reprod BioMed Online* 2003; **6**: 84–96.

38. **Jauniaux E, Hempstock J, Greenwold N** et al. Trophoblastic oxidative stress in relation to temporal and regional differences in maternal placental blood flow in normal and abnormal early pregnancies. *Am J Pathol* 2003; **162**: 115–25.

39. **Pijnenborg R, Bland JM, Robertson WB** et al. The pattern of interstitial trophoblastic invasion of the myometrium in early human pregnancy. *Placenta* 1981; **2**: 303–16.

40. **Jauniaux E, Watson AL, Hempstock J** et al. Onset of maternal arterial bloodflow and placental oxidative stress; a possible factor in human early pregnancy failure. *Am J Pathol* 2000; **157**: 2111–22.

41. **Watson AL, Skepper JN, Jauniaux E** et al. Changes in the concentration, localization and activity of catalase within the human placenta during early gestation. *Placenta* 1998; **19**: 27–34.

42. **Watson AL, Skepper JN, Jauniaux E** et al. Susceptibility of human placental syncytiotrophoblastic mitochondria to oxygen-mediated damage in relation to gestational age. *J Clin Endocrinol Metabol* 1998; **83**: 1697–705.

43. **Hempstock J, Jauniaux E, Greenwold N** et al. The contribution of placental oxidative stress to early pregnancy failure. *Hum Pathol* 2003; **34**: 1265–75.

44. **Hustin J, Jauniaux E, Schaaps JP.** Histological study of the materno-embryonic interface in spontaneous abortion. *Placenta* 1990; **11**: 477–86.

45. **Burton GJ, Jauniaux E.** Placental oxidative stress; from miscarriage to preeclampsia. *J Soc Gynecol Invest* 2004; **11**: 342–52.

46. **Hafner E**, **Metzenbauer M**, **Hofinger D** et al. Placental growth from the first to the second trimester of pregnancy in SGA-foetuses and pre-eclamptic pregnancies compared to normal foetuses. *Placenta* 2003; **24**: 336–42.

47. **Jauniaux E**, **Ramsay B**, **Campbell S**. Ultrasonographic investigation of placental morphologic characteristics and size during the second trimester of pregnancy. *Am J Obstet Gynecol* 1994; **170**: 130–7.

48. **Cindrova-Davies T**, **Yung HW**, **Johns J** et al. Oxidative stress, gene expression, and protein changes induced in the human placenta during labor. *Am J Pathol* 2007; **171**: 1168–79.

49. **Hung TH**, **Skepper JN**, **Burton GJ**. In vitro ischemia-reperfusion injury in term human placenta as a model for oxidative stress in pathological pregnancies. *Am J Pathol* 2001; **159**: 1031–43.

50. **Hubel CA**. Oxidative stress in the pathogenesis of preeclampsia. *Proc Soc Exp Biol Med* 1999; **222**: 222–35.

51. **Myatt L**, **Cui X**. Oxidative stress in the placenta. *Histochem Cell Biol* 2004; **122**: 369–82.

52. **Yung HW**, **Calabrese S**, **Hynx D** et al. Evidence of placental translation inhibition and endoplasmic reticulum stress in the etiology of human intrauterine growth restriction. *Am J Pathol* 2008; **173**: 451–62.

53. **Brewer JW**, **Hendershot LM**, **Sherr CJ** et al. Mammalian unfolded protein response inhibits cyclin D1 translation and cell-cycle progression. *Proc Natl Acad Sci USA* 1999; **96**: 8505–10.

54. **Widdows K**, **Drewlo S**, **Baczyk D** et al. Stereological evidence for reduced cytotrophoblast proliferation in severe IUGR. *Placenta* 2008; **29**: A.99.

55. **Hay N**, **Sonenberg N**. Upstream and downstream of mTOR. *Genes Dev* 2004; **18**: 1926–45.

56. **Yang Q**, **Guan KL**. Expanding mTOR signaling. *Cell Res* 2007; **17**: 666–81.

57. **Constancia M**, **Hemberger M**, **Hughes J** et al. Placental-specific IGF-II is a major modulator of placental and fetal growth. *Nature* 2002; **417**: 945–8.

58. **Inoki K**, **Li Y**, **Zhu T** et al. TSC2 is phosphorylated and inhibited by Akt and supresses mTOR signalling. *Nature Cell Biol* 2002; **4**: 648–57.

59. **Yang ZZ**, **Tschopp O**, **Hemmings-Mieszczak M** et al. Protein kinase B alpha/Akt1 regulates placental development and fetal growth. *J Biol Chem* 2003; **278**: 32124–31.

60. **Roos S**, **Powell TL**, **Jansson T**. Placental mTOR links maternal nutrient availability to fetal growth. *Biochem Soc Trans* 2009; **37**: 295–8.

61. **Roos S**, **Lagerlof O**, **Wennergren M** et al. Regulation of amino acid transporters by glucose and growth factors in cultured primary human trophoblast cells is mediated by mTOR signaling. *Am J Physiol Cell Physiol* 2009; **297**: C723–31.

62. **Jauniaux E**, **Nicolaides KH**. Placental lakes, absent umbilical artery diastolic flow and poor fetal growth in early pregnancy. *Ultrasound Obstet Gynecol* 1996; **7**: 141–4.

63. **Viero S**, **Chaddha V**, **Alkazaleh F** et al. Prognostic value of placental ultrasound in pregnancies complicated by absent end-diastolic flow velocity in the umbilical arteries. *Placenta* 2004; **25**: 735–41.

64. **Toal M**, **Keating S**, **Machin G** et al. Determinants of adverse perinatal outcome in high-risk women with abnormal uterine artery Doppler images. *Am J Obstet Gynecol* 2008; **198**: 330 e1–7.

65. **Chabes A**, **Pereda J**, **Hyams L** et al. Comparative morphometry of the human placenta at high altitude and at sea level. I. The shape of the placenta. *Obstet Gynecol* 1968; **31**: 178–85.

66. **Nordenvall M**, **Ullberg U**, **Laurin J** et al. Placental morphology in relation to umbilical artery blood velocity waveforms. *Eur J Obstet Gynecol Reprod Biol* 1991; **40**: 179–90.

67. **Thompson RS**, **Trudinger BJ**. Doppler waveform pulsatility index and resistance, pressure and flow in the umbilical placental circulation; an investigation using a mathematical model. *Ultrasound Med Biol* 1990; **16**: 449–58.

68. **Burton GJ**, **Charnock-Jones DS**, **Jauniaux E**. Regulation of vascular growth and function in human placenta. *Reproduction* 2009; **138**: 895–902.

69. **Gudmundsson S**, **Lindblad A**, **Marsal K**. Cord blood gases and absence of end-diastolic blood velocities in the umbilical artery. *Early Hum Dev* 1990; **24**: 231–7.

70. **Burton GJ**, **Jauniaux E**, **Charnock-Jones DS**. The influence of the intrauterine environment on human placental development. *Int J Dev Biol* 2010; **54**: 303–12.

Discussion

ROBERT BOYD: I'd like us to begin with a more general discussion, and then to discuss Graham Burton's paper more specifically. Hilary Critchley, could you share some of your thoughts about what we have discussed so far to kick things off with?

CRITCHLEY: The thoughts I have were beautifully set up by Graham's paper. If you have a stressor, such as a hypoxia–reperfusion event, he brought us to that stage by saying that there is a problem with the spiral artery conversion. All the time we are looking downstream at the effect on the placenta and the effect of the placenta on the fetus, but I think we have to go back to the local uterine environment and ask, what causes the spiral artery conversion? We know immune cells such as NK cells and macrophages are very important. David Barker began the meeting by making an important observation about placental shape, and it sounds that if you don't get the right spiral artery conversions you don't get the right placental shape. Then Alan and Tom's papers touched on maternal diet determining expression of key genes. Hormone receptors are expressed in immune cells and vascular cells. Could diet be studied preconceptually to see how that affects the maternal side? Ashley Moffett's talk touched on the ligand/receptor phenotype that is bad. What effect would an AA NK phenotype have on spiral artery conversion? We need functional assays; we are really not looking at how the maternal stromal cells or maternal vessels or maternal immune cells interact with the trophoblast. The talk on hypoxia and hyperoxia showed that they are very important in understanding ischaemia–reperfusion. As someone who is here with slightly lateral interests (I work on the endometrium in both the pregnant and non-pregnant state). I think we have a huge area that isn't looked at sufficiently, in the maternal interface with the trophoblast. It is interesting that NK cells 'hug' the vasculature and also 'hug' the glands. If NK cells are knocked out, the result is an extraordinarily thick vascular smooth muscle cell layer. In the human we have shown that if we give an agent that completely modifies the vascular smooth muscle layer so it is thick, we have no NK cells. We do have an opportunity to study the interface between the immune system, the vascular system and vascular smooth muscle conversion. If we get a handle on what is important in the normal, we might be able to follow this through to some of these effects of programming agents.

THORNBURG: The idea that oxidative stress may regulate the shape of the placenta is fascinating. If you look in embryos, there is always more than one tissue that has to interact with another to get the signal for apoptosis. For example, in the early embryo there is a solid limb with rays that will become the digits and then the cells between the presumptive digits die with the appropriate signalling between epithelial and mesenchymal cells. Following this line, we might think about the interaction of the chorion with the surface of the endometrium to understand the cross-talk underlying chorionic regression. Endometrial signals and chorionic signals may integrate in a way that would tell the chorionic cell to initiate the apoptotic pathway. Perhaps the cells from the chorion in the peripheral areas are 'preprogrammed' to die when they receive specific signals.

LOKE: The main difference between what Kent Thornburg is talking about (the formation of fingers) is that this is in the same individual, while the chorion–decidua interaction is between two individuals, so there must be a dramatic difference in the cellular signalling.

THORNBURG: There are places in the uterus that are preferred for the implantation of the blastocyst because of the signalling that goes back or forth. This might be another example of how these two kinds of tissues interact.

BURTON: Absolutely. I was trying to find out what the difference in density of the spiral arteries is in different parts of the literature. It would make sense for there to be regional variation, but I couldn't find any data.

CRITCHLEY: Depending on where the blastocyst implants, does this determine the shape of the placenta? This could be studied by ultrasound.

BURTON: How the placenta grows is an interesting question. I can't see how it could recruit spiral arteries after the period of the main trophoblast invasion, because there would not be a chance to convert those arteries and you'd be tapping into a high-pressure system straight away. We need to think again about the early conceptus growing and the uterus enlarging, and the growth of the two together.

MOFFETT: This was my question. Early in pregnancy we have the placenta around the whole uterus. Are you saying that the arteries are only plugged, and that there is not actually destruction of the media of those arteries, all around the uterine wall?

PIJNENBORG: There is an early conversion of the media, the smooth muscle area around the spiral arteries, because we have decidualization.

MOFFETT: Is that all around the uterus, then?

PIJNENBORG: Yes. But mainly in the implantation site.

MOFFETT: So when the superficial villi regress, those arteries repair themselves. We never considered that. We always thought that there were some invaded arteries and some not invaded, depending on where the placenta is. This is a different way of thinking about it. As you regress down, those that come under the basal plate are invaded more, then, as we get to 12–15 weeks?

PIJNENBORG: Yes.

MOFFETT: If that is the case, in the case of miscarriage, we should see some repair of arteries. I don't know if we do see that.

HUNT: I have really enjoyed your work on the ER stress. In thinking about placental growth and function, early on, the cells of the placenta (whether you are looking at the trophoblast or mesenchymal cells) will show differences in the sophistication of their spectrum of receptors or ligands. Do you think there is variation over time in the effect of stressors or the ability of the placenta to react to stressors with increased protein synthesis?

BURTON: I am sure there would be. Any cell will adapt to ER stress to a certain extent by increasing chaperone proteins. Most people who have looked at ER stress have done so in terms of a single insult, such as in the brain or in ischaemic heart disease. Here we are hypothesizing that there is a low-grade but repetitive injury, probably from the time of onset of the circulation, at 11–12 weeks. The ultrasound evidence is that these placentas are smaller than normal even at 11 weeks. Especially with IUGR and pre-eclampsia, they grow at a slower rate throughout pregnancy. It is a chronic stress, and different cells may respond in different ways.

PIJNENBORG: I'd like to return to the spiral arteries, and especially the story about the velocity of the blood coming into the intravillous space. I always think about the earliest observations of bloodflow in the pregnant uterus by Elizabeth Ramsey. She mentioned that maternal blood enters the intravillous space in spurts. I always questioned that. She did all her work on rhesus monkeys, which are not deep invaders and therefore have a highly contractile myometrial segment in the spiral arteries. I kept asking several of my clinical colleagues whether or not such spurt-like flow into the intervillous space might occur in the human as well. I never got a satisfactory answer.

CETIN: I don't think it is clear.

BURTON: In the rhesus monkey you can see it quite clearly. Ramsey did show that if she put radiotracers in the femoral artery at different stages of uterine relaxation, she saw different areas of the placenta perfused at different times. The arterial constrictions I showed were from the rhesus monkey, but there are old angiographic data from the 1950s showing constrictions in the human.

THORNBURG: We see the spurting in the rhesus monkey under MRI.

BURTON: In the old literature, women with phaeochromocytoma have very high rate of miscarriage in the second trimester. This is what you'd expect through intense vasoconstriction of that segment.

CETIN: I am a little confused. In the first weeks the best thing is not to have too much blood flow. Then all of a sudden you need blood flow. You said this is after 12 weeks. But we have clinical data showing that intrauterine blood flow at 11 weeks is predictive of pre-eclampsia. What is going on at this stage, exactly?

BURTON: One informative study was that of Schaaps et al. (Schaaps JP, Tsatsaris V, Goffin F et al. Shunting the intervillous space: new concepts in human uteroplacental vascularization. *Am J Obstet Gynecol* 2005; **192**: 323–32). The idea has always been that uterine artery bloodflow goes through the spiral arteries into the placenta. The placenta is a low-resistance shunt, and therefore that circulation dictates the Doppler waveform seen in uterine arteries. They did scanning of the uterine artery resistance before delivery and 2–3 days after delivery and the resistance index did not change. Clearly, bloodflow into the placenta has very little impact on the uterine artery Doppler waveforms that clinicians are measuring. They showed that there are large arteriovenous shunts beneath the placenta in the myometrium, and the placenta is tapping into these. So we have to rethink the concept that the spiral artery is a direct route to the placenta. I don't know what the determinants of the spiral artery waveform are. This is a big area that needs to be looked at in terms of the resistance of the vessel.

SIBLEY: We recently published a paper (Hutchinson E, Brownbill P, Jones N et al. Uteroplacental haemodynamics in the pathogenesis of pre-eclampsia. *Placenta* 2009; **30**: 634–41) in which we set out to try to model turbulence in the intervillous space using the perfused cotyledon. I think this is relevant to this discussion. We modelled turbulence by altering flow into the intervillous space. We measured the composition of fluid coming out, and found that increased turbulence not only altered the amount of soluble factors coming out, but also the quality was different. If you then put this perfusate on to cultured endothelial cells or leucocytes, you get different responses depending on whether turbulent or non-turbulent flow has been used. This all fits with the pre-eclampsia model. If you are getting both solubles and non-solubles being altered by turbulence, then it must have some effect on placental development.

GIUSSANI: I am trying to think of it the other way round. I can understand how early-onset blood flow will lead to increased oxidative stress and proapoptotic pathways, in turn leading to excessive regression of the villi, which will lead to the elliptical placental shape. But what about the other way around? Is there any evidence that you get incomplete regression under some circumstances, leading to an overlarge placenta?

BURTON: Yes, there is the situation of circumvallate placenta, where there is a larger placental base and then the membranes are tucked over the chorionic surface. One could imagine that this is an incomplete regression.

LOKE: Do spiral arteries behave differently with the second and subsequent pregnancies as opposed to the first? In other words, is there education, memory or tolerance?

PIJNENBORG: T. Y. Khong *et al.* (*Placenta* 2003; **24**: 348–353) made an observation on this some years ago. They found that converted spiral arteries never reconvert to the original shape after pregnancy. There is no complete repair to the normal smooth muscle layer, and the elastic membrane remains fragmented and often shows very odd structures.

LOKE: Does it matter if the women change husband?

PIJNENBORG: Not as far as I know.

ROBERT BOYD: I have a question for the epidemiologists. We have not discussed drugs at all. They are usually replicates of metabolites or receptor interactions. In terms of programming, what is the general health of those who have been exposed to thalidomide or other drugs *in utero*?

BARKER: The study from which I showed data was one where none of the subjects were taking medication because this was between the wars, when only simple remedies were available. I'd like to sidestep your question and remind people of the first slide I showed, which was the amazing change in shape of the placenta in Finland over 20 years. It became more oval. These were huge changes.

COOPER: I am not aware of pregnancy or prepregnancy use of drugs linked to outcomes in the offspring. Let's take the general practice database, where a lot of evidence linking drug use to later outcome has come from. Most of those don't have long enough records to link later life outcomes with fetal drug exposure.

CETIN: The problem here is that women taking drugs usually have a disease, and this disease would likely also have an effect.

Chapter 14
Glucocorticoids and placental programming

Owen R. Vaughan, Alison J. Forhead and Abigail L. Fowden

Introduction

Size at birth is critical in determining life expectancy. It affects not only neonatal viability but also adult rates of mortality and morbidity. Human epidemiological observations have shown that the smaller the neonate the less likely it is to survive at birth, and the more likely it is to develop adult-onset degenerative diseases such as hypertension and glucose intolerance [1]. The associations between low birth weight and the risks of developing adult disease have now been demonstrated in several populations of different ages, genders and ethnicities, and shown to occur independently of current weight and exercise level [1]. When intrauterine growth is restricted in experimental animals, there are changes in cardiovascular and metabolic function in the adult offspring consistent with the human epidemiological data [2]. Together the animal and human studies have led to the concept that tissues and organ systems can be programmed *in utero* by suboptimal conditions during critical stages of development. Because size at birth is determined primarily by the placental supply of nutrients to the fetus [3] the placenta may have an important role in the programming of adult phenotype and susceptibility to disease [4]. Indeed, in human populations the size and shape of the placenta have been related to the risk of developing particular diseases in later life (Chapter 2). However, compared to fetal somatic tissues, relatively little is known about programming of the placenta *per se*.

Glucocorticoids are known to cause intrauterine programming of adult physiological phenotype in a number of species, including rats, sheep and non-human primates [5]. They alter cardiovascular, endocrine and metabolic function both before and after birth, and lead to progressive hypertension and glucose intolerance as the offspring age postnatally [6,7]. They may also act as a common link between adverse environmental conditions and the programming of fetal tissues during intrauterine development, particularly if they alter the capacity of the placenta to supply nutrients to the fetus. However, the effects of glucocorticoids on placental development are not well established. This chapter examines the role of glucocorticoids as signals in the developmental programming of the placenta, with particular emphasis on the specific actions of these steroids on placental structure and function.

Glucocorticoids as developmental signals

Environmental signals

Glucocorticoids act as signals of environmental conditions during intrauterine development [7]. Their concentrations increase in both mother and fetus in response to most of the adverse environmental conditions known to cause intrauterine programming, including undernutrition, hypoxaemia, manipulation of dietary content, placental insufficiency, and maternal restraint and other stresses, such as injection and handling [5,7–9]. In late gestation, concentrations also vary with the time of day [10]. Glucocorticoid concentrations therefore act as signals of maternal stress and photoperiod as well as of the fetal availability of glucose and oxygen during late gestation (Figure 14.1).

Maturational signals

Glucocorticoids also act as maturational signals in the fetus near term. In all species studied to date, fetal glucocorticoid concentrations rise naturally towards term as a result of prepartum activation of the fetal

The Placenta and Human Developmental Programming, ed. Graham J. Burton, David J. P. Barker, Ashley Moffett and Kent Thornburg. Published by Cambridge University Press. © Cambridge University Press 2011.

Chapter 14. Glucocorticoids and placental programming

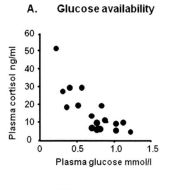
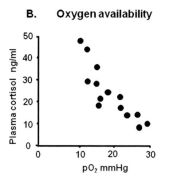
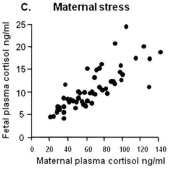
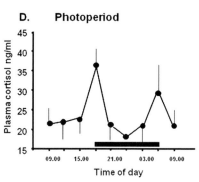
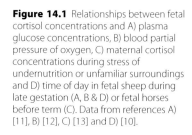

Figure 14.1 Relationships between fetal cortisol concentrations and A) plasma glucose concentrations, B) blood partial pressure of oxygen, C) maternal cortisol concentrations during stress of undernutrition or unfamiliar surroundings and D) time of day in fetal sheep during late gestation (A, B & D) or fetal horses before term (C). Data from references A) [11], B) [12], C) [13] and D) [10].

hypothalamopituitary–adrenal (HPA) axis [6]. This rise in basal glucocorticoid concentrations is responsible for maturation of a wide range of fetal tissues in preparation for birth [6]. Glucocorticoids switch the cell cycle from proliferation to differentiation by altering the expression of growth factors, enzymes, ion channels, transporters and hormone receptors in key tissues essential for neonatal survival, such as the lung, liver, gut and kidney [14]. Prepartum activation of the fetal HPA axis also increases the fetal adrenocortical response to adverse intrauterine conditions as term approaches. Basal and stress-induced increments in fetal glucocorticoid levels, therefore, normally provide a signal of the proximity to delivery [7]. However, early exposure to glucocorticoids during adverse environmental conditions is not a signal of imminent delivery but can induce many of the maturational changes prematurely, with consequences for tissue development both before and long after birth [7,8].

Fetoplacental glucocorticoid exposure

The placenta is exposed to glucocorticoids from both the maternal and the fetal circulations and thus receives environmental and maturational signals. Glucocorticoid concentrations are higher in the maternal than the fetal circulation, even in normal conditions, so there is a transplacental concentration gradient for glucocorticoid transfer from mother to fetus for most of gestation [14]. These steroids readily cross the placenta, and even in species such as the sheep, with a multilayered epitheliochorial placenta, 80% of the cortisol in the fetal circulation is of maternal origin before activation of the fetal HPA axis close to term [15]. Consequently, the placenta is exposed to glucocorticoids not only at the maternal and fetal facing surfaces of the trophoblast but also intracellularly at all cell layers between the maternal and fetal circulations. Receptors for the glucocorticoids are also found in most cell types in human, ovine and rodent placentas [16–19].

Increased fetoplacental exposure to natural glucocorticoids can occur in three main ways. First, maternal glucocorticoid concentrations can rise as a result of activation of the maternal HPA axis by nutritional and other stressful stimuli during pregnancy. In turn, this will increase transplacental transport of glucocorticoids and their fetal concentrations. Fetal glucocorticoid levels are known to rise in parallel with maternal concentrations in rats, sheep and horses at periods of gestation when the fetal HPA is unresponsive to stimuli (Figure 14.1C) [13,15,20]. Furthermore, the programming effects of maternal stress and protein deprivation on their adult offspring are prevented when the maternal adrenocortical response is abolished by adrenalectomy [21,22]. Clinically, increased fetoplacental exposure to maternally administered synthetic glucocorticoids can occur in asthmatics inhaling these steroids to improve ventilation during pregnancy, and in women at risk of preterm delivery given glucocorticoids antenatally to improve the viability of their infants [25,26]. Second, once the fetal HPA axis is activated in late gestation, fetoplacental glucocorticoid exposure can be increased by elevating fetal glucocorticoid concentrations independently of maternal levels, either in response to reduced fetal nutrient and oxygen availability (Figure 14.1A and B) or as part of the normal sequence of prepartum maturational events [6]. Finally, fetoplacental glucocorticoid exposure can be altered independently of any change in maternal or fetal glucocorticoid concentrations by changes in the placental activity of 11β-hydroxysteroid dehydrogenase (11βHSD). This enzyme occurs in two isoforms, 11βHSD type 1 (11βHSD1) which interconverts active and inactive glucocorticoids, and 11βHSD type 2 (11βHSD2), which converts active glucocorticoids to their inactive keto metabolites [5]. Placental 11βHSD2 therefore helps protect fetoplacental tissues from overexposure to the higher maternal glucocorticoid concentrations [5,9]. Across species, placental 11βHSD2 activity is inversely related to the magnitude of the transplacental glucocorticoid gradient [14] and in rats and sheep is downregulated by maternal undernutrition [23,24].

Glucocorticoids and placental development

The placental supply of nutrients and oxygen to the fetus occurs primarily by diffusion and transporter-mediated transport. In turn, these processes depend on placental size, morphology, transporter abundance and the synthesis and metabolism of nutrients and hormones by the placenta itself [4]. All of these factors can be influenced by glucocorticoids.

Placental size

In all species studied to date, maternal glucocorticoid administration during the last third of gestation leads to reduced placental weight (Table 14.1). The degree of placental growth restriction depends on the type of glucocorticoid administered, the dose and duration of treatment, and gestational age at the time of treatment (Table 14.1). The synthetic glucocorticoid dexamethasone appears to be more potent than other synthetic or natural glucocorticoids (Table 14.1). The decrease in placental weight induced by maternal administration of dexamethasone at a clinically relevant dose of 200 μg/kg per day was less when given intermittently by daily subcutaneous injections than when administered continuously via osmotic minipumps or orally in the drinking water (Table 14.1). Growth restriction also occurs in the rat placenta overexposed to glucocorticoids by reducing 11βHSD2 activity, either by deletion of the gene or by maternal treatment with carbenoxolone [43,44]. Conversely, reducing maternal glucocorticoid levels during pregnancy by the administration of metyrapone in vivo increases the weight of the rat placenta by 10–15% at term [43]. In human choriocarcinoma cell lines and cultured term explants, glucocorticoid administration in vitro leads to cytotrophoblast differentiation and reduced cell proliferation and migration [45,46], consistent with the growth-inhibitory action of glucocorticoids on the placenta in vivo (Table 14.1).

In the majority of studies, the fetus is also growth restricted by maternal glucocorticoid administration, but to a lesser extent than the placenta (Table 14.1). These observations suggest that placental efficiency may be increased by glucocorticoid treatment, despite the reduction in placental weight. In rats, sheep and non-human primates, specific measurements of the fetal to placental weight ratio show that this ratio increases after maternal glucocorticoid treatment in late gestation [34,38,41]. Similarly, when fetoplacental glucocorticoid exposure is increased by deletion of the *11βHSD2* gene, the placenta is more efficient, despite both placental and fetal growth restriction by late gestation [44]. This glucocorticoid-induced increase in efficiency may be due to morphological and/or functional changes in the placenta.

Placental morphology

In sheep, glucocorticoids affect the gross morphology of the placenta when given either maternally or fetally [38,40,47]. Increased cortisol exposure alters the frequency distribution of the ovine placentomes by reducing the numbers of the most everted placentomes. Tagging individual placentomes before cortisol administration showed that this was due to a decrease in the normal rate of placentome eversion with increasing gestational age [47]. In contrast,

Table 14.1 Effects of maternal administration of glucocorticoids during late gestation on placental and fetal weights in different species.

				Age (days)		% normal wt		
Species	Glucocorticoid	Dose	Route	At treatment	At delivery	Placenta	Fetus	Reference
Mouse	Dex	3×500 µg kg	Injection ip	15, 16, 17	20	93	81	27
	Dex	5×200 µg kg	Injection sc	11–15	16	90	100	28
	Dex	5×200 µg kg	Injection sc	14–18	19	90	95	28
Rat	TA	1×380 µg kg	Injection ip	16	21	53	67	29
	TA	5×20 mg kg	Injection im	15–19	20	82	85	30
	Beta	6×100 µg kg	Injection sc	15–20	21	80	88	31
	Beta	6×200 µg kg	Injection sc	15–20	21	68	80	31
	Dex	5×200 µg kg	Injection sc	15–19	20	76	89	32
	Dex	200 µg kg day	Minipump sc	13–20	20	48	77	33
	Dex	100 µg kg day	Minipump sc	15–21	21	73	93	34
	Dex	200 µg kg day	Minipump sc	15–21	21	50	71	34
	Dex	2000 µg kg day	Orally in water	15–22	22	44	35	35
	Dex	200 µg kg day	Orally in water	13–22	22	66	73	36
Rabbit	Beta	2×200 µg kg	Injection im	24, 25	26	77	53	37
Sheep	Cortisol	80 mg day	Continuous iv	118–128	128	75	No Δ	38
	Beta	3×500 µg kg	Injection im	104,111,118	121	63	68	39
	Dexa	3×120 µg kg	Injection im	103,110,117	119	83	84	40
Monkey	Beta	3×2 mg	Injection im	129–131	132	80	No Δ	41
	Beta	13×2 mg	Injection im	118–131	132	65	91	41
	Beta	2×3 mg	Injection im	24–48h before delivery	134–150	89	98	42

Note: Beta, Betmethasone; TA, triamcinoline; Dex, Dexamethasone; ip, intraperitoneal; sc, subcutaneous; im, intramuscular; iv, intravenous. Δ = change.

maternal dexamethasone administration at a dose of 200 µg/kg per day for 5 days in late gestation appears to have little effect on the gross morphology or volume fractions of the different zones of the rodent placenta [28,33]. However, at higher doses the dexamethasone-treated mouse placenta appears necrotic and has increased expression of several apoptotic genes [27]. In the $11\beta HSD2$-null placenta, the labyrinthine zone, responsible for nutrient transport, accounts for a smaller proportion of the total volume than in the wild-type placenta close to term [44].

At the ultrastructural level, glucocorticoid administration to either the mother or fetus reduces the number of binucleate cells (BNC) in the ovine placentomes [39,48]. These cells are formed in the fetal trophectoderm and migrate across the fetomaternal microvillous junction to form a syncytium by fusing with the maternal epithelium [49]. Glucocorticoid-induced changes in BNC frequency may therefore alter the morphological remodelling of the placenta that normally occurs with BNC migration. In preliminary experiments injecting pregnant mice with dexamethasone for 5 days before delivery, there was an increase in the surface area for nutrient exchange and a decrease in the thickness of the diffusion barrier in the labyrinthine zone at day 16 but not day 19 of pregnancy, compared to saline-injected controls [28]. Increasing glucocorticoid exposure by deleting the $11\beta HSD2$ gene has little effect on the volume densities of the blood vessels in the placenta at day 15, but reduces the surface density and length of the fetal capillaries in mutant relative to wild-type placentas nearer to term [44]. Similarly, continuous dexamethasone treatment of rat dams during late gestation reduces the volume densities of maternal and fetal blood spaces and the relative surface area of the fetal capillaries in the labyrinthine zone of the term placenta [36]. In addition, hypovascularization of the fetal villi is seen in term placentas of asthmatic women taking high doses of glucocorticoids during pregnancy [50]. However, the consequences for blood flow in the rodent placenta of this glucocorticoid-induced inhibition of

Chapter 14. Glucocorticoids and placental programming

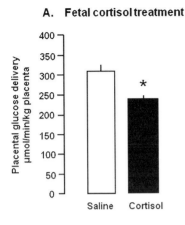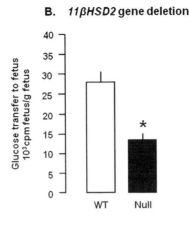

Figure 14.2 Mean (±SE) weight specific rates of placental glucose delivery to A) fetal sheep at 130 days of gestation (term 145 days) infused with saline (open columns) or cortisol (filled columns) for 5 days [51] and B) fetal mice either wild type (WT, open columns) or completely null (filled columns) for the 11β-hydroxysteroid dehydrogenase type 2 (*11βHSD2*) gene [44] at day 18 of pregnancy (term 21 days). *Significantly different from control groups, saline infused in A) and WT in B) $p < 0.02$.

angiogenesis during late gestation remains unclear. In sheep, less prolonged glucocorticoid administration to either the mother or the fetus appears to have little effect on uterine or umbilical blood flow during and immediately after treatment, except when the fetus is hypoxaemic, growth restricted, or has raised blood pressure [51–54]. In women at risk of preterm delivery there is also little evidence for changes in umbilical blood flow in response to a standard course of antenatal glucocorticoid treatment [55].

Placental nutrient transport and utilization

In sheep, maternal dexamethesone administration causes maternal hyperglycaemia and a greater transplacental glucose concentration gradient, which is likely to increase placental glucose delivery to the fetus [56]. However, short-term administration of dexamethasone or cortisol to fetal sheep has little effect on placental consumption or fetal delivery of glucose, although umbilical lactate uptake is increased [57,58]. On the other hand, fetal administration of cortisol for 5 days has been shown to reduce placental delivery of glucose and lactate, but not oxygen, to the fetus (Figure 14.2). This effect was more pronounced in placentas with a higher frequency of the more everted placentome types [47]. Decreased eversion of the placentomes when fetal cortisol levels are high may therefore help maintain a placental glucose supply to the sheep fetus during adverse intrauterine conditions that raise glucocorticoid concentrations [47]. The reduced umbilical uptake of glucose in response to more long-term fetal cortisol infusion was due, in part, to increased glucose consumption by the uteroplacental tissues [51]. Fetal cortisol may therefore alter glucose partitioning between the fetoplacental tissues in favour of the placenta. In mice, fetoplacental overexposure to glucocorticoids by deletion of the *11βHSD2* gene also reduces placental glucose delivery to the fetus (Figure 14.2), in association with reduced placental expression of the *Slc2a3/GLUT3* but not the *Slc2a1/GLUT1* isoform of the glucose transporters [44]. Altered expression of both these transporters has been observed in human term trophoblast cultured *in vitro* with triamcinolone, and in term placentas from rat dams treated either with a single dose of triamcinolone or continuously with dexamethasone for 6 days [27,29,59]. However, in sheep, no change in placental expression of GLUT1 or GLUT3 was observed after fetal cortisol administration for 5 days [51].

Total placental amino acid delivery to the sheep fetus was unaffected by either fetal cortisol or dexamethasone treatment [58]. However, fetal amino acid deamination was increased and umbilical uptake of the key gluconeogenic amino acid alanine was reduced by fetal glucocorticoid treatment [57,58]. The primary effect of glucocorticoids on amino acid metabolism may therefore be on fetal not placental tissues in the sheep. *In vitro* studies of human choriocarcinoma cell lines and term explants show that cortisol increases uptake of methyl amino-isobutyric acid (MeAIB), an amino acid analogue transported by the system A amino acid transporters [46,60]. Preliminary observations in mice suggest that, relative to controls, the small dexamethasone-treated placenta has a higher weight-specific clearance of MeAIB at day 16 but not at day 19 of pregnancy (Figure 14.3). Similar increases in maternofetal MeAIB transfer per gram of

Chapter 14. Glucocorticoids and placental programming

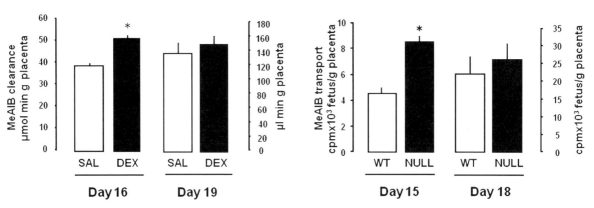

Figure 14.3 Mean (±SE) weight specific rates of A) clearance or B) transport of ^{14}C-methyl amino-isobutyric acid (MeAIB) across the mouse placenta overexposed to glucocorticoids by A) daily maternal administration of dexamethasone treatment (200 μg/kg, DEX, filled columns) for 5 days before measurement at days 16 and 19 compared to saline-injected animal (SAL, open columns) [28] or B) deletion of the 11β-hydroxysteroid dehydrogenase type 2 (*11βHSD2*) gene for 15 or 18 days of pregnancy (Null, filled columns) compared to wild-type littermates (WT, open columns) [44]. * Significantly different from control groups, saline infused in A) and WT in B) $p < 0.01$.

growth-restricted *11βHSD2*-null placenta are seen at day 15 of pregnancy (Figure 14.3), in association with normal fetal weight and increased expression of the *Slc38a2* and *Slc38a4* isoforms of the system A amino acid transporters [44]. However, by day 18, upregulation of placental amino acid transport and system A transporter expression is not maintained and both the *11βHSD2*-null placenta and the fetus weigh less than their wild-type counterparts [44]. In common with mouse placentas growth restricted by maternal undernutrition [61], the small glucocorticoid-overexposed placenta appears to adapt its phenotype to help maintain fetal growth before term, but cannot sustain an increased weight-specific supply of nutrients into late gestation, when the fetal demands for nutrients are rising most rapidly in absolute terms.

Placental endocrine function

The placenta produces a wide range of hormones, including steroids, glycoproteins, eicosanoids and peptides [3,7]. The secretion of several of these is sensitive to glucocorticoids. In sheep, glucocorticoid treatment reduces maternal concentrations of placental lactogen owing to the decreased number of trophoblast BNC cells that synthesize this hormone [39]. Concentrations remained depressed for up to 10 days after cessation of treatment, which suggest that glucocorticoids have long-term effects on BNC formation and migration in ovine placentomes. During late pregnancy, maternal dexamethasone treatment also causes dysregulated synthesis of the prolactin-like family of hormones in the rat placenta [33]. In general, expression of these hormones was enhanced in the labyrinth zone and suppressed in the junctional zone primarily responsible for producing these and other hormones in the rodent placenta. There are also decreases in placental leptin expression and increased transplacental leptin transfer in the dexamethasone-exposed rat placenta [34,43]. However, dexamethasone increases leptin expression and secretion in human trophoblast cultured *in vitro* [62].

In sheep, glucocorticoid administration to either the fetus or the mother reduces placental progesterone (P4) synthesis and increases oestrogen production, which can activate the myometrium and lead to the onset of myometrial contractions, as occurs during the prepartum period in response to the natural rise in fetal cortisol concentrations [63,64]. These changes in sex steroid synthesis are mediated in part by increased placental prostaglandin (PG) E_2, which in turn is due to glucocorticoid-induced upregulation of placental PGH synthase (PGHS) activity [18,61]. Increased activity of this enzyme and the rise in placental oestrogen availability leads to enhanced uteroplacental production of $PGF_{2\alpha}$, which is responsible for the increased uterine contractility along with the changes in placental P4 and oestrogen production [18,64]. These glucocorticoid-induced changes in placental steroid and PG production lead to changes in the circulating levels of these hormones, primarily in the mother, which influence the development of other tissues, such as the mammary glands. In the fetus, the glucocorticoid-stimulated rise in plasma PGE_2 level can act on the fetal HPA axis to stimulate more cortisol secretion via a positive feedback loop [65]. Similar increases in PGE_2 production in association with enhanced PGHS activity are seen in human term trophoblast cultured with dexamethasone *in vitro* [66].

Glucocorticoids alter not only the production but also the metabolism of key hormones in the placenta. They activate the prostaglandin dehydrogenases, which increase production of the inactive keto PG metabolites in parallel with the increased synthesis of the primary PGs. The net effect of glucocorticoid overexposure is therefore determined by the balance between PG synthesis and metabolism, which is also influenced by gestational age and maternal nutritional state [67]. Glucocorticoids also regulate their own metabolism. Placental 11βHSD2 activity is downregulated in sheep by both maternal and fetal glucocorticoid administration *in vivo* [68,69]. Similar decreases in 11βHSD2 activity are observed in perfused rat placentas after *in vitro* treatment with either dexamethasone or betamethasone, although 11βHSD2mRNA and protein abundance were unaffected in these experiments [70]. When dexamethasone was given orally to rats for 5 days, there was no change in 11βHSD2 activity or mRNA expression in the term placenta [16,35]. In contrast, 11βHSD2mRNA expression is reduced in term placentas from rats following restraint stress during late gestation, which suggests that natural corticosterone may be more effective than synthetic glucocorticoids in regulating placental 11βHSD2 in rats [71]. In human and non-human primates, maternal treatment with synthetic glucocorticoids increases placental 11βHSD2mRNA, protein and activity close to term [25,72]. Glucocorticoid treatment has also been shown to downregulate glucocorticoid receptor (GR) abundance in the human placenta *in vitro*, but to upregulate GR expression in ovine uninucleate trophoblast cells *in vivo* [18]. In addition, in rats, dexamethasone reduces labyrinthine expression of the multidrug resistance P-glycoprotein that restricts transport of glucocorticoids and other drugs across the maternoplacental interface [73]. Taken together, these observations suggest that glucocorticoids affect their own bioavailability, but in a manner that depends on species and the dose, duration and type of administration.

Mechanisms of glucocorticoid action

The mechanisms by which glucocorticoids act to alter placental phenotype remain unclear but may involve the *Igf2* gene and/or oxidative stress [74,75]. Glucocorticoid treatment may disturb the balance between pro-oxidant and antioxidant factors in tissues by altering the activity of important antioxidant enzymes [75]. However, little is known about the activity of these enzymes, or their glucocorticoid dependence, in the placenta. In contrast, the *Igf2* gene is known to be important in placental development and its expression is downregulated in both the junctional and the labyrinthine zones of the rat placenta by maternal glucocorticoid treatment [33,74]. Deletion of the *Igf2* gene either from all fetoplacental tissues (complete null) or specifically from the labyrinthine zone (*Igf2P0* null) in mice also leads to placental and fetal growth retardation and to morphological changes in the placenta, which are more severe in the complete null [75,76]. These changes include increased barrier thickness and reductions in the length, volume and surface area of the fetal capillaries in both mutant placentas relative to their respective wild-type counterparts [78]. In addition, the small *Igf2P0* null placenta is more efficient in supporting fetal growth and transports more MeAIB per gram than do wild-type placentas, despite its reduced size and passive permeability [77,78]. These changes in transport capacity are accompanied by enhanced expression of the *Slc38a4* isoform of the system A amino acid transporters and are more pronounced at day 16 than day 19 of pregnancy [74,77], consistent with the findings in the *11βHSD2* knockout and dexamethasone-treated mice (Figure 14.3).

Conclusion

Glucocorticoids can alter placental development in several different ways (Figure 14.4). They can influence the growth and morphological development of the placenta, particularly of the exchange surface and fetal blood vessels. They can also affect the abundance of the transporter molecules required for facilitated diffusion and active transport of key nutrients across the placenta. In part, these changes in morphology and nutrient transporter expression appear to compensate for the reduced size of the placenta overexposed to glucocorticoids before term (Table 14.1). In addition, glucocorticoids can modify the partitioning of nutrients between fetal and placental tissues and,

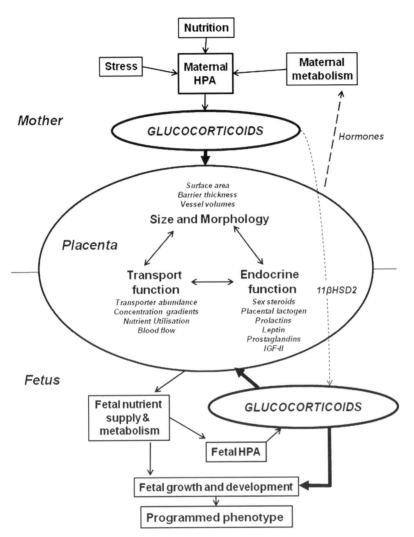

Figure 14.4 Schematic diagram showing the role of glucocorticoids in the maternal and fetal circulations in controlling placental phenotype and the consequences of these regulatory actions for fetal development.

through their actions on placental hormone production, may also influence the allocation of maternal nutrient resources to the gravid uterus (Figure 14.4). Overall, these actions of the glucocorticoids alter the placental capacity to supply nutrients to the fetus for development, with long-term consequences for the phenotype of the offspring (Figure 14.4). However, to date, the majority of the studies examining the effects of glucocorticoids on placental phenotype have concentrated on the period during or shortly after glucocorticoid administration, so the extent to which these effects persist to programme placental morphology and function in the long term remain largely unknown. Nor is it clear yet whether the actions of the glucocorticoids on placental development are sex-linked and thus explain the gender specificity of the intrauterine programming induced by particular environmental challenges.

Acknowledgements

We would like to thank all the members of the Department of Physiology, Development and Neuroscience who have helped with these studies over the years. We are also indebted to the BBSRC, the Royal Society, the Wellcome Trust and the Horserace Betting Board for their funding of the research, and the Centre for Trophoblast Research in the Department of Physiology, Development and Neuroscience for the studentship awarded to Owen Vaughan.

References

1. **Barker DJP**. *Mother, Babies and Disease in later Life*. London: BMJ Publishing Group, 1994.
2. **McMillen IC, Robinson JC**. Developmental origin of the metabolic syndrome; prediction, plasticity and programming. *Physiol Rev* 2005; **85**: 571–633.
3. **Fowden AL, Ward JW, Wooding FPB, Forhead AJ, Constancia M**. Programming placental nutrient transfer capacity. *J Physiol* 2006; **572**: 5–15.
4. **Fowden AL, Forhead AJ, Coan PM, Burton GJ**. The placenta and intrauterine programming. *J Neuroendocrinol* 2008; **20**: 439–50.
5. **Seckl JC**. Glucocorticoids, developmental programming and the risk of affective dysfunction. *Prog Brain Res* 2008; **167**: 17–34.
6. **Fowden AL, Li J, Forhead AJ**. Glucocorticoids and the preparation for life after birth: are there long term consequences of the life insurance? *Proc Nutr Soc* 1998; **57**: 113–22.
7. **Fowden Al, Forhead AJ**. Hormones as epigenetic signals in developmental programming. *Exp Physiol* 2009; **94**: 607–25.
8. **Fowden AL, Giussani DA, Forhead AJ**. Endocrine and metabolic programming during intrauterine development. *Early Hum Dev* 2005; **81**: 723–34.
9. **Seckl JC**. Glucocorticoids, fetoplacental 11β-hydroxysteroid dehydrogenase type 2 and early life origins of adult disease. *Steroids* 2001; **62**: 89–94.
10. **McMillen IC, Thorburn GD, Walker DW**. Diurnal variations in plasma concentrations of cortisol, prolactin, growth hormone and glucose in the fetal sheep and pregnant ewe during late gestation. *J Endocrinol* 1987; **114**: 65–72.
11. **Fowden AL, Mundy L, Silver M**. Developmental regulation of glucogenesis in the sheep fetus during late gestation. *J Physiol* 1998; **508**: 937–47.
12. **Fletcher AJW, Gardner DS, Edwards CMB, Fowden Al, Giussani, DA**. Cardiovascular and endocrine responses to acute hypoxaemia during and following dexamethasone infusion in the ovine fetus. *J Physiol* 2003; **549**: 271–87.
13. **Fowden, AL, Forhead AJ, Ousey, JC**. The endocrinology of equine parturition. *Exp Clin Endocrinol Diabetes* 2008; **116**: 393–403.
14. **Fowden AL, Forhead AJ**. Endocrine mechanisms of intrauterine programming. *Reproduction* 2004; **127**: 515–26.
15. **Hennessy DP, Coghlan JP, Hardy KJ, Scoggins BN, Wintour EM**. The origin of cortisol in the blood of fetal sheep. *J Endocrinol* 1982; **95**: 71–9.
16. **Waddell BJ, Benediktsson R, Brown RW, Seckl JR**. Tissue-specific messenger ribnucleic acid expression of 11β-hydroxysteroid dehydrogenase types 1 and 2 and the glucocorticoid receptor within the rat placenta suggest exquisite local control of glucocorticoid action. *Endocrinology* 1998; **139**: 1517–23.
17. **Speirs HJL, Seckl JR, Brown RW**. Ontogeny of glucocorticoid receptor and 11β-hydroxysteroid dehydrogenase type-1 gene expression identifies potential critical periods of glucocorticoid susceptibility during development. *J Endocrinol* 2004; **181**: 105–16.
18. **Whittle WL, Holloway AC, Lye S, Challis JR, Gibb W**. The pattern of glucocorticoid and estrogen receptors may explain differences in steroid dependency of intrauterine prostaglandin production at parturition in sheep. *J Soc Gynecol Invest* 2006; **13**: 506–11.
19. **Johnson RF, Rennie N, Murphy V** et al. Expression of glucocoticoid receptor messenger ribonucleic acid transcripts in the human placenta at term. *J Clin Endocrinol Metab* 2008; **93**: 4887–93.
20. **Ward IL, Weisz J**. Differential effects of maternal stress on circulating levels of corticosterone, progesterone and testosterone in male and female rat fetuses and their mothers. *Endocrinol* 1984; **114**: 1635–44.
21. **Barbazanges A, Piazza PV, Le Moal M, Maari S**. Maternal glucocorticoid secretion mediates long-term effects of prenatal stress. *J Neurosci* 1996; **16**: 3943–49.
22. **Gardner DS, Jackson AA, Langley-Evans SC**. The effect of prenatal diet and glucocorticoids on growth and systolic blood pressure in the rat. *Proc Nutr Soc* 1997; **57**: 235–40.
23. **Langley-Evans SC, Benediktsson R, Gardner DS, Jackson AA, Seckl JR**. Protein intake in pregnancy, placental glucocorticoid metabolism and the programming of hypertension. *Placenta* 1997; **17**: 169–72.
24. **McMullen S, Ogersby JC, Thurston LM** et al. Alterations in 11β-hydroxysteroid dehydrogenase (11βHSD) activities and fetal cortisol: cortisone ratios induced by nutritional restriction prior to conception and at defined stages of gestation. *Reproduction* 2004; **127**: 17–25.
25. **Clifton VL, Rennie N, Murphy VE**. Effects of inhaled glucocorticoid treatment on placental 11β-hydroxysteroid dehydrogenase type 2 activity and neonatal birth weight in pregnancies complicated by asthma. *Aust NZ J Obstet Gynaecol* 2006; **46**: 136–40.
26. **American College of Obstericians and Gynecologists**. ACOG Committee Opinion 402. Antenatal corticosteroid therapy for fetal maturation. *Obstet Gynecol* 2008; **111**: 805–7.

27. **Baisden B, Sonne S, Joshi RM, Ganapathy V, Shekhaurat PS**. Antenatal dexamethasone treatment leads to changes in gene expression in a murine late placenta. *Placenta* 2007; **28**: 1082–90.

28. **Vaughan OR, Sferruzzi-Perri AN, Coan PM, Burton GJ, Fowden AL**. Effects of dexamethasone treatment on amino acid clearance and diffusional exchange capacity of the mouse placenta depends on gestational age. *Placenta* 2009; **30**: A11.

29. **Hahn T, Barth S, Graf R** et al. Placental glucose transporter expression is regulated by glucocorticoids. *J Clin Endocrinol Metab* 1999; **84**: 1445–52.

30. **Shafrir E, Barash V, Zederman R, Kisselevitz R, Diamant TZ**. Modulation of fetal and placental metabolic pathways in response to maternal thyroid and glucocorticoid hormone excess. *Isr J Med Sci* 1994; **30**: 32–41.

31. **McDonald TJ, Franko KL, Brown JM** et al. Betamethasone in the last week of pregnancy causes fetal growth retardation but not adult hypertension in rats. *J Soc Gynecol Invest* 2003; **10**: 469–73.

32. **Franko KL, Forhead AJ, Fowden AL**. Differential effects of prenatal stress and glucocorticoid administration on postnatal growth and glucose metabolism in rate. *J Endocrinol* 2010; **204**: 319–29.

33. **Ain R, Canham LN, Soares MJ**. Dexamethasone-induced intrauterine growth restriction impacts the placental prolactin family, insulin-like growth factor-II and the AKT signalling pathway. *J Endocrinol* 2005; **185**: 253–63.

34. **Sugden MC, Langdown ML, Munns MJ, Holness MJ**. Maternal glucocorticoid treatment modulates placental leptin and leptin receptor expression and maternofetal leptin physiology during late pregnancy, and elicits hypertension associated with hyperleptinaemia in the early-growth-retarded adult offspring. *Eur J Endocrinol* 2001; **145**: 529–39.

35. **Burton BJ, Waddell BJ**. 11β-hydroxysteroid dehydrogenase in the rat placenta: developmental changes and the effects of altered glucocorticoid exposure. *J Endocrinol* 1994; **143**: 505–13.

36. **Hewitt DP, Mark PJ, Waddell BJ**. Glucocorticoids prevent the normal increase in placental vascular endothelial growth factor expression and placental vascularity during late pregnancy in the rat. *Endocrinology* 2006; **147**: 5568–74.

37. **Barrada MI, Blomquist CH, Kotts MS**. The effects of betamethasone on fetal development in the rabbit. *Am J Obstet Gynecol* 1980; **136**: 234–8.

38. **Jensen EC, Gallaher BW, Breien BH, Harding JE**. The effect of chronic maternal cortisol infusion on the late-gestation fetal sheep. *J Endocrinol* 2002; **174**: 27–36.

39. **Braun T, Li S, Moss TJM** et al. Maternal betamethasone administration reduces binucleate cell number and placental lactogen in sheep. *J Endocrinol* 2007; **194**: 337–47.

40. **Kutzler MA, Ruane EK, Coksaygan T, Vincent SE, Nathanielsz PW**. Effects of three couses of maternally administered dexamethasone at 0, 7, 0.75 and 0.8 of gestation on prenatal and postnatal growth in sheep. *Pediatrics* 2004; **113**: 313–19.

41. **Johnson JWC, Mitzner W, London WT, Palmer AE, Scott R**. Betamethasone and the rhesus fetus: multisystemic effects. *Am J Obstet Gynecol* 1979; **133**: 677–84.

42. **Epstein MF, Farrell PM, Sparks JW** et al. Maternal betamethasone and fetal growth and development in the monkey. *Am J Obstet Gynecol* 1977; **27**: 1261–3.

43. **Smith JT, Waddell BJ**. Leptin distribution and metabolism in the pregnant rat: transplacental leptin passage increases in late gestation but is reduced by excess glucocorticoids. *Endocrinology* 2003; **144**: 3024–30.

44. **Wyrwoll CS, Seckl JC, Megan MC**. Altered placental function of the 11β-hydroxysteroid dehydrogenase 2 knockout mice. *Endocrinology* 2009; **150**: 1287–93.

45. **Mandl M, Ghaffari-Tabrizi N, Haas J, Nohammer G, Desoye G**. Differential glucocorticoid effects on proliferation and invasion of human trophoblast cell lines. *Reproduction* 2006; **132**: 159–67.

46. **Audette MC, Greenwood SL, Sibley CP** et al. Dexamethasone stimulates placental system A transport and trophoblast differentiation in term villous explants. *Placenta* 2010; **31**: 97–105.

47. **Ward JW, Forhead AJ, Wooding FBP, Fowden AL**. Functional significance and cortisol dependence of the gross morphology of ovine placetomes during late gestation. *Biol Reprod* 2006; **74**: 137–45.

48. **Ward JW, Wooding FBP, Fowden AL**. The effect of cortisol on the binucleate cell population in the ovine placenta during late gestation. *Placenta* 2002; **23**: 451–8.

49. **Wooding FPB, Burton GJ**. *Comparative Placentation; Structure, Functions and Evolution*. Heidelberg: Springer-Verlag, 2008.

50. **Mayhew TM, Jenkins H, Todd B, Clifton VL**. Maternal asthma and placental morphology: effects of severity, treatment and fetal sex. *Placenta* 2008; **29**: 366–73.

51. **Ward JW, Wooding FPB, Fowden AL**. Ovine fetoplacental metabolism. *J Physiol* 2004; **554**: 529–41.

52. **Jellyman JK, Gardner DS, Fowden AL, Giussani DA**. Effects of dexamethasone on the uterine and umbilical

vascular beds during basal and hypoxemic conditions in sheep. *Am J Obstet Gynecol* 2004; **190**: 825–35.

53. **Kutzler MA, Coksaygan TC, Ferguson AD, Nathanielsz PW**. Effects of maternally administered dexamethasone and acute hypoxemia at 0.7 gestation on blood pressure and placental perfusion in sheep. *Hypertens Pregnancy* 2004; **23**: 75–90.

54. **Miller SL, Supramaniam VG, Jenkin G, Walker DW, Wallace EM**. Cardiovascular responses to maternal betamethasone administration in the intrauterine growth-restricted ovine fetus. *Am J Obstet Gynecol* 2009; **201**: 613e1–8.

55. **Cohlen BJ, Stigter RH, Derks JB, Mulder EJ, Visser GH**. Absence of significant hemodynamic changes in the fetus following maternal betamethasone administration. *Ultrasound Obstet Gynecol* 1996; **8**: 252–5.

56. **Franko KL, Giussani DA, Forhead AJ, Fowden AL**. Effects of dexamethasone on the glucogenic capacity of fetal, pregnant and non-pregnant adult sheep. *J Endocrinol* 2007; **191**: 67–3.

57. **Milley JR**. Fetal substrate uptake during increased ovine cortisol concentration. *Am J Physiol* 1996; **271**: E186–E191.

58. **Timmerman M, Teng C, Wilkening RB** et al. Effect of dexamethasone on fetal hepatic glutamine-glutamate exchange. *Am J Physiol* 2000; **278**: E839–E845.

59. **Langdown ML, Sugden MC**. Enhanced placental GLUT1 and GLUT3 expression in dexamethasone induced fetal growth retardation. *Mol Cell Endocrinol* 2001; **185**: 109–17.

60. **Jones HN, Ashworth CJ, Page KR, McArdle HJ**. Cortisol stimulates system A amino acid transport and SNAT2 expression in human placental cell lines (BeWo). *Am J Physiol* 2006; **291**: E596–603.

61. **Coan PM, Vaughan OR, Sekita Y** et al. Adaptations in placental phenotype support fetal growth during undernutrition of pregnant mice. *J Physiol* 2010; **588**: 527–38.

62. **Coya R, Gualillo O, Pineda J** et al. Effect of cyclic 3′,5′-adenosine monophosphate, glucocoticoids, and insulin on leptin messenger RNA levels and leptin secretion in cultured human trophoblast. *Biol Reprod* 2001; **65**: 814–19.

63. **Whittle WL, Patel FA, Alfraidy N** et al. Glucocorticoid regulation of human and ovine parturition: the relationship between fetal hypothalamic-pituitary-adrenal axis activation and intrauterine prostaglandin production. *Biol Reprod* 2001; **64**: 1019–32.

64. **Challis JRG, Sloboda D, Matthews SC** et al. Prostaglandins and mechanisms of preterm birth. *Reproduction* 2002; **124**: 1–17.

65. **Liggins GC, Scroop GG, Haughey KG**. Comparison of the effects of prostaglandin E2, prostacyclin and 1–24 adrenocorticotrophin on plasma cortisol levels of fetal sheep. *J Endocrinol* 1982; **95**: 153–62.

66. **Mirazi N, Alfraidy N, Martin R, Challis JRG**. Effects of dexamethasone and sulfasalazine on prostaglandin E2 output by human placental cells *in vitro*. *J Soc Gynocol Invest* 2004; **11**: 22–6.

67. **Fowden AL, Ralph M, Silver M**. Nutritional regulation of uteroplacental prostaglandin production and metabolism in pregnant ewes and mares during late gestation. *Exp Clin Endocrinol* 1994; **102**: 212–21.

68. **Clarke K, Ward JW, Forhead AJ, Giussani DA, Fowden AL**. Regulation of 11β-hydroxysteroid dehydrogenase type 2 (11β-HSD2) activity in ovine placenta by fetal cortisol. *J Endocrinol* 2002; **172**: 527–34.

69. **Kerzner LS, Stonestreet BS, Wu K-Y, Sadowska G, Malee MP**. Antenatal dexamethasone: effect on ovine placental 11β-hydroxysteroid dehydrogenase type 2 expression and fetal growth. *Pediatr Res* 2002; **52**: 706–12.

70. **Vackova Z, Vagnernova K, Libra A**. Dexamethasone and betamethasone administration during pregnancy affects the expression and function of 11β-hydroxysteroid dehydrogenase type 2 in the rat placenta. *Reprod Toxicol* 2009; **28**: 46–51.

71. **Mairesse J, Lesage J, Breton C** et al. Maternal stress alters endocrine function of the fetoplacental unit in rats. *Am J Physiol* 2007; **292**: E1526–E1533.

72. **Ma XH, Wu WX, Nathanielsz PW**. Gestation-related and betamethasone-induced changes in 11β-hydroxysteroid dehydrogenase types 1 and 2 in the baboon placenta. *Am J Obstet Gynecol* 2003; **188**: 13–21.

73. **Mark PJ, Sheldon S, Lewis JL, Hewitt D, Waddell BJ**. Changes in the placental glucocorticoid barrier during rat pregnancy: Impact on placental corticosterone levels and regulation by progesterone. *Biol Reprod* 2009; **80**: 1209–15.

74. **Fowden AL, Sibley C, Reik C, Constancia M**. Imprinted genes, placental development and fetal growth. *Horm Res* 2006; **65**: 49–57.

75. **Verhaeghe J, van Bree R, Van Herck E**. Oxidative stress after antenatal betamethasone: acute downregulation of glutathione peroxidase-3. *Early Hum Dev* 2009; **85**: 767–71.

76. **Fowden AL**. The insulin-like growth factors and fetoplacental growth. *Placenta* 2003; **24**: 803–12.

77. **Constancia M, Angiolini E, Sandovici I** et al. Adaptation of nutrient supply to fetal demand in the mouse involves interaction between the *Igf2* gene and

placental transporter systems. *PNAS* 2005; **102**: 19219–24.
78. **Coan PM**, **Fowden AL**, **Constancia M** et al. Disproportionate effects of *Igf2* gene knockout on placental morphology and diffusional exchange characteristics in the mouse. *J Physiol* 2008; **586**: 5023–32.

Discussion

CRITCHLEY: Have you had a chance to look at vasculature in your morphology studies? Glucocorticoids have an angiostatic effect in the cardiovascular system.

FOWDEN: Yes, we have done a lot of vessel counting. Vessel numbers are changed. There is a higher vessel density in the early dexamethasone exposed, but a lower vessel density in those we have left to go on for another 4 days. Again, like many of these interventions, dexamethasone treatment seems to advance the normal maturation processes that would be going on in the mouse placenta, and then we can't get any further adaptation.

HUNT: I am interested in what the target cells would be for the glucocorticoids. There should be target cells in the decidua and uterus in general, but also in the placenta. Has anyone studied the distribution?

FOWDEN: In most instances, we produce a mini placenta. It looks as if we are affecting both the labyrinth and the spongiotrophoblast in terms of volumes and areas. The glucocorticoid receptors are present in all the various different cell types that are present in the placenta from day 16 onwards. In fact, there is some evidence to suggest that the number of glutocorticoid receptors in the placenta generally comes down with increasing gestational age. This might explain why we see different effects when we give the dexamethasone early as opposed to giving it late. We have not seen any specific changes, but we know that the spongiotrophoblast is losing glycogen cells.

MOFFETT: Private IVF clinics are giving large doses of steroids to women to suppress their NK cells. These are high doses: around 25 mg of prednisolone for 3 months (it differs with clinic). What might be happening to the early placentas in these women?

FOWDEN: It will be having an effect. Whether or not it is a good effect is debatable. What the women want in the end is a pregnancy. It may have long-term adverse outcomes.

ROBERT BOYD: What they want in the end is a healthy baby, and then a healthy adult in due course, not just a pregnancy.

FOWDEN: What I mean is that the mother wants a baby, and is not thinking about whether this child will have high blood pressure at age 50. There is good evidence that a dose of prednisolone such as this will affect placental size in late gestation.

SMITH: With prednisolone, they always say that 95% of it is metabolised in the placenta and the fetus isn't exposed to it. But if you have changes in the placenta itself, maybe our relative confidence about using prednisolone is slightly misplaced.

FOWDEN: If you give prednisolone in late gestation you produce fetal and placental growth retardation.

FLEMING: Have you tried adding the glucocorticoid earlier than day 11 in the mouse model?

FOWDEN: We start giving it at day 11; it's day 11–14 in our earliest challenge.

JANSSON: The case for increased fetal cortisol as a player in developmental programming is very strong. Another important part of the stress response is activation of the sympathetic nervous system, and this would involve catecholamines. I would guess that in both those perturbations you would actually see activation of the adrenergic system on the fetal side too. Is there any reason we should look more into the role of catecholamines?

FOWDEN: Yes. This is just the effect of giving the saline injection compared with untreated animals. This saline injection, to which we have been doing all the comparisons, is raising the maternal corticosteroid concentrations. This is just an indicator of maternal stress.

ALWASEL: I have a general question. It is strange to me that sex hormones are not secreted in significant amounts. However, sex has an effect in fetal programming. Do you think this goes back to X and Y chromosomes, or endocrine differences?

FOWDEN: I think it is a combination of growth rates, sex steroids and innate differences. The rate of growth is likely to be important. The male fetuses grow faster and therefore they will be influenced more drastically in late gestation. Here we have seen that the females are more adaptive than the males are. It is more difficult to put our finger on the cause of this. There will be differences in sex steroids, particularly in the early phases of development, which may have placental consequences. The male placenta will have been exposed to testosterone at very early stages which the female placentas won't.

SFERRUZZI-PERRI: It might also be related to the responsiveness of the placenta to these endocrine signals. There are differences in expression of glucocorticoid receptor between male and female placentas.

LAMPL: Which way does it go?

FOWDEN: It is the opposite way round to the human situation. It is the males who are more responsive in the humans; females in the mice. It is not consistent across species.

THORNBURG: How does this apply to humans in stressful conditions?

FOWDEN: This is one of those areas that we don't pay enough attention to. There is wide interindividual variation in the responsiveness of peoples' HPA axes. In the human context we are therefore looking at extremes: war zones, bereaved mothers and mothers divorced during their pregnancies where glucocorticoid levels are rising dramatically. There would be some mileage in collecting these data, but if you ask for a blood sample peoples' glucocorticoid levels go up.

ROBERT BOYD: What about those studies of jogging mothers that were popular some years back?

FOWDEN: Separating out the factors is difficult. Yes, the exercise might make glucocorticoid levels rise, but some runners rather like the endorphin hit, so their catecholamine and glucocorticoid levels may be on the low side. Their nutrient levels could be different because of the excessive levels of exercise.

ROBERT BOYD: Your slide of photoperiodicity made me think of another question: the diurnal variation of nutrients in maternal plasma, and whether this is, for example, identifiably different during Ramadan than different times of the year. If it is, does this have a programming effect on the offspring?

ALWASEL: We're in the early stages of looking at this.

SIBLEY: David Barker, the population that you have studied will have been stressed during the war. Is the change in growth of placentas you mentioned a function of decreasing stress in the post-war years?

BARKER: What we know about the absolute size of the placentas in the Dutch famine is that in both men and women the size fell in relation to birth weight, but it fell more in the men: was this about not enough food, or the result of stress, or both?

MAGNUS: There is one condition in early pregnancy where there is stress and undernourishment, hyperemesis. Are there any studies on the long-term consequences of this?

FOWDEN: I don't know of any.

Chapter 15
Clinical biomarkers of placental development

Gordon C. S. Smith

Introduction

Many of the serious complications of pregnancy, including placental abruption, pre-eclampsia, stillbirth, intrauterine growth restriction and preterm birth, are the ultimate manifestations, to a greater or lesser extent, of suboptimal placentation. In the case of abruption the role of placentation is self-evident, and in the case of pre-eclampsia there are many lines of evidence that link the disease to placentation [1]. Poor fetal growth, in the absence of other environmental causes, is assumed to reflect poor function of the placenta. Similarly, there is evidence that links the placenta to many cases of preterm birth [2] and stillbirth [3]. In many cases, clinical measurements related to placentation are abnormal prior to these disorders being clinically evident. Hence, many of these measurements have been evaluated as screening tests, i.e. used as a means to identify from the general population the women who are most likely to develop complications. In general, the rationale for developing these screening tests is that, having identified high-risk women, interventions may be attempted with the aim of reducing maternal or perinatal morbidity and mortality consequent upon the given complication. The primary aim of this chapter is to summarize some of the most important aspects of current knowledge in relation to clinical assessment of placentation.

Ultrasonic assessment of placentation
Placental volume and thickness

Ultrasound allows the assessment of some physical characteristics of the placenta, such as thickness and, more recently using 3D methods, volume. The maximum thickness increases with advancing gestational age, and the mean in millimetres is approximately equal to the gestational age in weeks. Increased placental thickness has been associated with a number of adverse outcomes. The risk of adverse outcome is very high when the placenta exhibits the combination of being very thick and having a jelly-like, heterogeneous appearance on ultrasound: in a series of 16 such cases, perinatal death occurred in 50% [4]. Placental size can also be assessed using 3D ultrasound to estimate placental volume. Serial assessment of placental volume in approximately 1200 singleton pregnancies, with measurements at 12, 16 and 22 weeks' gestation, demonstrated a complex pattern of associations. Among women who subsequently developed pre-eclampsia, the mean placental volume was greater at 12 and 16 weeks but exhibited a reduced rate of growth from 16 to 22 weeks. Among women who subsequently delivered a small for gestational age (SGA) infant the mean placental volume was small compared to the average at 12 weeks, but maintained a normal rate of growth (i.e. remained small, with the same proportional deviation from the mean). Among women who subsequently developed pre-eclampsia and delivered an SGA infant, the mean placental volume was persistently smaller in all three measurements [5]. Another way to assess placental volume in early pregnancy is to relate it to the size of the fetus. This is expressed as the ratio of the fetal crown–rump length to the placental volume, called the 'placental quotient'. A study of almost 2500 singleton pregnancies demonstrated that the first-trimester placental quotient was reduced in women who went on to develop complicated pre-eclampsia or who ultimately delivered an SGA infant. However, the association was too weak to be clinically useful [6]. Interestingly, a lower placental quotient at 11–13 weeks was associated with increased uterine

---

The Placenta and Human Developmental Programming, ed. Graham J. Burton, David J. P. Barker, Ashley Moffett and Kent Thornburg. Published by Cambridge University Press. © Cambridge University Press 2011.

artery mean pulsatility index and notching (see below) in the second trimester, indicating that pregnancies exhibiting the 'classic' markers of impaired trophoblast invasion in the middle of pregnancy are demonstrably abnormal even in the first trimester [7].

Placental site

Implantation of the placenta in the lower uterine segment – placenta previa – can be a life-threatening complication of pregnancy. The development of ultrasound has transformed the clinical management of women presenting with vaginal bleeding, the cardinal symptom of placenta previa. The use of ultrasound in this condition is reviewed in the RCOG guideline on placenta previa [8]. The placenta is best localized using transvaginal ultrasound, which has a sensitivity of 87.5% and a specificity of 98.8% in women clinically suspected of the condition. If the edge of the placenta is <2 cm from the internal os in late gestation, delivery will usually be by caesarean section. Low-lying placentas will be detected during the routine anomaly scan at 20 weeks. Where this is detected and the internal os is only covered by the margin of the placenta, 90% will resolve. However, the chance of resolution is less in women with a previous caesarean section, or where the placenta is more centrally implanted over the cervix. Ultrasound is also useful in assessing morbid invasion of the placenta, i.e. placenta accreta or percreta. The features suggestive of morbid invasion are the presence of placental lacunae (given the placenta a 'moth-eaten' appearance), loss of the normal echolucent space between the base of the placenta and the myometrium, and the presence of large vessels in the interface between the base of the placenta and the myometrium. In women with a suspected morbidly invasive placenta, ultrasound had a sensitivity of around 80% [8]. Hence, although the majority of cases will be detected, a morbidly invasive placenta cannot be completely excluded by ultrasound. This has led to evaluation by magnetic resonance imaging (MRI), but this has lower sensitivity than ultrasound [8].

Placental maturity

Placental maturity describes the appearances of the placenta in relation to echodense material in the intralobular septae and periphery of the placenta. When mature placentae were examined histologically, the appearance was found to be due to deposition of fibrin and calcium [9]. A grading system was proposed by Grannum *et al.*, scoring these changes on a semiquantitative scale from 0 to 3 [10]. A small-scale prospective study suggested that there was no relationship between the Grannum score and outcome [9]. However, a subsequent large-scale, single-centre prospective study examined this with a combined observational and interventional component [11]. In the observational analysis, grade 3 maturity at 34–36 weeks was associated with a two- to threefold risk of meconium-stained liquor, emergency caesarean section for fetal distress, depressed 5-minute Apgar score and low birth weight; but was strongly associated with perinatal death (odds ratio 7.5, 95% confidence intervals 2.1–26.6). Moreover, in the interventional element of the analysis, revealing the placental maturity result to the clinician was associated with reduced perinatal mortality. This analysis is somewhat undermined by the very high perinatal mortality in the control group, where there were 13 perinatal deaths in 1011 women, despite the fact that the group excluded extreme preterm births. Nevertheless, both elements of this analysis suggest that there may be clinical utility in the assessment of placental maturity, but this requires further study.

Placental lesions

Echolucent areas in the placenta are frequently observed, which may or may not exhibit blood flow depending on whether they are perfused by maternal arteries. These do not appear to be associated with an increased risk of adverse outcome [12]. In contrast to the benign prognosis of these findings, another appearance has been described where the echolucent area has an echodense border, termed echogenic cystic lesions. These were associated with histologically confirmed villous infarcts and their presence, in combination with high-resistance uterine artery Doppler flow velocimetry (see below), was strongly predictive of an adverse outcome in a high-risk population [13].

Doppler flow velocimetry of the uteroplacental circulation

Uterine artery Doppler

As discussed elsewhere in this book, invasion of the maternal uterine resistance vessels by the trophoblast is an essential process in normal placentation. Uterine artery Doppler flow velocimetry can be used to assess the resistance to blood flow in the uterine circulation.

The uterine circulation is high resistance in the non-pregnant state and the resistance to flow decreases with progressive placental invasion in the first half of pregnancy [14]. A high-resistance pattern of flow is an indicator of impaired implantation and, clinically, is associated with an increased risk of a range of placentally related complications. Resistance is assessed by quantitative measures of end-diastolic flow, such as the pulsatility index (PI; higher measures indicate increased resistance), and the presence of notching of the waveform. A number of studies have addressed the relationship between uterine artery Doppler resistance indices and the risk of adverse outcome in the first trimester. One of the earliest studies demonstrated that there was no association between measurements at 7–11 weeks and outcome, but being in the top quartile of PI at 12–13 weeks was associated with a fourfold risk of hypertensive disorders [15]. A study of over 3000 women assessed the screening properties of this measurement. A uterine artery mean PI at 11–14 weeks in the top 5% of the population was associated with an increased risk of both pre-eclampsia and delivery of an SGA infant. The sensitivity for the detection of all pre-eclampsia was 27%, but the sensitivity for pre-eclampsia requiring delivery prior to 34 weeks was 50%. The equivalent figures for delivery of an SGA infant were 12% and 24%, respectively [16].

The process of trophoblast invasion continues into mid-gestation, and the majority of studies have addressed the predictive ability of uterine artery Doppler flow velocimetry performed around 20–24 weeks, when trophoblast invasion is thought to be complete. One of the largest studies (>30 000 women) evaluating the ability of this technique to predict pre-eclampsia quantified resistance using the mean PI at 23 weeks' gestation and analysed the additional prediction obtained by the presence of notching and other maternal characteristics. This study reported that there was a log linear relationship between the mean PI and the risk of pre-eclampsia, and that the presence of bilateral, but not unilateral, notches was also significantly associated with an increased risk of the condition. Adding maternal characteristics (such as age, parity, race, etc.) had no effect on the prediction of early-onset disease but significantly enhanced prediction of later-onset disease. Uterine Doppler resistance measurements in the top 5% had a sensitivity for pre-eclampsia resulting in delivery before 34 weeks of 78%, but only 32% for the condition where delivery occurred at or beyond 34 weeks [17]. Analysis of stillbirth risk using the same dataset yielded similar results. Treating women in the top 5% of predicted risk on the basis of the Doppler results as 'screen positive', the sensitivity for stillbirth prior to 33 weeks was 58% and the positive likelihood ratio was 12.1. In contrast, being in the top 5% of predicted risk performed very poorly as a screening test for stillbirth at or after 33 weeks, with a sensitivity of 7% and a positive likelihood ratio of 1.3. When analysed by cause of stillbirth, this was explained by two facts. First, high-resistance patterns of uterine artery Doppler were much more strongly associated with stillbirth due to a placental cause, compared with non-placental causes. Second, stillbirths due to a placental cause (growth restriction, abruption and pre-eclampsia) tended to occur prior to 33 weeks, whereas other stillbirths occurred at later gestational ages. Hence, abnormal uterine artery Doppler flow velocimetry appears to be a much better predictor of placentally related complications of pregnancy occurring at preterm gestational ages.

Umbilical artery Doppler

Doppler flow velocimetry can also be used to assess the resistance to flow on the fetal side of the placenta. High-resistance patterns of flow in the umbilical arteries are associated with maldevelopment of the villous vascular tree on the fetal side of the placenta [18]. The waveform is characterized by the presence of end-diastolic flow. Normal is defined as flow present and indices of resistance in the normal range. Abnormality is characterized by three stages: (1) increased resistance indices but end-diastolic flow still present; (2) end-diastolic flow absent; and (3) reversed end-diastolic flow. Randomized controlled trials of umbilical artery Doppler flow velocimetry in the management of high-risk pregnancies have been performed and have demonstrated a strong trend towards decreased perinatal mortality [19]. This may be due to identifying infants at high risk of stillbirth and preventing the loss through elective delivery. However, it could also improve outcome by providing reassurance in cases where the infant is SGA, but being SGA does not reflect an abnormal process (i.e. the fetus is constitutionally small). This would then reduce the inappropriate induction of iatrogenic prematurity and thereby also the risk of neonatal death. However, studies have not shown that screening of unselected women using umbilical artery Doppler reduces perinatal mortality. This may be due to problems in the way these studies were designed (see below). Furthermore, in clinical practice these measures and not used in isolation. The decision to deliver on the basis of abnormal umbilical artery Doppler will depend on the gestational age, and other methods of assessing fetal wellbeing, such as amniotic fluid index, middle cerebral artery and ductus venosus Doppler, computerized analysis of fetal heart rate, and, in the USA in particular, ultrasound assessment of the biophysical profile [20].

Biochemical markers of placentation

Pregnancy-associated plasma protein A (PAPP-A)

First-trimester maternal serum levels of PAPP-A have been widely studied in the assessment of Down's syndrome risk. However, the biology of the protein is such that associations might be anticipated for other complications of pregnancy. PAPP-A is part of the system controlling the insulin-like growth factors (IGF) in trophoblast, being a protease for insulin-like growth factor-binding protein (IGFBP)4 [21]. IGFBPs bind IGF-I and IGF-II, inhibiting their interaction with cell surface receptors, and therefore have a key role in modulating IGF activity [22]. Because PAPP-A breaks down IGFBP [21], low levels of PAPP-A would be expected to be associated with high levels of IGFBP and hence low levels of free IGF. Messenger RNA for PAPP-A has been identified in syncytiotrophoblast and mononuclear cells of possible extravillous trophoblast origin, and the protein has been localized to placental septae, anchoring villi and chorionic villi [23]. The IGFs have a key role in regulating fetal growth [24], have also been shown to control uptake of glucose and amino acids in cultured trophoblast [25], and are thought to have an important role in the autocrine and paracrine control of trophoblast invasion of the deciduas [26]. Finally, mice which were null mutant for the gene encoding PAPP-A exhibited severe early-onset intrauterine growth restriction [27]. An association between PAPP-A and later outcome was, therefore, biologically plausible.

Early studies relating first-trimester measurement of PAPP-A to perinatal outcome reported inconsistent results. One study compared first-trimester PAPP-A levels in 73 babies ultimately born less than the fifth percentile for gestational age and 87 ultimately born preterm with matched controls. There was no statistically significant difference between the groups [28]. However, another study found a positive correlation between PAPP-A at 8–14 weeks and eventual birth weight [29], and an analysis of 60 IVF pregnancies described lower concentrations of PAPP-A in the first trimester in eight women who eventually delivered preterm [30]. A further study of 5297 women demonstrated lower PAPP-A at 10–14 weeks' gestation among women who miscarried, delivered babies SGA and developed pre-eclampsia [31]. However, PAPP-A was used in these women to estimate the risk of the fetus having Down's syndrome, which is a potential cause of confounding.

Two large-scale studies evaluated first trimester PAPP-A measurement in a non-interventional way, i.e. data were ascertained but the management of the pregnancy was not influenced by the early pregnancy assessment. Both studies demonstrated that low levels of PAPP-A at 10–13 weeks were associated with an increased risk of delivering an SGA infant, preterm delivery, pre-eclampsia and stillbirth [32,33]. The associations persisted in multivariate analysis. A further analysis of the earlier of the two cohorts assessed whether PAPP-A levels prior to 13 weeks' gestation were associated with fetal growth and the timing of labour among entirely uncomplicated pregnancies. A total of 4288 women who had PAPP-A assayed at 8–12 weeks of gestation (dated by ultrasound, equivalent to 6–10 weeks after conception), who ultimately had uncomplicated singleton pregnancies and delivered normal, live babies at full term, were studied [34]. This demonstrated that both the eventual birth weight (stratified by gestational age) and the timing of labour at term were correlated with the early pregnancy levels of PAPP-A.

Both of the large-scale cohort studies of early pregnancy PAPP-A demonstrated associations with stillbirth but lacked detailed information on the cause. In order to address this, data from the multicentre Scottish cohort study was linked to a national registry of perinatal deaths. Moreover, the number of participants in the cohort was sufficient to allow limitation of the analysis to women who were, by strict definition, in the first trimester (i.e. prior to 13 weeks gestational age). This analysis demonstrated that low first-trimester levels of PAPP-A were very strongly associated with stillbirth due to placental causes (i.e. associated with growth restriction, placental abruption or pre-eclampsia) but not stillbirths due to other causes [35]. Women with PAPP-A levels in the lowest 5% had a 40–50-fold risk of stillbirth related to a placental cause compared to the rest of the population. Moreover, when analysed by quintiles of PAPP-A, there were no placentally related stillbirths among women with maternal serum levels in the top 60%. When analysed by gestational age, the increased risk of stillbirth was apparent across the whole range of gestation. This observation raises the hope that effective clinical prediction of the most severe complications of late pregnancy may be possible in the first trimester. However, the positive predictive value of low PAPP-A on its own was too low to be clinically useful. The very high relative risk partly reflects the extreme rarity of the outcome in women with normal levels, rather than a strong predictive association among women with low PAPP-A.

α-fetoprotein

α-fetoprotein (AFP) is the major fetal oncotic protein and, when the fetus is structurally normal, high maternal circulating levels are thought to reflect a defect in placentation, leading to greater permeability to AFP across the fetomaternal barrier. It has long been appreciated that, in the absence of congenital abnormality, high maternal serum levels of AFP in the second trimester of pregnancy are associated with an increased risk of adverse perinatal outcome, including fetal death, preterm delivery and low birth weight [36]. More recent studies have addressed the gestational age dependence and causes of the associations. In a population of more than 70 000 nulliparous women having serum screening performed between 15 and 20 weeks in Scotland, women with AFP levels in the top 5% had an odds ratio for stillbirth of approximately 2.8. When analysed by cause of stillbirth, there were strong associations with losses due to pre-eclampsia (odds ratio 7.8), abruption (odds ratio 3.0), and stillbirths where the infant was SGA (odds ratio 3.6) [37]. Like high-resistance patterns of uterine artery Doppler, AFP was much more strongly associated with stillbirths at extreme preterm gestational ages, and there was no significant association between AFP levels at 15–20 weeks and the risk of stillbirth at term. Elevated levels of AFP were also associated with a two- to threefold risk of spontaneous preterm birth [2].

Human chorionic gonadotrophin (hCG)

Human chorionic gonadotrophin is a heterodimeric glycoprotein produced by the syncytiotrophoblast which, in the first trimester of pregnancy, maintains the corpus luteum. A number of forms of hCG have been studied in relation to predicting pregnancy outcome, specifically, levels of the free β subunit of hCG, total hCG and hyperglycosylated hCG. Down's syndrome screening employs the free β subunit of hCG in the first trimester and total hCG from 15–20 weeks gestational age. First-trimester levels of the free β subunit of hCG were not associated with the risk of delivering an SGA infant, preterm birth, pre-eclampsia or stillbirth [32]. A large-scale study demonstrated weak associations between total hCG levels in the second trimester and the risk of pregnancy complications [38]. However, further analysis demonstrated that, like uterine artery Doppler, elevated mid-gestational levels of hCG were strongly associated with complications at extreme preterm gestational ages, but were only weakly associated with risks in later pregnancy. Women with hCG levels in the top 5% were at increased risk of spontaneous preterm birth between 24–28 weeks' gestational age, but not at later gestations [2]. Similarly, elevated hCG was associated with an increased risk of antepartum stillbirth [37]. The association was strong for stillbirth related to pre-eclampsia (7–8-fold) and the risk of stillbirth was elevated at extreme preterm gestational ages, but not at term. Finally, there is some evidence that levels of hyperglycosylated hCG may be a better predictor of complications than total hCG [39].

Angiogenesis-associated proteins and pre-eclampsia

A number of studies over the last 5 years have provided strong evidence that maternal serum levels of proteins which are known to control angiogenesis differ among women who go on to develop pre-eclampsia [40–42]. The key proteins identified thus far are soluble fms-like tyrosine kinase 1 (sFlt-1), placental growth factor (PlGF) and soluble endoglin (sEng). sFlt-1 is a soluble form of the membrane-bound vascular endothelial growth factor receptor 1 (VEGFR-1), which is synthesized and released by endothelial cells and the placenta. It is produced by alternative splicing of VEGFR-1 and binds VEGF-A and PlGF, preventing them exerting their biological effects through their membrane-bound receptors. PlGF is a type of vascular endothelial growth factor expressed in both the placenta and normal endothelial cells which binds specifically to VEGFR-1. Endoglin is an 180 kDa transmembrane protein, which is part of a family of receptors for transforming growth factor β (TGF-β). It is expressed on vascular endothelium and syncytiotrophoblast and is known to play a key role in angiogenesis and in regulation of vascular tone through its interaction with endothelial nitric oxide synthase (eNOS) [43]. Soluble endoglin (sEng) circulates and prevents the binding of TGF-β to membrane-bound endoglin and the subsequent activation of eNOS.

Levine and colleagues demonstrated that, among women who subsequently developed pre-eclampsia, PlGF levels were lower at 13–16 weeks gestational age, sEng levels were elevated about 8–12 weeks prior to diagnosis, and sFlt-1 levels were elevated approximately 5 weeks prior to diagnosis [41,44]. Subsequently, several studies have looked at the ability of changes in the concentrations of angiogenic factors to predict the risk of pre-eclampsia. Women whose levels of sEng increased between the first and second trimesters had an odds ratio for preterm pre-eclampsia of 14.9 (compared to women whose levels of sEng decreased). This was a stronger association than seen for sFlt-1 or PlGF: an increase in sFlt-1 alone had an odds ratio of 3.9 and an increase in PlGF concentration less than the median had an odds ratio of 4.3. Women with the lowest increase in PlGF concentration from the first to the second trimester had an odds ratio of 13.8. Women in the highest quartile for an increase in sFlt-1 concentration from first to second trimester had an odds ratio of 9.2 compared to controls. The combination of low PlGF and high sFlt-1 gave an odds ratio of 35.3 compared to high PlGF and low sFlt-1 [45]. A key aim in much of the research into pre-eclampsia is to identify the sequence of events leading to the condition. A nested case–control study conducted within a large-scale prospective cohort study demonstrated that at 10–14 weeks low PlGF was associated with subsequent development of the disease, whereas there was no association with levels of sFlt-1 [46]. The results were similar when confined to samples obtained between 10 and 12 weeks and when confined to women with severe pre-eclampsia. This suggests that PlGF is low prior to development of the condition owing to decreased production from the placenta, rather than as a consequence of increased binding by sFlt-1, at least in early pregnancy.

Maternal serum levels of angiogenic factors were also associated with other complications of pregnancy. Romero et al. [47] found that pregnancies where the baby was ultimately born SGA had higher concentrations of sEng from 10 weeks' gestation. They also had lower concentrations of PlGF throughout gestation than those with normal pregnancies. There was no significant difference in the concentration of sFlt-1 between patients who ultimately delivered an SGA infant and controls at 25 or 40 weeks. In addition to being risk factors for pre-eclampsia, abnormal uterine artery Doppler velocimetry and a low plasma PlGF concentration were found to be independent risk factors for the occurrence of SGA without pre-eclampsia. In combination they were associated with an odds ratio of 2.7 for the delivery of an SGA infant in the absence of pre-eclampsia [48]. A nested case–control study conducted within a large-scale prospective cohort study demonstrated that at 10–14 weeks, low PlGF was associated with an increased risk of delivering an SGA infant, but there was no association with the risk of preterm birth [46]. The same study demonstrated that higher levels of sFlt-1 were associated with a reduced risk of delivering an SGA infant, delivering preterm, and with the risk of stillbirth. Moreover, the apparent protective effect of high levels of sFlt-1 was greater among women with low or average levels of PlGF. These findings again highlight the importance of understanding the changes that can be observed in the earliest stages of pregnancy.

Other placentally derived proteins

A number of other studies have reported the relationship between maternal serum concentrations of other placentally derived proteins and adverse outcome. A nested case–control study of about 50 women who went on to develop pre-eclampsia with approximately 100 controls (matched on maternal characteristics) confirmed the association with low PAPP-A and PlGF, but also found associations with high inhibin A and activin A. In contrast, there were no associations with pregnancy-specific β_1-glycoprotein, placental lactogen, leptin, interleukin (IL)-8 or C-reactive protein [49]. The associations with inhibin A and activin A have subsequently been confirmed. Moreover, the predictive ability of activin A was comparable to that of PAPP-A, and that of inhibin A was somewhat stronger [50]. Another IGFBP protease has been studied (ADAM12, a protease for IGFBP3 and 5). The risk of pre-eclampsia was increased among women with low levels of ADAM12 in the first trimester, whereas high levels in the second trimester were associated with an increased risk of the condition. In the first trimester, analysis of the protein added little to the prediction obtained from PAPP-A [51]. Finally, the placentally produced galectin, placental protein 13 (PP13), has been the focus of a number of studies for the first-trimester prediction of pre-eclampsia. Low levels were very strongly predictive of early-onset pre-eclampsia but less predictive of disease at term, whether mild or severe [52].

Combined ultrasonic and biochemical assessment of placentation

A number of studies have examined the relationships between the ultrasonic and biochemical measurements discussed above. Higher values of PAPP-A have been demonstrated where the placental quotient was higher in the first trimester [53]. Moreover, first-trimester levels of PAPP-A are positively correlated with second-trimester growth of femoral length and abdominal circumference [54]. Two studies have demonstrated that adding PAPP-A to uterine artery Doppler had a minimal effect on the prediction of pre-eclampsia [55,56]. However, adding PAPP-A to uterine artery Doppler has been shown to improve prediction of an SGA infant [55], and the same was found for maternal serum PlGF [57]. One study combined PAPP-A, PlGF and uterine artery mean PI and found that, although PAPP-A was associated with pre-eclampsia and was more strongly predictive of severe disease, adding it to a model with uterine artery mean PI and PlGF had a minimal effect on the predictive ability obtained by the latter two parameters on their own [58]. Espinoza et al. found that abnormal uterine artery Doppler velocimetry and a low plasma PlGF concentration were independent risk factors for pre-eclampsia [48]. A combination of the two gave an odds ratio (OR) of 43.8 for the development of early-onset pre-eclampsia. In contrast, sFlt-1 concentration was of limited predictive value. However, in a study of 63 women with abnormal mid-gestation uterine artery Doppler, both sFlt-1 and PlGF were found to be predictive of those who would subsequently develop complications [59]. Finally, a Chinese study used z-scores to quantify deviation of both PAPP-A and the first-trimester fetal crown–rump length (an indicator of early fetal growth) from expected values. Although both z-scores for both parameters were associated with low birth weight, they did not perform well as a screening test [60]. Hence, a complex pattern of association exists between ultrasonic measures of growth and the risk of different adverse outcomes. Large-scale prospective cohort studies would help resolve the details of these relationships and would also yield screening performance indices from an unselected population.

Clinical utility of assessing placentation

The studies described above clearly indicate that clinical manifestations of serious complications of pregnancy, such as growth restriction, pre-eclampsia and stillbirth, are preceded by abnormalities in clinical biomarkers of placentation. Although based on clinical observations, these findings are relevant to understanding the biology of normal pregnancy and the pathophysiology of complicated pregnancies. Apart from the biological significance of these observations, the primary clinical purpose of tests that discriminate women destined to have complications is that they may be used to screen the whole population. The necessity for such an approach is underlined by the fact that the majority of serious complications of pregnancy occur in women with no known risk factors. The exemplar of obstetric screening is the approach to detecting fetuses affected by Down's syndrome. Previously, the offer of an invasive diagnostic test (with its associated risk of inducing a loss) was conducted on the basis of specific clinical risk factors, principally advanced maternal age. However, this approach detected a small proportion of Down's syndrome cases (low sensitivity) and the personal risk of Down's syndrome in many of these women was low (low positive predictive value). In the last 20 years, a highly effective screening programme, based on ultrasonic and biochemical associations with Down's syndrome, has led to methods of screening that detect in the region of 90% of cases of Down's syndrome for a <5% rate of invasive testing [61].

When considering screening for other complications of pregnancy using early pregnancy markers there are two main considerations. First, how effectively do the measures discriminate between women who are going to have complicated or uncomplicated pregnancies? Second, if a woman is known to be at high risk, how might this information be used to inform management? Meta-analyses of current methods of fetal monitoring do not indicate any methods of fetal assessment that reduces perinatal mortality when used as a screening tool in an unselected population [3]. Interpretation of the negative results is, however, problematic, and the meta-analysis of umbilical artery Doppler in low-risk pregnancies is a good illustration of the problems of interpretation [62]. First, the trials in this meta-analysis were designed in the absence of reliable information on how the test performed as a predictor of stillbirth in a population of low-risk women. The adequate design of an interventional trial requires knowledge of how the test performs in identifying women at increased risk. In the case of stillbirth, this includes both the discriminative power of the test and the gestational age dependence [63]. Without this information, a trial cannot be adequately designed. Second, a trial of an effective prenatal screening tool may yield a negative result because there is no intervention which is effective at preventing the outcome among women who screen positive. Only one of the RCTs included in the Cochrane meta-analysis of umbilical artery Doppler had a protocol for the treatment of women who screened positive. It is impossible, therefore, to determine whether these trials yielded a negative result due to failure of the screening tool or failure of the intervention. Future studies of population-based screening for pregnancy complications related to abnormal placentation need to be preceded by high-quality, non-interventional prospective cohort studies characterizing the screening properties of novel methods of risk assessment in an unselected population. Having identified effective screening tools, candidate interventions could be evaluated in RCTs among women who screen positive.

Future research

Several large-scale prospective cohort studies are in progress which aim to improve on currently available methods for clinical detection of impaired placentation. Three of these will be compared to illustrate the strengths and weaknesses of different approaches, namely, the POP study [64], the SCOPE study [65] and the Southampton Women's Study (SWS) [66]. Two of the studies, POP and SCOPE, confined recruitment to women having their first ongoing pregnancy. There are two major reasons why this would be considered. First, primiparity is associated with increased risks of a number of adverse outcomes of pregnancy, including preterm birth, pre-eclampsia, stillbirth and intrapartum caesarean section. By focusing on primiparous women, there will be a greater proportion of the cohort who experience adverse events. Second, one of the best predictors of the outcome of pregnancy is the outcome of a woman's previous pregnancy. This information is clearly not available for primiparae, hence there is a particular need for tests to predict risk in this population.

An important feature of the POP study is that assessments were made through all three trimesters of pregnancy. The SWS study performed scans at 19 and 34 weeks and the SCOPE study at 20 and 24 weeks. The rationale for first-trimester assessment is based on the multiple associations, described above, between first-trimester measurements and the risk of complications in later pregnancy. The rationale for performing scans and phlebotomy at 28 and 36 weeks in the POP study is previous analyses of other datasets. As discussed above, elevated AFP, hCG and uterine artery Doppler measures of resistance in mid-gestation were all much more strongly associated with complications at extreme preterm gestations than complications in late gestation. Ideally, screening of the pregnant population would identify women at high risk of complications in late pregnancy as well as at extreme preterm gestations. This may be better achieved with serial assessment in the third trimester. Another key question is whether information obtained through a research scan should be used for *ad hoc* screening of research participants. In the POP study, with very few exceptions (undiagnosed congenital abnormality/placenta previa, presentation at term and profound oligohydramnios), the position was that the research scan data should not be so employed. The justification for concealing this information was recommendations from NICE in the UK. Their guideline on antenatal care, which is intended to inform all NHS activity in the UK, explicitly states that routine Doppler and growth studies should not be performed on unselected women [67]. The basis for their position is the result of meta-analyses of randomized controlled trials of these methods which failed to show any benefit [68]. Other studies have, however, taken different views. The SWS study revealed the results of a growth scan at 34 weeks. The SCOPE study reveals the following information: cervical length <15 mm at 20 weeks, absent or reversed end-diastolic velocity on umbilical artery Doppler at 24 weeks, and abdominal circumference <10th percentile at 24 weeks. The concern with revealing scan results is that this will bias care, which could in turn lead to iatrogenic outcomes, such as preterm delivery.

Conclusion

Multiple lines of evidence indicate an important role for abnormal placentation in the pathophysiology of many of the most important complications of pregnancy. However, most of the currently described methods of assessment of placentation do not provide sufficient discrimination of women who are likely to experience complications. Studies are currently in progress using multiple modalities to identify women at high risk of complications. Understanding the underlying biological mechanisms and the development of effective screening tools may allow the ultimate goal of interventions that reduce the maternal and perinatal morbidity and mortality related to abnormal placentation.

References

1. **Redman CW**, **Sargent IL**. Latest advances in understanding preeclampsia. *Science* 2005; **308**(5728): 1592–4.
2. **Smith GCS**, **Shah I**, **White IR** et al. Maternal and biochemical predictors of spontaneous preterm birth among nulliparous women: a systematic analysis in relation to the degree of prematurity. *Int J Epidemiol* 2006; **35**(5): 1169–77.
3. **Smith GCS**, **Fretts RC**. Stillbirth. *Lancet* 2007; **370**(9600): 1715–25.
4. **Raio L**, **Ghezzi F**, **Cromi A** et al. The thick heterogeneous (jellylike) placenta: a strong predictor of adverse pregnancy outcome. *Prenat Diagn* 2004; **24**(3): 182–8.
5. **Hafner E**, **Metzenbauer M**, **Hofinger D** et al. Placental growth from the first to the second trimester of pregnancy in SGA-foetuses and pre-eclamptic pregnancies compared to normal foetuses. *Placenta* 2003; **24**(4): 336–42.
6. **Hafner E**, **Metzenbauer M**, **Hofinger D** et al. Comparison between three-dimensional placental volume at 12 weeks and uterine artery impedance/notching at 22 weeks in screening for pregnancy-induced hypertension, pre-eclampsia and fetal growth restriction in a low-risk population. *Ultrasound Obstet Gynecol* 2006; **27**(6): 652–7.
7. **Hafner E**, **Metzenbauer M**, **Dillinger-Paller B** et al. Correlation of first trimester placental volume and second trimester uterine artery Doppler flow. *Placenta* 2001; **22**(8–9): 729–34.
8. **RCOG**. *Placenta Praevia and Placenta Praevia Accreta: Diagnosis and Management.* Guideline No. 27.1–12 2005.
9. **Hill LM**, **Breckle R**, **Ragozzino MW**, **Wolfgram KR**, **O'Brien PC**. Grade 3 placentation: incidence and neonatal outcome. *Obstet Gynecol* 1983; **61**(6): 728–32.
10. **Grannum PA**, **Berkowitz RL**, **Hobbins JC**. The ultrasonic changes in the maturing placenta and their relation to fetal pulmonic maturity. *Am J Obstet Gynecol* 1979; **133**(8): 915–22.
11. **Proud J**, **Grant AM**. Third trimester placental grading by ultrasonography as a test of fetal wellbeing. *Br Med J (Clin Res Ed)* 1987; **294**(6588): 1641–4.
12. **Gudmundsson S**, **Dubiel M**, Sladkevicius placental morphologic and functional imaging in high-risk pregnancies. *Semin Perinatol* 2009; **33**(4): 270–80.
13. **Viero S**, **Chaddha V**, **Alkazaleh F** et al. Prognostic value of placental ultrasound in pregnancies complicated by absent end-diastolic flow velocity in the umbilical arteries. *Placenta* 2004; **25**(8–9): 735–41.
14. **Lyall F**. The human placental bed revisited. *Placenta* 2002; **23**(8–9): 555–62.
15. **Van Den Elzen HJ**, **Cohen-Overbeek TE**, **Grobbee DE**, **Quartero RW**, **Wladimiroff JW**. Early uterine artery Doppler velocimetry and the outcome of pregnancy in women aged 35 years and older. *Ultrasound Obstet Gynecol* 1995; **5**(5): 328–33.
16. **Martin AM**, **Bindra R**, **Curcio P**, **Cicero S**, **Nicolaides KH**. Screening for pre-eclampsia and fetal growth restriction by uterine artery Doppler at 11–14 weeks of gestation. *Ultrasound Obstet Gynecol* 2001; **18**(6): 583–6.
17. **Yu CK**, **Smith GCS**, **Papageorghiou AT**, **Cacho AM**, **Nicolaides KH**. An integrated model for the prediction of preeclampsia using maternal factors and uterine artery Doppler velocimetry in unselected low-risk women. *Am J Obstet Gynecol* 2005; **193**(2): 429–36.
18. **Krebs C**, **Macara LM**, **Leiser R** et al. Intrauterine growth restriction with absent end-diastolic flow velocity in the umbilical artery is associated with maldevelopment of the placental terminal villous tree. *Am J Obstet Gynecol* 1996; **175**(6): 1534–42.
19. **Neilson JP**, **Alfirevic Z**. Doppler ultrasound for fetal assessment in high risk pregnancies. *Cochrane Database Syst Rev* 2000; **2**: CD000073.
20. **Turan S**, **Miller J**, **Baschat AA**. Integrated testing and management in fetal growth restriction. *Semin Perinatol* 2008; **32**(3): 194–200.
21. **Lawrence JB**, **Oxvig C**, **Overgaard MT** et al. The insulin-like growth factor (IGF)-dependent IGF binding protein-4 protease secreted by human fibroblasts is pregnancy-associated plasma protein-A. *Proc Natl Acad Sci USA* 1999; **96**(6): 3149–53.

22. **Clemmons DR**. Role of insulin-like growth factor binding proteins in controlling IGF actions. *Mol Cell Endocrinol* 1998; **140**(1–2): 19–24.

23. **Bonno M, Oxvig C, Kephart GM** et al. Localization of pregnancy-associated plasma protein-A and colocalization of pregnancy-associated plasma protein-A messenger ribonucleic acid and eosinophil granule major basic protein messenger ribonucleic acid in placenta. *Lab Invest* 1994; **71**(4): 560–6.

24. **van Kleffens M, Groffen C, Lindenbergh-Kortleve DJ** et al. The IGF system during fetal-placental development of the mouse. *Mol Cell Endocrinol* 1998; **140**(1–2): 129–35.

25. **Kniss DA, Shubert PJ, Zimmerman PD, Landon MB, Gabbe SG**. Insulinlike growth factors. Their regulation of glucose and amino acid transport in placental trophoblasts isolated from first-trimester chorionic villi. *J Reprod Med* 1994; **39**(4): 249–56.

26. **Irwin JC, Suen LF, Martina NA, Mark SP, Giudice LC**. Role of the IGF system in trophoblast invasion and pre-eclampsia. *Hum Reprod* 1999; 14 Suppl 2: 90–6.

27. **Conover CA, Bale LK, Overgaard MT** et al. Metalloproteinase pregnancy-associated plasma protein A is a critical growth regulatory factor during fetal development. *Development* 2004; **131**(5): 1187–94.

28. **Morssink LP, Kornman LH, Hallahan TW** et al. Maternal serum levels of free beta-hCG and PAPP-A in the first trimester of pregnancy are not associated with subsequent fetal growth retardation or preterm delivery. *Prenat Diagn* 1998; **18**(2): 147–52.

29. **Pedersen JF, Sorensen S, Ruge S**. Human placental lactogen and pregnancy-associated plasma protein A in first trimester and subsequent fetal growth. *Acta Obstet Gynecol Scand* 1995; **74**(7): 505–8.

30. **Johnson MR, Riddle AF, Grudzinskas JG** et al. Reduced circulating placental protein concentrations during the first trimester are associated with preterm labour and low birth weight. *Hum Reprod* 1993; **8**: 1942–7.

31. **Ong CY, Liao AW, Spencer K, Munim S, Nicolaides KH**. First trimester maternal serum free beta human chorionic gonadotrophin and pregnancy associated plasma protein A as predictors of pregnancy complications. *Br J Obstet Gynaecol* 2000; **107**(10): 1265–70.

32. **Smith GCS, Stenhouse EJ, Crossley JA** et al. Early pregnancy levels of pregnancy-associated plasma protein A and the risk of intra-uterine growth restriction, premature birth, pre-eclampsia and stillbirth. *J Clin Endocrinol Metab* 2002; **87**: 1762–7.

33. **Dugoff L, Hobbins JC, Malone FD** et al. First-trimester maternal serum PAPP-A and free-beta subunit human chorionic gonadotropin concentrations and nuchal translucency are associated with obstetric complications: a population-based screening study (the FASTER Trial). *Am J Obstet Gynecol* 2004; **191**(4): 1446–51.

34. **Smith GCS, Stenhouse EJ, Crossley JA** et al. Development Early-pregnancy origins of low birth weight. *Nature* 2002; **417**: 916.

35. **Smith GCS, Crossley JA, Aitken DA** et al. First-trimester placentation and the risk of antepartum stillbirth. *JAMA* 2004; **292**(18): 2249–54.

36. **Waller DK, Lustig LS, Smith AH, Hook EB**. Alpha-fetoprotein: a biomarker for pregnancy outcome. *Epidemiology* 1993; **4**(5): 471–6.

37. **Smith GCS, Shah I, White IR** et al. Maternal and biochemical predictors of antepartum stillbirth among nulliparous women in relation to gestational age of fetal death. *Br J Obstet Gynaecol* 2007; **114**(6): 705–14.

38. **Walton DL, Norem CT, Schoen EJ, Ray GT, Colby CJ**. Second-trimester serum chorionic gonadotropin concentrations and complications and outcome of pregnancy. *N Engl J Med* 1999; **341**(27): 2033–8.

39. **Cole LA**. Hyperglycosylated hCG. *Placenta* 2007; **28**(10): 977–86.

40. **Maynard SE, Min JY, Merchan J** et al. Excess placental soluble fms-like tyrosine kinase 1 (sFlt1) may contribute to endothelial dysfunction, hypertension, and proteinuria in preeclampsia. *J Clin Invest* 2003; **111**(5): 649–58.

41. **Levine RJ, Maynard SE, Qian C** et al. Circulating angiogenic factors and the risk of preeclampsia. *N Engl J Med* 2004; **350**(7): 672–83.

42. **Venkatesha S, Toporsian M, Lam C** et al. Soluble endoglin contributes to the pathogenesis of preeclampsia. *Nature Med* 2006; **12**(6): 642–9.

43. **Robinson CJ, Johnson DD**. Soluble endoglin as a second-trimester marker for preeclampsia. *Am J Obstet Gynecol* 2007; **197**(2): 174–5.

44. **Levine RJ, Lam C, Qian C** et al. Soluble endoglin and other circulating antiangiogenic factors in preeclampsia. *N Engl J Med* 2006; **355**(10): 992–1005.

45. **Vatten LJ, Eskild A, Nilsen TI** et al. Changes in circulating level of angiogenic factors from the first to second trimester as predictors of preeclampsia. *Am J Obstet Gynecol* 2007; **196**(3): 239–6.

46. **Smith GCS, Crossley JA, Aitken DA** et al. Circulating angiogenic factors in early pregnancy and the risk of preeclampsia, intrauterine growth restriction, spontaneous preterm birth, and stillbirth. *Obstet Gynecol* 2007; **109**(6): 1316–24.

47. **Romero R, Nien JK, Espinoza J** et al. A longitudinal study of angiogenic (placental growth factor) and

anti-angiogenic (soluble endoglin and soluble vascular endothelial growth factor receptor-1) factors in normal pregnancy and patients destined to develop preeclampsia and deliver a small for gestational age neonate. *J Matern Fetal Neonatal Med* 2008; **21**(1): 9–23.

48. **Espinoza J, Romero R, Nien JK** et al. Identification of patients at risk for early onset and/or severe preeclampsia with the use of uterine artery Doppler velocimetry and placental growth factor. *Am J Obstet Gynecol* 2007; **196**(4): 326–13.

49. **Zwahlen M, Gerber S, Bersinger NA**. First trimester markers for pre-eclampsia: placental vs. non-placental protein serum levels. *Gynecol Obstet Invest* 2007; **63**(1): 15–21.

50. **Spencer K, Cowans NJ, Nicolaides KH**. Maternal serum inhibin-A and activin-A levels in the first trimester of pregnancies developing pre-eclampsia. *Ultrasound Obstet Gynecol* 2008; **32**(5): 622–6.

51. **Spencer K, Cowans NJ, Stamatopoulou A**. ADAM12s in maternal serum as a potential marker of pre-eclampsia. *Prenat Diagn* 2008; **28**(3): 212–16.

52. **Romero R, Kusanovic JP, Than NG** et al. First-trimester maternal serum PP13 in the risk assessment for preeclampsia. *Am J Obstet Gynecol* 2008; **199**(2): 122.

53. **Metzenbauer M, Hafner E, Hoefinger D** et al. Three-dimensional ultrasound measurement of the placental volume in early pregnancy: method and correlation with biochemical placenta parameters 7. *Placenta* 2001; **22**(6): 602–5.

54. **Leung TY, Chan LW, Leung TN** et al. First-trimester maternal serum levels of placental hormones are independent predictors of second-trimester fetal growth parameters. *Ultrasound Obstet Gynecol* 2006; **27**(2): 156–61.

55. **Pilalis A, Souka AP, Antsaklis P** et al. Screening for pre-eclampsia and fetal growth restriction by uterine artery Doppler and PAPP-A at 11–14 weeks' gestation. *Ultrasound Obstet Gynecol* 2007; **29**(2): 135–40.

56. **Spencer K, Cowans NJ, Chefetz I, Tal J, Meiri H**. First-trimester maternal serum PP-13, PAPP-A and second-trimester uterine artery Doppler pulsatility index as markers of pre-eclampsia. *Ultrasound Obstet Gynecol* 2007; **29**(2): 128–34.

57. **Poon LC, Zaragoza E, Akolekar R, Anagnostopoulos E, Nicolaides KH**. Maternal serum placental growth factor (PlGF) in small for gestational age pregnancy at 11(+0) to 13(+6) weeks of gestation. *Prenat Diagn* 2008; **28**(12): 1110–15.

58. **Akolekar R, Zaragoza E, Poon LC, Pepes S, Nicolaides KH**. Maternal serum placental growth factor at 11 + 0 to 13 + 6 weeks of gestation in the prediction of pre-eclampsia. *Ultrasound Obstet Gynecol* 2008; **32**(6): 732–9.

59. **Stepan H, Unversucht A, Wessel N, Faber R**. Predictive value of maternal angiogenic factors in second trimester pregnancies with abnormal uterine perfusion. *Hypertension* 2007; **49**(4): 818–24.

60. **Leung TY, Sahota DS, Chan LW** et al. Prediction of birth weight by fetal crown–rump length and maternal serum levels of pregnancy-associated plasma protein-A in the first trimester. *Ultrasound Obstet Gynecol* 2008; **31**(1): 10–14.

61. **Malone FD, Canick JA, Ball RH** et al. First-trimester or second-trimester screening, or both, for Down's syndrome. *N Engl J Med* 2005; **353**(19): 2001–11.

62. **Bricker L, Neilson JP**. Routine doppler ultrasound in pregnancy. *Cochrane Database Syst Rev* 2000; (**2**): CD001450.

63. **Smith GC**. Estimating risks of perinatal death. *Am J Obstet Gynecol* 2005; **192**(1): 17–22.

64. **Pasupathy D, Dacey A, Cook E** et al. Study protocol. A prospective cohort study of unselected primiparous women: the pregnancy outcome prediction study. *BMC Pregnancy Childbirth* 2008; **8**(1): 51.

65. **The SCOPE study**. http://www.scopestudy.net 2008.

66. **Southampton Woman's Survey**. http://www.mrc.soton.ac.uk/sws 2008.

67. **NICE**. *Antenatal Care: Routine Care for the Healthy Pregnant Woman 2008*. London: RCOG Press, 2008.

68. **Bricker L, Neilson JP**. Routine ultrasound in late pregnancy (after 24 weeks gestation). *Cochrane Database Syst Rev* 2006; (**2**).

Discussion

ROBERT BOYD: What stands out for me is that none of your data were analysed by gender of the baby. Have you looked at this?

SMITH: In terms of male and female differences, we have looked at the risk of stillbirth in growth-restricted females. The interesting thing here is that growth restriction seems to be more strongly associated with stillbirth in male babies compared with female babies.

FALL: What about gendering the grandparents?

SMITH: There are similar associations for either the grandmother or grandfather on the maternal side.

BARKER: Another argument in favour of what you say is that we have looked in Helsinki at the offspring from preeclamptic pregnancies. They are at increased risk of stroke, but not of coronary heart disease.

LOKE: AFP is an embryonic protein. Why was it chosen in your screening?

SMITH: The first reason for choosing it is because it is measured in Down's syndrome screening. Virtually everything that relates biochemical measurements in early pregnancy to subsequent outcome is a spin-off of Down's syndrome screening. Down's-affected pregnancies have low levels of AFP. Therefore the data are collected on a massive scale and our analyses are somewhat opportunistic. In terms of the biology, AFP is thought to be the fetal equivalent of albumin. It is a protein in the fetal circulation that is produced in relatively stable amounts, and it indicates placental permeability. High AFP in the mother represents a defect of placental permeability.

LOKE: In your slide, you say AFP and hCG; do you mean AFP *or* hCG?

SMITH: We looked at them both independently. We looked at the different patterns of the different causes of stillbirth. We put everything into the model to see whether we could get clinically useful predictions of stillbirth. Even putting the two measurements together plus all the maternal characteristics we still don't get a clinically useful predictive test.

LOKE: There is no synergism between AFP and hCG?

SMITH: They provide separate additive information. They do provide independent information, and if both are elevated it is slightly synergistic, but it is still not clinically strong enough to be useful in discriminating people who are likely to have stillbirth.

JANSSEN: Isn't one implication of your data that the relationship between birth weight and later cardiovascular disease could be genetic rather than environmental?

SMITH: Stepping back, the things that cause complicated pregnancy and the things that cause cardiovascular disease are likely to be similar. One of the things known clearly is that there tends to be familial aggregation of risk factors. Heart disease and stroke tend to run in families. Some of those things are genetic, some could be epigenetic, and some could be environmental. We can't tease these out. The animal studies show intergenerational effects: an insult to one generation persists through generations, and we can't tease out the cause.

BURTON: I am trying to tie up what you are showing with the uterine artery Doppler. I don't really understand what physical characteristics of the arteries dictate that waveform. Is it possible that in these women with some sort of endothelial dysfunction, that the flow-mediated dilatation would give you that abnormal Doppler waveform, rather than downstream effects like spiral artery conversion?

SMITH: I guess it is possible. You could test this by looking at the prepregnancy uterine artery Doppler. This is not something that we have done.

POSTON: Women with pre-eclampsia have abnormal flow-mediated dilatation during pregnancy and some time afterwards, but no one has ever looked the other way round to see whether people have it before. The assumption is it is still there many years afterwards.

SMITH: My understanding is that the uterine artery Doppler around 22 weeks' gestation looks completely different from prepregnancy. The prepregnancy Doppler is very similar to the high-resistance pattern of flow seen in women who are ultimately going to develop pre-eclampsia.

ROBERT BOYD: The change you are referring to might be in pregnancy in the feeder arteries, rather than the spiral arteries. It doesn't have to be prepregnancy.

BURTON: Absolutely. There is the whole of the uterine artery. In the distributing network, I would have thought the flexibility of the walls of those vessels will be a more important determinant of that arterial wave form, rather than just the distal 2 mm.

SIBLEY: Going back to phenotyping the placenta non-invasively, can you comment on what John Kingdom is doing in Toronto with his placenta clinic? He has placental measurements and cord insertion. You didn't mention umbilical artery Doppler: can you comment on this also?

SMITH: Clinically, umbilical artery Doppler is one assessment that if you know the result of, the pregnancy outcome is likely to be better than if you don't. This is particularly in the context of growth restriction. Uterine artery Doppler is particularly valuable in terms of predicting early onset severe pre-eclampsia. In terms of identifying growth-restricted babies that are vulnerable to stillbirth, then umbilical artery Doppler is more useful. John has taken a slightly heterogeneous group. My feeling is that you have to go to unselected populations. The problem with John's population is that you have partly a bunch of people who are at very high risk, another bunch of people who are unselected but had low PAPP-A, and these are grouped together and you then you try to identify what predicts outcome. It would be even more powerful if you had an unselected group.

POSTON: Looking at these placental markers of pre-eclampsia, as we have done, they only really start differentiating in the second trimester. Is it possible that we'll ever find a placental marker that will be a good first-trimester screen? If not, what should we be looking for? Uterine artery Doppler is not very good in the first trimester.

SMITH: Uterine artery Doppler is slightly predictive in the first trimester. There is a three- to fourfold increased risk if you top 10% above the 10th centile. PP13 is a marker I haven't mentioned, but there are some papers suggesting striking associations of PP13 and later pre-eclampsia.

POSTON: The relationship of PP13 and gestational age shows a biphasic type of relationship, which does not facilitate interpretation.

SMITH: One of the things we have done with all our studies is convert everything to multiples and median. Things change with gestation age and then you get biases.

POSTON: Should we be looking for placental markers in the first trimester?

SMITH: The PP13 data suggest we should. We need to do the whole thing in a better way where markers are much better characterized and assays are standardized. The other potential is combinations of test. The combination of AFP and PAPP-A together is quite striking, but each on its own wasn't a great predictor. We might end up with ultrasonic tests combined with biochemical tests.

POSTON: The field is certainly moving towards multiple biomarkers, especially if they diverge with gestation.

SMITH: In our study we have done a combination of ultrasound and biochemistry.

ROBERT BOYD: Once you have a predictive test the thing to do is to tell the results to half the women and then you have your controlled trial for the effect of steroids in early gestation.

SMITH: There's a serious point there. In the past when people came up with a screening test they went straight to a randomized controlled trial, where they either do the test or don't. What you need to do is to identify what your screening test tells you and then do RCTs of an intervention. The trials of routine screening with umbilical artery Doppler all lack a standardized intervention.

MOFFETT: Lucilla's point is good, in that these makers are mainly for things that are stress responses to failure of invasion. We need to look for direct markers that reflect trophoblast invasion directly. At the moment these things are all being picked up as a sort of secondary effect.

POSTON: My current state of thinking is that if we look for a really good profile of cardiovascular risk biomarkers it may be possible to identify women at risk of pre-eclampsia before they become pregnant.

ROBERT BOYD: Graham Burton, it seems to me that the appropriate death of trophoblastic tissue and its resorption in the non-placental areas is quite an important thing. Are there any proteins coming out from that that you could look for in maternal circulation?

BURTON: I did think of this to try to quantify villous regression. Proteins are difficult, because I don't think they are specific enough. What I did think of was fetal DNA. Interestingly, when Dennis Low first described free fetal DNA in the maternal serum, in many of the women there was a little blip at 10–12 weeks of gestation in the amount of free fetal DNA in the maternal blood, which then comes down again. It probably doesn't meet the requirements for a screening test as such, but it may be a way of quantifying the amount of villous regression and damage at the 10–12-week period.

ROBERT BOYD: It might be degraded or apoptotic DNA.

BURTON: I think that approach might be more profitable than looking for a protein.

Chapter 16

The placental roots of cardiovascular disease

Kent Thornburg, Perrie F. O'Tierney, Terry Morgan and Samantha Louey

Introduction

Over the past 20 years the scientific literature has reported important relationships between placental size and augmented risk of cardiovascular disease in the offspring. The foundational paper that led to this idea was published by Barker and colleagues in 1989 [1]. They showed an inverse relationship between the mortality of men and women who lived in Hertfordshire, UK, and their weight at birth (Figure 16.1). This report was a landmark discovery; it showed clearly what many developmental biologists had previously noted – that the plastic nature of development underlies the lifelong health of the offspring. But this finding by Barker *et al.* was more specific [1]. Birth weight was discovered to be a clear, statistically reliable predictor of death from one chronic disease condition.

One exciting part of the birth weight story is the detective work from which it arose. Barker's team sought the culprit responsible for high death rates from ischaemic heart disease in the north of England. For them, the geographical map was as powerful as electrophoretic gel is for the molecular biologist. The team noticed that regions in England and Wales with high mortality rates from ischaemic heart disease overlapped perfectly regions with high rates of neonatal death (reviewed by Barker [2]).

In observing the regional correspondence of adult cardiovascular mortality with infant mortality, the question arose: why did the babies who were strong enough to survive infancy go on to die of cardiovascular disease as adults? The most compelling idea was that the same intrauterine conditions that predisposed to neonatal death also predisposed to cardiovascular death in later life. Thus Figure 16.1 clearly shows that the more slowly a fetus grew before birth, the more likely it was to suffer from ischaemic heart disease some 60 years later. The figure also shows a clear

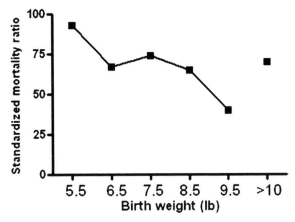

Figure 16.1 Relationship between birth weight and mortality from ischaemic heart disease, expressed as a standardised mortality ratio in 15 000 men and women in Hertfordshire, UK. Data from [1].

graded decrease in mortality for cardiovascular disease over the range of birth weights between 5 and 9 lb (2.3–4.1 kg). Interestingly, for babies above 9.5 lb, mortality rates rose again. Thus, even this 20-year-old study, which was completed before there was much notice of the increased prevalence of obesity in western populations, showed that babies born at the extreme high end of the birth weight scale are also at increased risk for death from cardiovascular disease in adult life.

These studies by Barker and colleagues stimulated many groups to investigate the mechanisms underlying intrauterine 'programming' of adult disease. Programming is the change in gene expression patterns that alters the structure and function of developing organs. Most organs are 'plastic' during development and have a narrow array of biological options from which they can respond to a biological insult. Of the many types of perinatal stress that have been identified, glucocorticoid excess, hypoxia and over- or undernutrition are the most common. All three appear to underlie adult-onset cardiovascular disease.

The Placenta and Human Developmental Programming, ed. Graham J. Burton, David J. P. Barker, Ashley Moffett and Kent Thornburg. Published by Cambridge University Press. © Cambridge University Press 2011.

Chapter 16. Placental roots of cardiovascular disease

The role that the placenta plays in the programming process is now coming to light.

The placenta connection

All nutrients are transferred to the fetus through the placenta. Thus, the organ itself must be complicit if not causative in the nutritional inadequacies that suppress fetal growth or in excesses that augment rates of fetal growth. It is now reasonable to surmise that the same suboptimal placental growth and function that underlies poor fetal growth also leads to a high risk for adult-onset disease. For example, maternal hypertension and pre-eclampsia, which independently predispose the offspring to cardiovascular disease, can now be added to the list of conditions that are strongly associated with a specific placental size and shape. Some of the recent evidence for the involvement of the placenta in the development of hypertension and cardiovascular disease are covered in Chapter 2.

Current evidence shows that several cardiovascular diseases are associated with mothers and placentas with specific phenotypes. It has been known for over a decade that placental enlargement in relation to the growth of the fetus is related to the risk for heart disease in later life. Martyn *et al.* in Barker's group [3] studied 13 000 men from birth cohorts in the UK. Cardiovascular disease-related deaths were higher among men born with low birth weight. However, these investigators found that death rates due to ischaemic heart disease were highest when the placental weight, expressed as a percentage of fetal body weight, was at extremes of the 'U'-shaped relationship. When the placenta weighed between 18% and 19% of fetal weight the rate of death from coronary disease was lowest, but the disease rate rose steeply with increasing or decreasing placental ratios. The highest death rates for both diseases were in men whose placental ratios exceeded 22% (reviewed by Godfrey [4]). In other words, the largest percentage of men died of heart disease if their placentas were large compared to their body weights.

Risnes *et al.* studied 31 000 people born in Norway between 1934 and 1959 [5]. Of 382 men and women who died from cardiovascular disease (median age, 51.3 years), the placental ratios were positively associated with cardiovascular disease mortality. Like the Martyn study above, this Norwegian study showed that a large placenta relative to birth weight was associated with an increased risk of death from cardiovascular causes. Both studies suggest that whereas small placentas were inadequate for growth, large ones were often inefficient at meeting the needs of the fetus for normal growth and development as well. One must conclude that the environmental conditions that led to placental inadequacy were associated with biological changes in the fetus that manifest themselves as cardiovascular disease in later life.

Increasingly, investigators are finding that several anatomic features of the delivered placenta have biological importance [4,6–8]. There is a positive correlation between placental weight, maternal BMI and fetal birth weights [9–12]. Placental weight predicts a person's risk for adult hypertension [13], coronary heart disease and stroke [3], and this relationship depends on maternal factors that reflect her childhood growth, including height and hip size [3].

Placental shape in pre-eclampsia

A birth cohort of over 13 000 men and women born in maternity hospitals between 1934 and 1944 in Helsinki, Finland, has made it possible to relate adult-onset disease with anthropometric measurements taken at birth [14]. Careful records of mother and baby were made at birth. The weight of the placenta and the lengths of its major axis (length) and perpendicular minor axis (width) were measured. From the length and width, the surface area of the placenta was estimated as the length × width × π/4. The mean thickness of the placenta – weight divided by area – was also determined.

The Finnish team sought to determine whether pre-eclampsia was associated with a particular size and shape of the placenta. Maternal characteristics, including blood pressures and urinary proteins, were measured after 20 weeks of pregnancy. From these data four groups of mothers were defined: group 1, severe pre-eclampsia with proteinuria and a systolic blood pressure of 160 mmHg or more, or a diastolic pressure of >110 mmHg; group 2, mild pre-eclampsia with proteinuria and a systolic blood pressure of 140–160 mmHg, or a diastolic pressure of 90–110 mmHg; group 3, hypertension without proteinuria and a systolic pressure of >140 mmHg, or a diastolic pressure of >90 mmHg; and group 4, normotensive.

From these groups, the relationship between the severity of pre-eclampsia in the mother and the shape of the placenta that she delivered was determined [14]. Although many studies have examined the role of

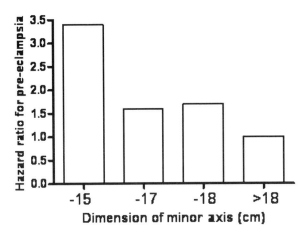

Figure 16.2 Hazard ratios for mild and severe pre-eclampsia according to the dimensions of the minor placental axis. p for trend <0.0001. Modified from [14].

placental features as they relate to pre-eclampsia [15–17], no previous studies had examined the association of placental axial dimensions with maternal pre-eclampsia. In the Helsinki study, some 5% of the mothers had pre-eclampsia. Hypertension without proteinuria was associated with a small placenta, but not with alterations in thickness or in the aspect ratio of the two axes (length/width). However, placentas from women who had pre-eclampsia were different: their width, area and aspect ratios were reduced. The effect was most profound in women with severe pre-eclampsia, where the condition was associated with reduced placental length and width and an increased thickness [14].

In women with severe pre-eclampsia, hazard ratios increased with reductions in both major and minor axes. Hazard ratios for pre-eclampsia also increased with increasing placental thickness. Figure 16.2 shows the trend in hazard ratios for pre-eclampsia with the minor axis dimension. The interesting finding was that in a simultaneous regression there was a strong trend of increasing risk as the length of the minor axis became smaller ($P < 0.0001$). There was no such trend with the maximal diameter ($P = 0.9$). This finding was surprising. How could the absolute dimension of the minor axis of the placenta be associated with pre-eclampsia in the mother without the longer dimension of the placenta also being important? These findings suggest that the major and minor axial dimensions of the placenta are driven by different forces and have unique biological roles to play. We speculate that the placenta grows along a biological axis that is established in conjunction with the rostrocaudal axis of the embryo soon after implantation. At present, it is not clear whether disproportionate deviations from the normal placental aspect ratio and thickness are a causative contributor to later disease risk, or whether it is merely a reflection of adaptations made by the placenta to minimize the negative effects of a stressor that affects fetal development. Whereas these studies did not test the hypothesis that the offspring were destined for cardiovascular disease, another study showed that the offspring of mothers with pre-eclampsia are at considerable for lethal thrombotic stroke [18].

Human placental insufficiency

The processes that regulate the ultimate shape of the placenta are not well understood (see Chapters 1 and 13). Because placental shape is an important predictor of chronic disease in offspring, we sought to determine whether it was related to poor neonatal outcomes in pregnancies at Oregon Health and Science University (OHSU, Portland, Oregon, USA). Dr Terry Morgan and his pathology team performed a retrospective study (2005–2006) of 247 consecutive singleton placentas referred for examination to determine potential relationships of placental characteristics (including placental dimensions and pathology) to an abnormal condition of the newborn. Elective terminations and spontaneous abortions before 22 weeks' gestation were excluded. Clinical records were reviewed and pregnancy outcomes recorded, including neonatal sex and age, neonatal weight (intrauterine growth restriction (IUGR) and small for gestational age (SGA) calculated by routine methods [19,20], trimmed placental weight, placental dimensions (length, width, thickness), aspect ratio (length/width), numbers of gross placental infarctions and additional clinical features, including race, gravida, hospital length of stay, and APGAR scores. Histologic sections were scored by placental pathologists who were blinded to the clinical diagnosis; they also quantified features of uteroplacental insufficiency, including advanced villous maturation (AVM), chorangiosis and microscopic infarctions. The presence of AVM is a sensitive feature of relative uteroplacental insufficiency (UPI). Figure 16.3 shows the histological features of placental insufficiency. When AVM is present, female placentas are more oval than male placentas. The group found no association of shape with intra-amniotic infection, meconium or other features of UPI.

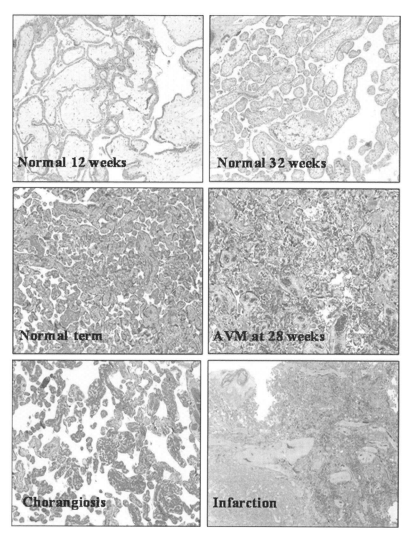

Figure 16.3 Histologic features of uteroplacental insufficiency. As the placenta ages it arborizes like a tree with increasing architectural complexity (i.e. tertiary villi are like the leaves). Accelerated villous maturation (AVM) is premature hypermaturation of the chorionic villi with syncytial knotting, and perivillous fibrosis. Chorangiosis is excessive angiogenesis in the villi. Infarction is the consequence of ischaemia and failed maternofetal compensation. (See colour plate section.)

In this cohort, IUGR was slightly more common in females (30/129, 23%) than males (21/118, 18%). The sex of offspring was not independently associated with conventional features of uteroplacental insufficiency, including placenta weight, infarctions, chorangiosis, AVM or placental thickness. However, the shape of the placenta was significantly less round for females than for males. Significant differences in placental shape were related to advanced villous maturation, which is the most sensitive measure of placental insufficiency. Placental shape did not vary by gestational age at birth, evidence of infection, maternal age, parity or APGAR scores. Placental shape did not specifically associate with maternal pregnancy outcomes (Figure 16.4), but the degree of elongation was more pronounced in female fetuses (Figure 16.4), especially if the mother had complications associated with placental insufficiency (Figure 16.5).

Surprisingly, the placenta was more symmetrical in the presence of advanced villous maturation than in age-matched placentas without AVM. The meaning of this finding is not clear. It could mean that the process that leads to placental symmetry is also associated with the compensation for uteroplacental insufficiency. Alternatively, discordant growth of the major and minor placental axes may occur in stressed placentas that are unable to compensate for the insufficiency by advancing the maturation of the placenta. These data also suggest that in this stressed population, males are more likely to grow a symmetrical placenta than are females.

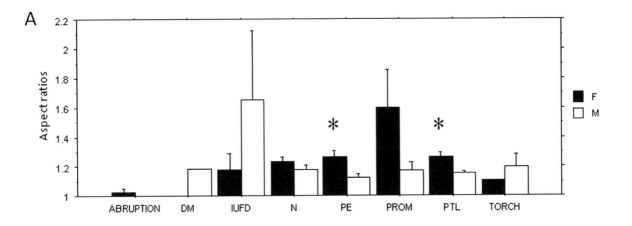

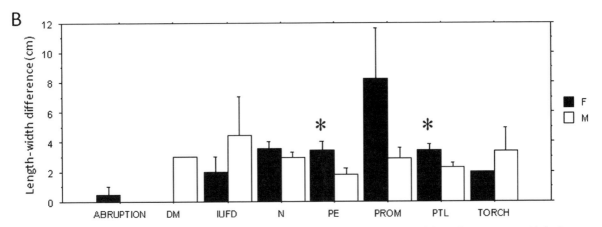

Figure 16.4 Differences in placental shape were not related to maternal pregnancy outcome, but did reveal an association with fetal sex. On average, the placenta was less round in female progeny, especially if the mother had complications associated with relative placental insufficiency (e.g. abruption, pre-eclampsia, preterm labour). Bars are mean ± SEM. Aspect Ratio = length/width; DM = diabetes mellitus; IUFD = intrauterine fetal demise; N = normal; PE = pre-eclampsia; PROM = premature rupture of membranes; PTL = preterm labour; TORCH = maternally derived infections that cross the placenta (toxoplasmosis, other infections, rubella, cytomegalovirus, herpes simplex virus).

These data are important, not because they bring predictive value for offspring, but because they show that there may be a biological pattern that underlies the pathological characteristics of the insufficient placenta and that its shape is sex dependent. Ironically, the ability of the male to maintain symmetrical placental growth during placental insufficiency and grow a large placenta does not necessarily indicate a good outcome for the baby [21]. Fetal growth restriction is more common in boys than girls, and male neonates are more likely to die from relative uteroplacental insufficiency [22,23]. During times of famine, more boys than girls succumb before birth [24], and maternal complications are more common if the baby is a boy [22,23].

Thus, boys are known to live more dangerously in the womb: they appear to prioritize linear growth over building placental mass [21]. Thus, we have come to expect that placentas from boys will be different from those of girls, and that these differences will play out as predictors of cardiovascular disease.

The Helsinki cohort data illustrate that placental size and shape at delivery are predictive of adult chronic illness, including cardiovascular disease [14,25]. It has been known for some time that placental growth is sensitive to the mother's nutritional attributes, such as diet and body composition. Not surprisingly, poor maternal nutrition is also predictive of chronic illness in the offspring. Rather than behaving

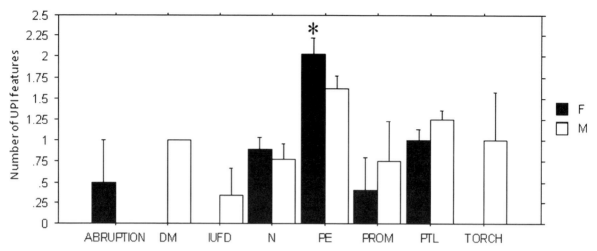

Figure 16.5 Features of uteroplacental insufficiency (UPI) are common in pre-eclampsia (PE). Compared to 'normal' placentas (N) that were submitted to pathology for meconium, maternal diabetes and infection (PROM, TORCH), placentas from mothers who had pre-eclampsia had more features of relative insufficiency in PE (P < 0.05). Bars are mean ± SEM. DM = diabetes mellitus; IUFD = intrauterine fetal demise; PROM = premature rupture of membranes; PTL = preterm labour; TORCH = maternally derived infections that cross the placenta (toxoplasmosis, other infections, rubella, cytomegalovirus, herpes simplex virus).

simply as a conduit for nutrients between mother and baby, the placenta is an active monitor and modulator of the transport and metabolism of the nutrients that the baby receives.

Pre-conceptional maternal nutrition affects future cardiovascular disease risk in the offspring

Fleming *et al.* demonstrated that the growth patterns of the rodent embryo are sensitive to its 'nutritional milieu' even before implantation [26]. This is a phase of development when most women are not aware that they are pregnant and are unlikely to be overly concerned about diet, or even alcohol use. In fact, alterations in maternal diet restricted to the preimplantation period have profound effects on the future cardiovascular health of the embryo (as reviewed by Watkins and Fleming [27]). Children born to undernourished or poorly nourished (high protein and fat intake) mothers have a significantly higher incidence of coronary heart disease, hypertension and insulin resistance in adulthood [28–30]. Maternal body mass index (BMI), one marker of nutritional history before pregnancy [31], has a significant impact on a child's risk of developing the metabolic syndrome and dying from coronary heart disease (Figure 16.6) [32–35]. A child born to an obese mother (>30 kg/m^2) has an increased probability of becoming obese later in life

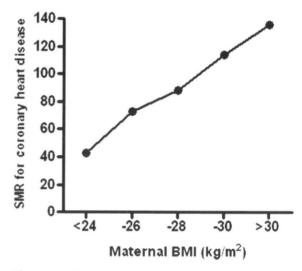

Figure 16.6 Standardized mortality ratios (SMR) for coronary heart disease in adult offspring according to maternal BMI at term. P = 0.0004 for mother's height ≤ 1.6 m. Data from [35].

[36,37]. Once a woman becomes obese she will perpetuate the cycle, generation after generation. Thus, the consequences to the offspring of a poor maternal diet, high maternal BMI and metabolic abnormalities may be severe.

Analyses of maternal diets in relation to fetal outcomes in the Southampton Women's Survey indicate that birth weight is inversely related to energy intake in early pregnancy [38,39], although there are exceptions.

Interestingly, women of low prepregnancy weight have smaller babies than women of average weight, despite similar weight gains during pregnancy [31]. This suggests that prepregnancy nutritional status may be more important to placental and fetal growth than maternal diet and weight gain during pregnancy.

The placenta mediates and modifies effect of maternal nutrition on the future health of offspring

A mother's diet and body composition both contribute to the 'nutrient pool' available to the placenta, which in turn controls nutrient transport and supply to the fetus [40]. Sufficient macro- and micronutrient provisions during gestation are critical for proper fetal growth and development. Accordingly, neonates exposed to deficiencies or excesses of nutrients *in utero* due to maternal diet, disease or placental dysfunction are at a greater risk of developing chronic disease later in life.

In the last decade a significant increase in maternal BMI and fetal and placental weights has been observed in normal, uncomplicated full-term singleton pregnancies [11]. In the case of obesity, however, bigger placentas do not mean more efficient function. Obese women have an elevated plasminogen activator inhibitor (PAI)-1: PAI-2 ratio which is indicative of compromised placental function [41]. Thus, the placenta is sensitive to lifelong maternal nutrition, not only the diet during pregnancy. Indices of placental growth (weight, linear dimensions) act as markers of placental adaptations or altered function associated with challenges to fetal cardiovascular development.

Maternal nutrition modifies placental pathways involved in nutrient transport, consistent with its known effect on fetal growth. Depending upon the timing of the insult, the placenta may be small and insufficient or large and inefficient. Either leads to increased cardiovascular disease risk. The stimulation of placental and fetal growth by early gestational undernutrition in sheep is associated with increases in the placental glucose transporter GLUT-1 [42]. A decrease in placental amino acid transporter activity precedes the IUGR in offspring of rats fed a low-protein diet during pregnancy [43]. Interestingly, the growth-restricted placentas of mice undernourished throughout pregnancy (fed 80% of an *ad lib* diet) had a greater nutrient transport capacity than the larger placentas of controls [44]. Despite these placental adaptations to maternal undernutrition, however, offspring were growth restricted. Placentas of obese women have increased inflammation [45] and nitrative stress [46], profiles which are associated with placental dysfunction, as seen in pre-eclampsia and gestational diabetes [47,48]. Maternal high-fat diet-induced obesity in a non-human primate model was associated with fetal hyperlipidaemia, severe fatty fetal liver and increased juvenile adiposity in the offspring [49]. McCurdy *et al.* suggested that the fatty livers were due to elevated placental lipid transfer [49]. It remains unknown how placental nutrient transport is affected by maternal obesity. A high maternal BMI is associated with neonatal omega-3 fatty acid deficiency despite normal maternal omega-3 levels [50]. This condition is associated with cardiovascular disease. The non-human primate findings are consistent with animal data demonstrating that placental transport is sensitive to maternal diet and body type and modulates fetal nutrient delivery, rather than merely acting as a passive conduit between the maternal and fetal nutrient pools.

Potential mechanisms for placental sensitivity to maternal nutrition

The mechanisms by which the placenta senses maternal nutrition are not understood. Roos *et al.* described a potential role for the mammalian target of rapamycin (mTOR) as a 'nutritional sensor' in placental trophoblast cells [51]. Placental mTOR is an attractive candidate, considering its capacity to activate amino acid transporters, stimulate cell growth and differentiation, respond to amino acid levels and its association with disturbances to fetal growth [51,52]. The maternal environment and placental uptake and metabolism of nutrients probably also affects epigenetic mechanisms within the placenta [53]. The proinflammatory cytokines tumour necrosis factor (TNF)-α, interleukin (IL)-1β and IL-6 alter the activity and expression of nutrient transporters in several tissues, including the placenta [54,55]. The overall effect of a proinflammatory environment – as occurs in obese [41] and diabetic women [56] – on placental nutrient uptake and metabolism remains controversial; however, it may serve as another mechanism by which the placenta is sensitive and responsive to the maternal environment.

Programming targets for cardiovascular disease

The development of a muscular pump that is sufficiently constructed to last a century requires a number of interlocking processes that occur in perfect temporal sequence. The anatomical structures of the heart must be precisely manufactured and placed. The chambers of the heart must be able to hold a volume of blood that will properly stretch the contractile sarcomeres for optimal performance and provide a stroke volume appropriate to the needs of the growing body. The muscle itself must be composed of cardiomyocytes containing their full complement of contractile proteins, mitochondria, reticular calcium storage sites and many other features and organelles. The myocardium must also be built on an extracellular matrix/scaffold that is properly compliant for filling and emptying. The working muscle must contain a network of vascular elements that deliver oxygen and fresh plasma at a rate sufficient to provide a chemical working environment within which the contractile machinery will deliver optimal performance.

The systemic arterial tree must be anatomically designed to provide a relatively low-resistance flow to all body organs while simultaneously offering a systolic pulsatile resistance to flow that stimulates the myocardium to generate an optimal force with each beat. The continuous cyclic work load of the heart must be of sufficient magnitude to stimulate the maintenance of cardiomyocyte size, contractile protein turnover and matrix integrity. The construction of the venous system is equally important. The overall resistance to flow in the venous circulation, along with its total container volume, must be constructed to regulate the mechanical pressure force in the growing heart during the diastolic portion of the cardiac cycle.

The endothelium is a specialized tissue that lines the vascular tree and must be properly primed to interact with the underlying vascular smooth muscle to regulate tone throughout the circulation. The microcirculation must be properly constructed with an adequate number of elements to provide for a range of resistances within each organ, to regulate flow that ranges from low at rest to enormous under stressful conditions. For example, blood flow through the fetal heart can increase from about $250\,\text{mL}/(\text{min} \times 100\,\text{g})$ at rest to levels that exceed $1.5\,\text{L}/(\text{min} \times 100\,\text{g})$ under extremely hypoxic conditions [57]. The microcirculation must also be properly innervated.

Diseases of the cardiovascular system are the leading cause of death worldwide. More than 17 million people die annually of cardiovascular disease, and no other disease is a close contender on a global scale [58]. However, the variety of events that lead to cardiovascular death can be reduced to a handful. The primary causes include occlusion of blood vessels serving the myocardium (myocardial infarction or heart attack), loss of ability to fill or provide contractile force by the myocardium (heart failure), and death due to electrical abnormalities that lead to fibrillation and/or erratic pump function (sudden death). Vascular deaths are most often the result of thrombotic or haemorrhagic stroke. Tracking the mechanisms by which early life conditions lead to these delayed lethal outcomes provides a noble opportunity for investigators. It is our expectation that these three lethal processes will have links with placental development that will bring new discoveries regarding the aetiology of this unnecessary disease.

The cardiovascular system is vulnerable to disruption by stressful events throughout development from conception onwards. However, there are two periods when stressors can have an inordinate effect. Over the first 6 weeks of embryonic life the heart is particularly vulnerable to stressors, chemical or mechanical, that alter the formation of its structural relationships. In the last weeks of pregnancy the heart is also vulnerable because it is during that plastic period when the final numbers of cardiomyocytes appear to be set, and when the final architecture of the coronary tree is determined at the microvascular level. These two periods are critical windows of vulnerability and represent times when placental inadequacies will have their most devastating lifelong effects on the heart.

Haemodynamic signals

Haemodynamic signals are important in guiding the gene expression patterns of the developing heart, even during the early embryonic period when the heart is a single tube and chambers have not yet formed [59]. Poelmann and colleagues showed that alterations in flow patterns through the heart of the chicken embryo lead to abnormal anatomic arrangements that are permanent [60]. In these embryos a disruption of the flow patterns in the vitelline vein, which returns blood from the yolk sac to the heart, led to cardiac defects, often in the form of incomplete chamber walls. These experiments show that even in the embryo, blood flow patterns in the heart are determined at least in part by the architecture of the vascular bed in the early yolk sac placenta. Whereas the arterial pressures generated by the 1 mm heart of the embryo are about 100 times lower than those found in the adult, even the small early haemodynamic forces are nonetheless sensed by the nascent cardiac structures.

When embryonic cardiomyocytes are grown in a non-flow environment, such as the anterior chamber of the eye of the rat, the cardiac chambers do not form properly [61]. There is increasing evidence that flow patterns reflect altered haemodynamic forces that impart signals to heart structures that regulate chamber growth and valve formation.

The fetal heart senses haemodynamic forces

A placenta may have one or more defects that affect the growth of the heart. It may be 'insufficient' – a global term indicating that the transport of oxygen and nutrients is inadequate to support robust fetal growth. The insufficiency might reflect a low membrane surface area for substrate diffusion, inadequate numbers of transport proteins, a thickened exchange barrier, or an inadequate vascular tree on the fetal side or poor maternal vascularization. In some cases of placental insufficiency, especially when the placenta is relatively small, it offers an increase in pulsatile blood flow resistance (impedance). Increases in resistance, as indicated by reduced diastolic flow velocities in human umbilical arteries, are countered with adaptations by the developing heart. This cardiac response is similar to that seen in the adult heart in cases of systemic hypertension. With each beat the heart must eject a stroke volume against the instantaneous resistance of the placental bed. The genetic machinery of the cells in the myocardium is set in motion by mechanical sensors to accommodate changing vascular conditions. In humans, severe placental insufficiency can be severe enough that blood flows backwards from the placenta during diastole. This reversed diastolic flow in the umbilical artery is indicative of increased loading of the right and left ventricles and portends a bad outcome for the fetus [62].

In a recent study of *HOXA13* knockout mice [63], placental vascular development was found to be very important to heart growth. In the placenta *HOXA13* is important for normal vascular development. When the gene is not expressed (*HOXA13* -/-), the placental endothelium does not develop normally but the underlying layers of trophoblast appear to be normal. However, the knockout condition is lethal: most die on embryonic day 14. Most fascinating is the fact that the *HOXA13* gene is not expressed in the embryonic heart, yet the primary lethal defect is found in the heart. The thickness of the ventricular wall of the embryonic heart is reduced by some 43% [63]. These data suggest that a poorly formed placental vascular tree leads to haemodynamic changes which are likely to alter both systolic and diastolic forces in the embryonic heart. It appears, as found in the chick experiments mentioned above, that mechanical forces, which can arise from faulty vascular architecture in the placenta, may lead to cardiac defects in the early embryo, some of which are embryo lethal.

Heart cell growth

Once cardiomyocytes are firmly committed to the cardiac phenotype in the embryo, they grow continuously by cell proliferation. Although these aspiring cardiomyocytes enlarge to some small degree during embryonic and early fetal life, they remain relatively small and maintain their cigar shape. During this early phase of development, they maintain an immature phenotype, with sarcomeres neatly arranged in a series along the sarcolemma and mitochondria and glycogen particles randomly scattered around the nucleus. The cardiomyocytes have no t-tubules at this stage. The hormones and growth factors that regulate the growth of cardiomyocytes up to this point have not been thoroughly studied. In precocious species, the myocardium goes through a maturation phase that begins gradually at about two-thirds gestation but then gains momentum until birth. The most obvious external feature of this phase is the switch from mononucleated to binucleated and enlargement of the myocytes. As the cells gain a second nucleus and go through their so-called terminal differentiation phase, they change their appearance, become fatter, generate t-tubules and place contractile units throughout the cytoplasm.

It is generally accepted that binucleated cardiomyocytes cannot divide, but can readily enlarge. Therefore, over the last weeks of gestation the myocardium relinquishes its generative capacity in favour of cardiomyocyte maturation and enlargement. At the time of birth, and certainly within weeks after birth, the population of cardiomyocytes that can produce new, functional myocytes is minimal.

The placenta may be important in the determination of cardiac cell numbers. Abnormal numbers may be the result of hypoxaemia, abnormal haemodynamic forces or changes in the growth factor environment. There is evidence that the 'set point' for cardiomyocyte number is established around the time of birth. Every organ 'decides' how many cells it should have and how large the organ should be to meet body demands. How such decisions are made is not known. One of the most telling studies in the developing heart comes from Zhang's group [64], which subjected pregnant rats to a low-oxygen environment over most of the second half of gestation. As adults, the males were found to have lower numbers of myocytes for life. The cardiomyocytes were larger also. The reason why the once hypoxic hearts have a low endowment of cardiomyocytes is not clear. As an altricial species, rat cardiomyocytes do not ordinarily undergo terminal differentiation until 1–2 weeks after birth [65], so presumably there were plenty of cells that could have generated a normal cardiomyocyte population postnatally but did not. A more likely explanation is that the set point for the number of cells was influenced by the chemical and haemodynamic environment at some crucial time point in the perinatal period.

Large-animal models of placental insufficiency

Several important models of placental insufficiency have been developed in sheep. The placental exchange area can be reduced by infusing microspheres into the supply arteries of the placenta or by carunclectomy. The first method, often called umbilicoplacental embolization, uses 50 μm diameter mucopolysaccharide microspheres which are slowly infused into the umbilical artery. As the spheres randomly lodge in the placental microcirculation nutritional exchange is reduced and placental resistance to flow is temporarily increased [66,67]. The second method of inducing placental insufficiency in sheep takes advantage of the fact that placental units (placentomes) are formed only at anatomically prepared sites in the uterus called caruncles. Ordinarily a sheep uterus can contain up to a hundred of these sites. If many of these sites are removed before mating, the number of placentation sites is reduced, as is the surface area of the developing placenta. These two models have several features in common. They slow fetal growth and cause fetal hypoxaemia, hypercapnia, hypoglycaemia and mild acidaemia – conditions seen in human intrauterine growth restriction.

In experiments where the near-term fetal heart is faced with an increased systolic load, the myocardium is stimulated to grow by hyperplasia and hypertrophy; the increased load also stimulates an increased rate of maturation and binucleation of cardiomyocytes [68]. Evidently these outcomes do not apply directly to models of placental insufficiency. In these models, cell cycle activity and proliferation become depressed. Perhaps more interesting is the fact that the cardiomyocytes of these fetuses cease their maturation process, which is normally seen in near-term hearts. Thus, a fetus with placental insufficiency is characterized by a small heart and a reduced number of cardiomyocytes that remain immature during fetal life [69,70]. One big piece of the puzzle has yet to be determined. If a fetus is born with an underendowed myocardium populated with immature myocytes, will it recover completely? Zhang's rat experiments suggest that it will not.

These animal models show in detail what was surmised from human data: that the placenta is key to the development of the heart. Experimental data have also shown that the placenta is sensitive to the nutritional status of the mother. Early gestation undernutrition stimulates an increase in expression of placental IGF-binding proteins [71] and placental glucose transporter-1 [42]. Although placental insufficiency can lead to lower circulating fetal IGF levels [72], not all models behave similarly. The hyperthermia model of placental insufficiency in sheep leads to an increased expression of placental IGF-binding proteins [73]. In addition, placental expression levels of VEGF, angiopoietins (ang-1, ang-2) and the angiopoietin receptor Tie-2, are increased for a period early in gestation but are not maintained at high levels throughout gestation [74,75]. Thus, there are biological mysteries that remain unsolved with regard to the types of insult that lead to abnormal cardiovascular development.

Chemical regulators of heart development

There is no question that disruptions to healthy placental growth (through either maternal influences or placental insufficiency) can have serious effects on placental morphology, vascular growth, transport functions, and downstream effects on the developing heart. Although genetic, nutritional, haemodynamic and structural modifications can also affect fetal heart growth, the intrauterine role of chemical regulation in long-term cardiac outcome cannot be discounted. Growth of the fetal heart is stimulated by IGF-1, IGF-2, cortisol and angiotensin II [76–78]; each of their associated hormonal systems can act directly on the placenta and heart, hence abnormalities in placental hormone and growth factor regulation also affect the heart. In contrast to the growth promoters, thyroid hormone can cross the human placenta [79] and can affect the growth of the heart by inhibiting cardiomyocyte proliferation [80]. It is not known whether these chemical regulators will have the same effects on heart growth in compromised pregnancies. In our sheep model of placental insufficiency (umbilicoplacental embolization), cell cycle activity is suppressed in cardiomyocytes despite elevated circulating cortisol [69] which, in unstressed fetuses, stimulates cardiomyocyte cell cycle activity [78]. Similarly, whereas excess thyroid hormone might suppress proliferation in a 'healthy' fetus [80], IUGR fetuses are characterized by low circulating thyroid hormone levels [81] and growth hormone levels which may lead to suppressed cardiomyocyte maturation and poor brain development. Undoubtedly several other growth factors that have dual effects on the placenta and heart have not yet been characterized. Thus, the balance between the pro- and anti-proliferative chemical regulators is influenced by placental actions, but the details of how these chemicals are integrated await further research.

Conclusion

There are many avenues of evidence suggesting that the placenta is complicit in the growth of the fetus and hence the programming of disease. Such findings include studies from Finland relating placental size and shape to hypertension in offspring. In these studies, the programming effect was conditioned by maternal phenotype. Either a small or a large placenta in relation to fetal body weight predicts disease in later life. A small placenta is associated with increased pulsatile resistance that stimulates abnormal growth and septation in the embryo and premature cardiomyocyte maturation in the fetus. A large placenta is the result of excess, apparently compensatory, growth that leads to inefficient nutrient exchange and is associated with coronary heart disease mortality. The mystery uncovered in these studies relates to the shape of the placenta. The width, defined as the dimension of the minor axis of the placenta, seems to be a particularly powerful predictor of cardiovascular disease, including heart failure (see Chapter 1). The growth of the heart is affected by the regulation and transport of hormones that regulate cardiomyocyte proliferation by the placenta. Thus, the placenta is the key to the development of the body's muscle pump and its most vulnerable organ. One thing is clear: the roots of heart failure, coronary artery disease and vascular disease will never be solved without solid systematic investigation of placental processes that lead to a poorly constructed cardiovascular system.

References

1. **Barker DJ, Winter PD, Osmond C, Margetts B, Simmonds SJ**. Weight in infancy and death from ischaemic heart disease. *Lancet* 1989; **2**(8663): 577–80.
2. **Barker DJP**. *Mothers, Babies and Health in Later Life*. Edinburgh: Churchill Livingstone, 1998.
3. **Martyn CN, Barker DJ, Osmond C**. Mothers' pelvic size, fetal growth, and death from stroke and coronary heart disease in men in the UK. *Lancet* 1996; **348**(9037): 1264–8.
4. **Godfrey KM**. The role of the placenta in fetal programming – a review. *Placenta* 2002; **23** Suppl A: S20–S27.
5. **Risnes KR, Romundstad PR, Nilsen TI, Eskild A, Vatten LJ**. Placental weight relative to birth weight and long-term cardiovascular mortality: findings from a cohort of 31,307 men and women. *Am J Epidemiol* 2009; **170**: 622–31.
6. **Yampolsky M, Salafia CM, Shlakhter O et al.** Centrality of the umbilical cord insertion in a human placenta influences the placental efficiency. *Placenta* 2009; **30**: 1058–64.
7. **Misra DP, Salafia CM, Miller RK, Charles AK**. Non-linear and gender-specific relationships among placental growth measures and the fetoplacental weight ratio. *Placenta* 2009; **30**: 1052–7.
8. **Toal M, Keating S, Machin G et al.** Determinants of adverse perinatal outcome in high-risk women with abnormal uterine artery Doppler images. *Am J Obstet Gynecol* 2008; **198**: 330–7.
9. **Thame M, Osmond C, Wilks RJ et al.** Blood pressure is related to placental volume and birth weight. *Hypertension* 2000; **35**: 662–7.
10. **Taricco E, Radaelli T, Nobile de Santis MS, Cetin I**. Foetal and placental weights in relation to maternal characteristics in gestational diabetes. *Placenta* 2003; **24**: 343–7.

11. **Swanson LD, Bewtra C.** Increase in normal placental weights related to increase in maternal body mass index. *J Matern Fetal Neonatal Med* 2008; **21**: 111–13.

12. **Perry IJ, Beevers DG, Whincup PH, Bareford D.** Predictors of ratio of placental weight to fetal weight in multiethnic community. *Br Med J* 1995; **310**: 436–9.

13. **Barker DJ, Bull AR, Osmond C, Simmonds SJ.** Fetal and placental size and risk of hypertension in adult life. *Br Med J* 1990; **301**: 259–62.

14. **Kajantie E, Thornburg K, Eriksson JG, Osmond L, Barker DJ.** In preeclampsia, the placenta grows slowly along its minor axis. *Int J Dev Biol* 2009; **54**: 469–73.

15. **Costa SL, Proctor L, Dodd JM** et al. Screening for placental insufficiency in high-risk pregnancies: is earlier better? *Placenta* 2008; **29**: 1034–40.

16. **Egbor M, Ansari T, Morris N, Green CJ, Sibbons PD.** Pre-eclampsia and fetal growth restriction: how morphometrically different is the placenta? *Placenta* 2006; **27**: 727–34.

17. **Nordenvall M, Sandstedt B, Ulmsten U.** Relationship between placental shape, cord insertion, lobes and gestational outcome. *Acta Obstet Gynecol Scand* 1988; **67**: 611–16.

18. **Kajantie E, Eriksson JG, Osmond C, Thornburg K, Barker DJ.** Pre-eclampsia is associated with increased risk of stroke in the adult offspring: the Helsinki birth cohort study. *Stroke* 2009; **40**: 1176–80.

19. **Roberts DJ, Post MD.** The placenta in pre-eclampsia and intrauterine growth restriction. *J Clin Pathol* 2008; **61**: 1254–60.

20. **Naeye RL.** Do placental weights have clinical significance? *Hum Pathol* 1987; **18**: 387–91.

21. **Eriksson JG, Kajantie E, Osmond C, Thornburg K, Barker DJ.** Boys live dangerously in the womb. *Am J Hum Biol* 2009; **22**: 330–5

22. **Di Renzo GC, Rosati A, Sarti RD, Cruciani L, Cutuli AM.** Does fetal sex affect pregnancy outcome? *Gender. Med* 2007; **4**: 19–30.

23. **Ingemarsson I.** Gender aspects of preterm birth. *Br J Obstet Gynaecol* 2003; **110**(supp 20): 34–8.

24. **Ravelli AC, Der Meulen JH, Osmond C, Barker DJ, Bleker OP.** Obesity at the age of 50 y in men and women exposed to famine prenatally. *Am J Clin Nutr* 1999; **70**: 811–16.

25. **Barker DJ, Thornburg KL, Osmond C, Kajantie E, Eriksson JG.** The surface area of the placenta and hypertension in the offspring in later life. *Int J Dev Biol* 2009; **54**: 525–30.

26. **Kwong WY, Wild AE, Roberts P, Willis AC, Fleming TP.** Maternal undernutrition during the preimplantation period of rat development causes blastocyst abnormalities and programming of postnatal hypertension. *Development* 2000; **127**: 4195–202.

27. **Watkins AJ, Fleming TP.** Blastocyst environment and its influence on offspring cardiovascular health: the heart of the matter. *J Anat* 2009; **215**: 52–9.

28. **Painter RC, de Rooij SR, Bossuyt PM** et al. Early onset of coronary artery disease after prenatal exposure to the Dutch famine. *Am J Clin Nutr* 2006; **84**: 322–7.

29. **Shiell AW, Campbell DM, Hall MH, Barker DJ.** Diet in late pregnancy and glucose-insulin metabolism of the offspring 40 years later. *Br J Obstet Gynaecol* 2000; **107**: 890–5.

30. **Shiell AW, Campbell-Brown M, Haselden S** et al. High-meat, low-carbohydrate diet in pregnancy: relation to adult blood pressure in the offspring. *Hypertension* 2001; **38**: 1282–8.

31. **Neggers Y, Goldenberg RL.** Some thoughts on body mass index, micronutrient intakes and pregnancy outcome. *J Nutr* 2003; **133**: 1737S–40S.

32. **Boney CM, Verma A, Tucker R, Vohr BR.** Metabolic syndrome in childhood: association with birth weight, maternal obesity, and gestational diabetes mellitus. *Pediatrics* 2005; **115**: e290–e296.

33. **Loos RJ, Phillips DI, Fagard R** et al. The influence of maternal BMI and age in twin pregnancies on insulin resistance in the offspring. *Diabetes Care* 2002; **25**: 2191–6.

34. **Mi J, Law C, Zhang KL** et al. Effects of infant birth weight and maternal body mass index in pregnancy on components of the insulin resistance syndrome in China. *Ann Intern Med* 2000; **132**: 253–60.

35. **Forsen T, Eriksson JG, Tuomilehto J** et al. Mother's weight in pregnancy and coronary heart disease in a cohort of Finnish men: follow up study. *Br Med J* 1997; **315**: 837–40.

36. **Gale CR, Javaid MK, Robinson SM** et al. Maternal size in pregnancy and body composition in children. *J Clin Endocrinol Metab* 2007; **92**: 3904–11.

37. **Hull HR, Dinger MK, Knehans AW, Thompson DM, Fields DA.** Impact of maternal body mass index on neonate birth weight and body composition. *Am J Obstet Gynecol* 2008; **198**: 416.

38. **Godfrey K, Robinson S, Barker DJ, Osmond C, Cox V.** Maternal nutrition in early and late pregnancy in relation to placental and fetal growth. *Br Med J* 1996; **312**: 410–14.

39. **Godfrey KM, Barker DJ, Robinson S, Osmond C.** Maternal birth weight and diet in pregnancy in relation to the infant's thinness at birth. *Br J Obstet Gynaecol* 1997; **104**: 663–7.

40. **Godfrey KM**. Maternal regulation of fetal development and health in adult life. *Eur J Obstet Gynecol Reprod Biol* 1998; **78**: 141–50.

41. **Stewart FM, Freeman DJ, Ramsay JE** et al. Longitudinal assessment of maternal endothelial function and markers of inflammation and placental function throughout pregnancy in lean and obese mothers. *J Clin Endocrinol Metab* 2007; **92**: 969–75.

42. **Dandrea J, Wilson V, Gopalakrishnan G** et al. Maternal nutritional manipulation of placental growth and glucose transporter 1 (GLUT-1) abundance in sheep. *Reproduction* 2001; **122**: 793–800.

43. **Jansson N, Pettersson J, Haafiz A** et al. Downregulation of placental transport of amino acids precedes the development of intrauterine growth restriction in rats fed a low protein diet. *J Physiol* 2006; **576**: 935–46.

44. **Coan PM, Vaughan OR, Sekita Y** et al. Adaptations in placental phenotype support fetal growth during undernutrition of pregnant mice. *J Physiol* 2010; **588**: 527–38.

45. **Challier JC, Basu S, Bintein T** et al. Obesity in pregnancy stimulates macrophage accumulation and inflammation in the placenta. *Placenta* 2008; **29**: 274–81.

46. **Roberts VH, Smith J, McLea SA** et al. Effect of increasing maternal body mass index on oxidative and nitrative stress in the human placenta. *Placenta* 2009; **30**: 169–75.

47. **Radaelli T, Varastehpour A, Catalano P, Hauguel-de Mouzon S**. Gestational diabetes induces placental genes for chronic stress and inflammatory pathways. *Diabetes* 2003; **52**: 2951–8.

48. **Sitras V, Paulssen RH, Gronaas H** et al. Differential placental gene expression in severe preeclampsia. *Placenta* 2009; **30**: 424–33.

49. **McCurdy CE, Bishop JM, Williams SM** et al. Maternal high-fat diet triggers lipotoxicity in the fetal livers of nonhuman primates. *J Clin Invest* 2009; **119**: 323–35.

50. **Wijendran V, Bendel RB, Couch SC** et al. Fetal erythrocyte phospholipid polyunsaturated fatty acids are altered in pregnancy complicated with gestational diabetes mellitus. *Lipids* 2000; **35**: 927–31.

51. **Roos S, Powell TL, Jansson T**. Placental mTOR links maternal nutrient availability to fetal growth. *Biochem Soc Trans* 2009; **37**: 295–8.

52. **Roos S, Jansson N, Palmberg I** et al. Mammalian target of rapamycin in the human placenta regulates leucine transport and is downregulated in restricted fetal growth. *J. Physiol* 2007; **582**: 449–59.

53. **Gheorghe CP, Goyal R, Holweger JD, Longo LD**. Placental gene expression responses to maternal protein restriction in the mouse. *Placenta* 2009; **30**: 411–17.

54. **Jones HN, Jansson T, Powell TL**. IL-6 stimulates system A amino acid transporter activity in trophoblast cells through STAT3 and increased expression of SNAT2. *Am J Physiol Cell Physiol* 2009; **297**: C1228–35.

55. **Thongsong B, Subramanian RK, Ganapathy V, Prasad PD**. Inhibition of amino acid transport system a by interleukin-1beta in trophoblasts. *J Soc Gynecol Invest* 2005; **12**: 495–503.

56. **Ategbo JM, Grissa O, Yessoufou A** et al. Modulation of adipokines and cytokines in gestational diabetes and macrosomia. *J Clin Endocrinol Metab* 2006; **91**: 4137–43.

57. **Thornburg KL, Reller MD**. Coronary flow regulation in the fetal sheep. *Am J Physiol* 1999; **277**: R1249–60.

58. **MacKay J, Mensah GA**. *The Atlas of Heart Disease and Stroke*. Geneva: World Health Organization, 2004.

59. **Liu A, Wang R, Thornburg KL, Rugonyi S**. Efficient postacquisition synchronization of 4-D nongated cardiac images obtained from optical coherence tomography: application to 4-D reconstruction of the chick embryonic heart. *J Biomed. Opt* 2009; **14**: 044020.

60. **Verberne ME, Gittenberger-De Groot AC, Van Iperen L, Poelmann RE**. Contribution of the cervical sympathetic ganglia to the innervation of the pharyngeal arch arteries and the heart in the chick embryo. *Anat Rec* 1999; **255**: 407–19.

61. **Bishop SP, Anderson PG, Tucker DC**. Morphological development of the rat heart growing in oculo in the absence of hemodynamic work load. *Circ Res* 1990; **66**: 84–102.

62. **Kiserud T, Ebbing C, Kessler J, Rasmussen S**. Fetal cardiac output, distribution to the placenta and impact of placental compromise. *Ultrasound Obstet Gynecol* 2006; **28**: 126–36.

63. **Shaut CA, Keene DR, Sorensen LK, Li DY, Stadler HS**. HOXA13 is essential for placental vascular patterning and labyrinth endothelial specification. *PLoS Genet* 2008; **4**: e1000073.

64. **Li G, Bae S, Zhang L**. Effect of prenatal hypoxia on heat stress-mediated cardioprotection in adult rat heart. *Am J Physiol Heart Circ Physiol* 2004; **286**: H1712–19.

65. **Clubb FJ Jr, Bishop SP**. Formation of binucleated myocardial cells in the neonatal rat. An index for growth hypertrophy. *Lab Invest* 1984; **50**: 571–7.

66. **Trudinger BJ, Stevens D, Connelly A** et al. Umbilical artery flow velocity waveforms and placental

resistance: the effects of embolization of the umbilical circulation. *Am J Obstet Gynecol* 1987; **157**: 1443–8.

67. **Morrow RJ, Adamson SL, Bull SB, Ritchie JW.** Effect of placental embolization on the umbilical arterial velocity waveform in fetal sheep. *Am J Obstet Gynecol* 1989; **161**: 1055–60.

68. **Barbera A, Giraud GD, Reller MD** et al. Right ventricular systolic pressure load alters myocyte maturation in fetal sheep. *Am J Physiol Regul Integr Comp Physiol* 2000; **279**: R1157–64.

69. **Louey S, Jonker SS, Giraud GD, Thornburg KL.** Placental insufficiency decreases cell cycle activity and terminal maturation in fetal sheep cardiomyocytes. *J Physiol* 2007; **580**: 639–48.

70. **Morrison JL, Botting KJ, Dyer JL** et al. Restriction of placental function alters heart development in the sheep fetus. *Am J Physiol Regul Integr Comp Physiol* 2007; **293**: R306–13.

71. **Osgerby JC, Wathes DC, Howard D, Gadd TS.** The effect of maternal undernutrition on the placental growth trajectory and the uterine insulin-like growth factor axis in the pregnant ewe. *J Endocrinol* 2004; **182**: 89–103.

72. **Langford K, Blum W, Nicolaides K** et al. The pathophysiology of the insulin-like growth factor axis in fetal growth failure: a basis for programming by undernutrition? *Eur J Clin Invest* 1994; **24**: 851–6.

73. **de Vrijer B, Davidsen ML, Wilkening RB, Anthony RV, Regnault TR.** Altered placental and fetal expression of IGFs and IGF-binding proteins associated with intrauterine growth restriction in fetal sheep during early and mid-pregnancy. *Pediatr Res* 2006; **60**: 507–12.

74. **Hagen AS, Orbus RJ, Wilkening RB, Regnault TR, Anthony RV.** Placental expression of angiopoietin-1, angiopoietin-2 and tie-2 during placental development in an ovine model of placental insufficiency-fetal growth restriction. *Pediatr Res* 2005; **58**: 1228–32.

75. **Regnault TR, Orbus RJ, de Vrijer B** et al. Placental expression of VEGF, PlGF and their receptors in a model of placental insufficiency-intrauterine growth restriction (PI-IUGR). *Placenta* 2002; **23**: 132–44.

76. **Sundgren NC, Giraud GD, Schultz JM** et al. Extracellular signal-regulated kinase and phosphoinositol-3 kinase mediate IGF-1 induced proliferation of fetal sheep cardiomyocytes. *Am J Physiol Regul Integr Comp Physiol* 2003; **285**: R1481–9.

77. **Sundgren NC, Giraud GD, Stork PJ, Maylie JG, Thornburg KL.** Angiotensin II stimulates hyperplasia but not hypertrophy in immature ovine cardiomyocytes. *J Physiol* 2003; **548**: 881–91.

78. **Giraud GD, Louey S, Jonker S, Schultz J, Thornburg KL.** Cortisol stimulates cell cycle activity in the cardiomyocyte of the sheep fetus. *Endocrinology* 2006; **147**: 3643–9.

79. **Burrow GN, Fisher DA, Larsen PR.** Maternal and fetal thyroid function. *N Engl J Med* 1994; **33**: 1072–8.

80. **Chattergoon NN, Giraud GD, Thornburg KL.** Thyroid hormone inhibits proliferation of fetal cardiac myocytes in vitro. *J Endocrinol* 2007; **192**: R1–R8.

81. **Kilby MD, Verhaeg J, Gittoes N** et al. Circulating thyroid hormone concentrations and placental thyroid hormone receptor expression in normal human pregnancy and pregnancy complicated by intrauterine growth restriction (IUGR). *J Clin Endocrinol Metab* 1998; **83**: 2964–71.

Discussion

ROBERT BOYD: There was a controlled trial of T3 about 10 years ago in the UK. Have you accessed that to see whether there is any evolving change in cardiac function?

THORNBURG: No, I have never found anyone who has tried to connect thyroid treatment with cardiac function or health in later life. The number of heart cells in a heart can't be detected by a paediatrician with ultrasound. All babies, no matter what their thyroid levels, are being sent home as perfectly normal because their ultrasound measurements will all be normal.

FERGUSON SMITH: On the T3 story, I wanted to raise a possibility. I'm not sure about human, but in mouse the type 3 diodinase is expressed at very high levels in the placenta. Dio3 is an imprinted gene.

THORNBURG: They are actually expressed in placenta before they are in the fetal body. We clicked on the work of Abby Fowden and Alison Forhead showing that cortisol stimulates the deiodinases in the heart, and we have data on this now.

FERGUSON SMITH: The type 3 diodinase is of course the negative regulator of thyroid hormones. In the mouse fetus its expression is quite restricted.

BURTON: To pick up on your point on oxidative stress, whatever we do to stress the placenta, the two tissues that always light up are the trophoblast and the fetal endothelial cells. They are highly sensitive to oxidative stress. With chronic exposure I would expect to see vascular regression in those placentas. This would increase placental resistance.

RICHARD BOYD: I was interested in the comment you made that the arrhythmia is a blank. Two things you said suggested that it is less than a total blank. One is the T3 story in terms of the β-adrenoreceptor development, and the other is the ventricular muscle development in terms of excitation contraction coupling.

THORNBURG: That's a good hypothesis. But at this time there is not a shred of evidence that arrhythmias are programmed. However, I think this is because no one has ever looked.

MOFFETT: You looked at the heart. But what about the spiral arteries in these fetuses? I was asking Graham Burton to count spiral arteries in the Boyd collection, but I think he needs to count spiral arteries in fetal uteri and newborn uteri.

THORNBURG: I agree. No one has ever done this.

SIBLEY: We need to discuss how we collect some of these data. Most delivery units aren't even weighing placentas any more, let alone measuring them. I don't know how we reverse this.

BARKER: We shouldn't overestimate how difficult it is. It is still routine in northern Finland.

SIBLEY: In the UK, trying to change the culture is tricky. I have a plug for another model. It is the teenage mothers. We have talked a bit about their nutrients and outcomes with tendency to small size. We made an interesting observation that teenagers as a whole have lower system A activity in their placentas than adult mothers. The teenagers system A is like that of an adult mother if she was having an SGA baby. If you split the teenagers' system A activity placentas into growing and non-growing, then it is only the teenagers that are not growing that have the low system A activity. As an adjunct to these data, Christina Heywood (personal communication) who did the study also did the morphology to ask whether this is a gross placental effect. She couldn't find any other changes, and so it seems to be a transporter defect, not a morphology effect. Under what conditions do placental phenotypes represent transport capacity? In the teenagers there is some evidence that it might be a pure transport effect rather than a surface area or morphology effect.

THORNBURG: Someone made the point earlier, does it matter how many transporters there are if you don't have enough of the molecule that needs to go across.

JACKSON: Iodine deficiency and thyroid disorders are common, widely spread and geographically distinct. From what you said, you'd expect to have a fairly sharp geographical distinction in cardiac problems mapped against that.

THORNBURG: If you look at the map of coronary artery disease around the world it is the highest in the Soviet countries by far. No one has ever tried to look at the micro level: we would need to see whether or not those relationships exist.

ROBERT BOYD: What about soft water as a confounder? Does that co-correlate with the iodine map?

BARKER: No.

THORNBURG: Even though we are interested in the heart, there are other organs such as the kidney and pancreas when it comes to programming. I hope other groups will be interested in following up the relationship between the placenta and the outcomes of those organs. These are the kinds of connections that will help us go back to the biological roots of programming. This is one thing that we as biologists can offer in partnerships with the epidemiologists.

Chapter 17
Placental function and later risk of osteoporosis

Cyrus Cooper, Laura Goodfellow, Nicholas Harvey, Susie Earl, Christopher Holroyd, Zoe Cole and Elaine Dennison

Introduction

Osteoporosis is a skeletal disorder characterized by low bone mass and microarchitectural deterioration of bone tissue, with a consequent increase in bone fragility and susceptibility to fracture [1,2]. It is a widespread condition, often unrecognized in clinical practice, which may have devastating health consequences through its association with fragility fractures. The risk of osteoporotic fracture ultimately depends on two factors: the mechanical strength of bone and the forces applied to it. Bone mass (a composite measure that includes contributions from bone size and from its volumetric mineral density) is an established determinant of bone strength, and the bone mass of an individual in later life depends upon the peak attained during skeletal growth and the subsequent rate of bone loss. Several longitudinal studies attest to the tracking of bone mass through childhood and adolescence, and mathematical models suggest that modifying peak bone mass will have biologically relevant effects on skeletal fragility in old age. There is evidence to suggest that peak bone mass is inherited, but current genetic markers are able to explain only a small proportion of the variation in individual bone mass or fracture risk [3]. Environmental influences during childhood and puberty have been shown to benefit bone mineral accrual, but the relatively rapid rate of mineral gain during intrauterine and early postnatal life, coupled with the plasticity of skeletal development *in utero*, offer the possibility of profound interactions between the genome and early environment at this stage in the life course. There is a strong biological basis for

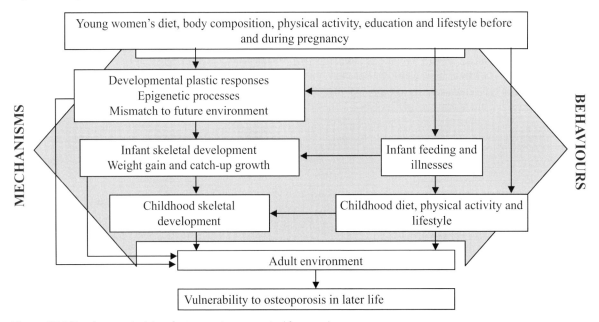

Figure 17.1 Developmental origins of osteoporosis: a conceptual framework.

The Placenta and Human Developmental Programming, ed. Graham J. Burton, David J. P. Barker, Ashley Moffett and Kent Thornburg. Published by Cambridge University Press. © Cambridge University Press 2011.

such a model of disease pathogenesis (Figure 17.1). Researcher have repeatedly demonstrated that minor alterations to the diet of pregnant animals can produce lasting changes in the body build, physiology and metabolism of the offspring [4]. This is one example of a ubiquitous phenomenon (phenotypic or developmental plasticity) that enables one genotype to give rise to a range of different physiological or morphological states in response to different prevailing environmental conditions during development. Its essence lies in the critical period during which a system is plastic and sensitive to the environment, followed by a loss of that plasticity and a fixed functional capacity. The evolutionary benefit of the phenomenon is that in a changing environment it maximizes phenotypic diversity and enables the production of phenotypes that are better matched to their environment than would be possible with the production of the same phenotype in all environments. This chapter will address the role played by influences during intrauterine or early postnatal life in establishing the risk of osteoporosis in later years. It will cover the patterns of normal skeletal growth during intrauterine life, infancy and childhood; the epidemiological evidence linking the risk of low bone density and fracture to environmental influences during early development; the impact of maternal nutrition and lifestyle on intrauterine bone mineral accrual; and the mechanisms underlying the relationship between developmental plasticity and osteoporosis.

Normal skeletal growth

Peak bone mass

At any age, the amount and quality of an individual's skeleton reflect their experiences from intrauterine life through the years of growth into young adulthood. The skeleton grows as the body grows, in length, breadth, mass and volumetric density. For men and women of normal body weight, total skeletal mass peaks a few years after fusion of the long bone epiphyses. The exact age at which bone mineral accumulation reaches a plateau varies with skeletal region and with how bone mass is measured. Areal density, the most commonly used measurement with dual energy X-ray absorptiometry (DXA), peaks earliest (before age 20 years) at the proximal femur, whereas total skeletal mass peaks 6–10 years later [5]. However, total skeletal mass does not reflect the considerable heterogeneity in mineral accrual at various skeletal sites [6,7]. Thus, the growth velocity of total body length is high immediately after birth, but slows rapidly during infancy; it accelerates again at 12 months of age, due to more rapid longitudinal growth of the legs, but not of the spine. Growth velocity of the legs continues to remain around twice that of the axial skeleton until puberty. Thus, there is a disassociation between the peak velocities for height gain and bone mineral accrual in both genders [8]. Sex steroid production slows long bone growth as the epiphyses begin to fuse, whereas axial growth accelerates. Adverse environmental exposures during the prepubertal period therefore produce greater deficits in the dimensions of the axial skeleton, whereas those in infancy and during puberty lead to greater restriction of appendicular growth. Gender differences in bone width are established during the peripubertal period: cortical width increases by periosteal bone formation in males, with relatively greater endocortical apposition observed in females. Thus, males build a longer and wider long bone with only a slightly thicker cortex than in females. The cortical mass is placed further from the neutral axis of the long bone in males, conferring greater resistance to bending by the correspondingly larger muscle mass.

The importance of peak bone mass for bone strength during later life was initially suggested by cross-sectional observations that the dispersion of bone mass does not widen with age [9]. This led to the proposition that bone mass tracks throughout life, and that an individual at the high end of the population distribution at age 30 is likely to remain at that end at age 70. Recent longitudinal studies have confirmed this tracking, at least across the pubertal growth spurt [10].

Bone growth *in utero*

The fetal skeleton develops in two distinct components, intramembranous (the skull and facial bones) and endochondral (the remainder of the skeleton) ossification. Intramembranous ossification begins with a layer or membrane of mesenchymal cells which becomes highly vascular; the mesenchymal cells then differentiate into isolated osteoblasts, which begin to secrete osteoid. The osteoid matrix is mineralized at the end of the embryonic period to form bony spicules, which are precursors of the lamellae of the haversian systems. There is no cartilage model preceding ossification in this type of bone development. Endochondral ossification is responsible for the formation of the bones that are the main sites of fragility fracture in later life. This form of ossification depends on a pre-existing cartilaginous model that undergoes invasion by osteoblasts and is only subsequently mineralized. The development of this cartilage model can be seen by 5 weeks' gestation, with the migration and condensation of mesenchymal cells in areas destined to form the bone [11]. These precartilaginous anlagen reflect the shape, size, position and number of skeletal elements that will be present in the mature skeleton. There is then an ordered differentiation of mesenchymal stem cells into chondrocyte precursors, proliferative chondrocytes, prehypertrophic chondrocytes and hypertrophic chondrocytes. During these stages of differentiation there is expansion of the bony template and the production of an extracellular matrix rich in cytokines, which facilitate vascular invasion and mineralization. The major regulator of the proliferation of chondrocytes is PTHrP [12], which is secreted by the perichondral cells; other proliferative stimuli include cytokines of the GH/IGF axis [13]. 1,25 $(OH)_2$ vitamin D_3 [14] and tri-iodothyronine [15] are stimuli for the differentiation of the chondrocytes through different stages. Once the cartilage model has been formed, vascular growth factors embedded in the matrix are released by chondrocyte metalloproteinases. This stimulates angiogenesis and, under the influence of Cbfa1 [16], osteoblasts from the perichondrium invade and lay down matrix, which is then mineralized.

During the period of a normal human pregnancy the fetus accumulates approximately 30 g of calcium, the majority of which is accrued during the third trimester. To supply this demand, there is a requirement for an adequate maternal supply of calcium to the placenta and increased placental calcium transfer to maintain a higher fetal serum calcium concentration than in the mother [17]. This maternofetal gradient emerges as early as 20 weeks of gestation [18]. It is mainly influenced by low levels of fetal parathyroid hormone activity; lack of fetal parathyroids in mice leads to low fetal calcium levels and decreased mineralization [19]. PTH does not cross the placenta, and maternal hypo- and hyperparathyroidism appear to affect the fetus via reducing or increasing the calcium load presented to the fetal circulation. Although both fetal PTH and PTHrP activity contribute to fetal plasma calcium concentration, the action of PTH appears to predominate. The main effect of PTHrP seems to be on placental calcium transport [19]. There is also evidence that PTH and PTHrP differentially affect mineralization of cortical and trabecular bone [20,21], and thus are attractive candidates for the physiological mediation of intrauterine skeletal programming.

The rate of maternofetal calcium transfer increases dramatically after 24 weeks, such that in a healthy term human fetus around two-thirds of total body calcium, phosphorus and magnesium are accumulated during this period. Factors that increase placental calcium transport capacity as gestation proceeds are only partly genetically controlled, and are achieved through regulatory hormones including 1,25 $(OH)_2$ vitamin D_3, parathyroid hormone, PTHrP and calcitonin. As the majority of fetal bone is gained during the last trimester, one of the major variables affecting bone mass at birth is gestational age. Other factors known to influence neonatal BMC include environmental variables such as season of birth and maternal lifestyle. Newborn total body bone mineral content has been demonstrated to be lower in winter births than in infants born during the summer [22]. This observation is concordant with lower cord serum 25 $(OH)_2$ vitamin D concentrations observed during the winter months, consequent upon maternal vitamin D deficiency. Other postulated contributors to impaired bone mineral acquisition during intrauterine life include maternal smoking, alcohol consumption, caffeine intake and diabetes mellitus [23].

Developmental origins of osteoporotic fracture

Epidemiological evidence that the risk of osteoporosis might be modified by the intrauterine and early postnatal environment has emerged from two groups of studies, first, retrospective cohort studies in which bone mineral measurements were undertaken, and in which fracture risk was ascertained, among adults whose detailed birth and/or childhood records were preserved; and second, mother–offspring cohorts relating the nutrition, body build and lifestyle of pregnant women to the bone mass of their offspring.

Birth weight, growth in infancy and adult bone mass

The association between weight in infancy and adult bone mass was shown in a cohort study of men and women aged 60–75 years who were born, and still lived in, Hertfordshire, England [24,25]. These studies showed highly significant relationships between

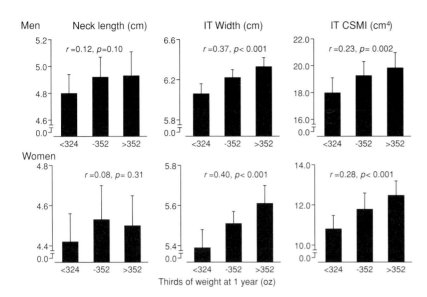

Figure 17.2 Weight in infancy and femoral geometry. IT, intertrochanteric; CSMI, cross-sectional moment of martia.

weight at 1 year and adult bone area at the spine and hip ($p < 0.005$); the relationships with bone mineral content (BMC) at these two sites were weaker but remained statistically significant ($p < 0.02$). The relationships also remained after adjusting for known genetic markers of osteoporosis risk, such as polymorphisms in the gene for the vitamin D receptor (VDR) [26], and after adjustment for lifestyle characteristics in adulthood that might have influenced bone mass (physical activity, dietary calcium intake, cigarette smoking and alcohol consumption). These findings confirmed previous observations in studies performed in the USA, Australia, Sweden and the Netherlands [27].

Bone density, geometry and strength

Hip structure analysis in the Hertfordshire Cohort Study (Figure 17.2) has demonstrated that poor growth *in utero* and during the first year of life is associated with disproportion of the proximal neck of the femur in later life (narrower neck but preserved axis length), with a corresponding reduction in mechanical strength of the region, over and above that attributable to reduced BMC *per se* [28]. In addition, the use of peripheral computed tomography (pQCT) within the same cohort demonstrated strong associations between birth weight, weight at 1 year, and each of bone width, length, area, fracture load and strength–strain index, at the tibia in both men and women (Figure 17.3), with less marked associations in a similar direction for the proximal radius [29]. These data add to those from hip structure analysis; thus poor growth *in utero* and during the first year of postnatal life is associated with alterations in bone architecture, cortical size and geometry, in addition to a deficit in densitometrically measured BMC, resulting in compromised bone strength and increased fracture risk in later life. These studies also complement data from the Helsinki Cohort Study, which directly linked growth rates *in utero* and during childhood with the risk of hip fracture. Three independent determinants of hip fracture risk were observed, including tall maternal stature, shortness at birth and low rate of childhood growth [30].

Maternal nutrition, body composition and neonatal skeletal development

Studies in mother–offspring cohorts have shown that body composition and lifestyle of mothers during pregnancy are related to the bone mass of their offspring at birth and in childhood. These influences on skeletal growth and mineralization were determined partly by the umbilical venous concentrations of IGF-1 and leptin [31]. In addition, further studies have confirmed that independent predictors of greater neonatal whole body bone area and bone mineral content include greater maternal birth weight, height, parity, fat stores (triceps skinfold thickness) and lower physical activity in late pregnancy. Maternal smoking was statistically significantly (and independently) associated with lower neonatal bone mass. These

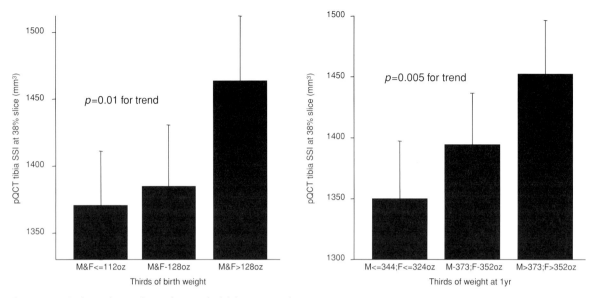

Figure 17.3 Birth weight, weight in infancy and adult bone strength.

relationships were observed in both male and female offspring [32].

In another study using DXA to assess the body composition of 198 children at age 9, reduced maternal height, lower pre-conception maternal weight, reduced maternal fat stores during late pregnancy, a history of maternal smoking during pregnancy and lower maternal social class were all associated with reduced whole-body BMC in the child at age 9 years [33]. In addition, lower ionized calcium concentration in umbilical venous serum also predicted reduced childhood bone mass. Around 31% of the mothers had insufficient and 18% had deficient circulating concentrations of 25(OH)-vitamin D during late pregnancy (11–20 and <11 μg/l, respectively). Lower concentrations of serum 25(OH)-vitamin D in mothers during late pregnancy were associated with reduced whole-body and lumbar spine BMC in children at age 9 (Figure 17.4). Maternal vitamin D status was also statistically significantly associated with childhood bone area and areal BMD. Adjunctive evidence supporting a role for maternal vitamin D status was obtained in the Southampton Women's Survey, where maternal vitamin D concentrations again correlated with neonatal bone mass [34]. These findings suggested that vitamin D supplementation of pregnant women, especially during winter months, could lead to long-lasting reductions in the risk of osteoporotic fracture in the offspring.

Further evidence that maternal calcium homoeostasis might play a role in the trajectory of intrauterine and early postnatal skeletal development emerged from a mother–offspring cohort study in Pune, India, which confirmed that children of mothers who had a higher intake of calcium-rich foods during pregnancy had higher total and lumbar spine BMC and areal BMD, independent of parental size and DXA measurements [35]. Circulating maternal 25(OH)-vitamin D concentrations in this cohort were relatively high, and were not associated with childhood skeletal measures. Thus, in populations in nutritional transition, where maternal sunlight exposure is sufficient to maintain adequate vitamin D status, the availability of calcium becomes a more critical determinant of fetal and childhood bone mineral accrual.

In most studies, maternal diet has been considered in terms of intake of specific nutrients, such as calcium and vitamin D. However, these nutrients are part of a broader dietary pattern, and one recent study has explored maternal diet in more detail in relation to skeletal health in the offspring, showing that a high maternal prudent diet score (high intakes of fruit and vegetables, wholemeal bread, rice and pasta, yoghurt and breakfast cereals, and low intakes of chips and roast potatoes, sugar, white bread, processed meat, crisps, tinned vegetables and soft drinks) was found to be associated with greater bone size and areal BMD

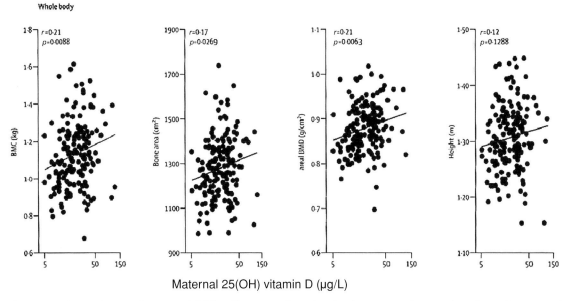

Figure 17.4 Maternal vitamin D status and childhood bone mass. Reproduced with permission from Javaid MK, Crozier SR, Harvey NC, Gale CR, Dennison EM, Boucher BJ, Arden NK, Godfrey KM, Cooper C, Princess Anne Hospital Study Group. Maternal vitamin D status during pregnancy and childhood bone mass at age 9 years: a longitudinal study. *Lancet* 2006; **367**(9504): 36–43. © Elsevier.

in the offspring [36]. The observed effect was independent of social class, education, maternal height, maternal smoking status and late pregnancy vitamin D levels, as well as childhood height, weight and exercise.

Animal models for the developmental origins of osteoporosis

Animal models for the developmental origins of osteoporosis have been established. In the first such model, the feeding of a low-protein diet to pregnant rats produced offspring that exhibited a reduction in bone area and BMC, with altered growth plate morphology in adulthood [37,38]. This study also examined whether maternal protein restriction affected the proliferation and differentiation of bone marrow stromal cells [39]. The results suggested that normal proliferation and differentiation were suppressed in offspring from mothers on low-protein diets as assessed by alkaline phosphatase-positive fibroblast colony formation at 4 and 8 weeks. In a further study, dams were given a low-protein diet (LPD) during pregnancy and 135 offspring were studied at different ages. Serum alkaline phosphatase concentrations reached peak levels earlier, and serum IGF-1 and 25(OH)-vitamin D levels were lower, in the offspring of protein-restricted dams, confirming the important role of the nutritional environment during intrauterine development [37]. Using micro-CT on samples of bone removed in late adulthood (75 weeks), it was observed that the offspring from LPD dams had femoral heads with thinner, less dense trabeculae; mechanical testing showed these samples to be structurally weaker [37].

These data further support the need for a programme of interventional research aimed at improving general dietary quality among women before conception and during pregnancy, in addition to studies that evaluate the targeting of specific micronutrients such as vitamin D and calcium.

Epigenetic mechanisms

Epigenetics refers to changes in phenotype or gene expression caused by mechanisms other than changes in the underlying DNA sequence. These changes are stable and heritable, and may last through multiple generations [40]. The two most-studied forms of epigenetic marking are DNA methylation and histone modification. DNA methylation involves the addition of a methyl group to cytosine residues at the carbon-5 position of CpG dinucleotides. DNA methylation is generally associated with gene repression, either by decreased binding of transcription factors or by attracting methyl-CpG-binding proteins that act as transcriptional repressors [41,42]. There is usually an inverse relationship between the extent of DNA methylation of regulatory CpGs and gene expression. Histone modification refers to post-translational modification of histone tails. Histones are small proteins involved in the packaging of DNA into chromatin, and if the way that DNA is wrapped around the histones changes, gene expression can also change. Histone modification can occur either by methylation or acetylation. These two types of epigenetic modification are mechanistically linked and work together to affect chromatin packaging, which in turn determines which gene or gene set is transcribed. The enzymes controlling these processes have recently been indentified and include DNA methyltransferases [43].

DNA methylation patterns differ through the phases of development. After conception, and with the exception of imprinted genes, gamete methylation patterns are erased during early blastocyst formation. During the implantation stage, methylation patterns become established via *de novo* methylation by the activities of DNA methyltransferases (Dnmt) 3a and 3b. Patterns of DNA methylation are maintained through mitosis by Dnmt1 activity [44]. In adulthood, there are variations in the amount and pattern of methylation depending on cell and tissue type. During embryonic and fetal development, maternal or environmental factors can disrupt these patterns of DNA methylation: examples of this process have been shown in animal models and will be discussed in further detail in this chapter. This disregulation of developmental programming via abnormal DNA methylation may permit specific genes to undergo inappropriate expression during adult life, resulting in disease development [43]. Emerging evidence strongly suggests that these epigenetic mechanisms underlie the processes of developmental plasticity.

Experimental data from animal models

Numerous studies in animals involving prenatal nutrient imbalance have provided important information regarding the biological basis for developmental plasticity. For example, the embryos of pregnant rats fed a low-protein diet during the preimplantation period showed altered development in multiple organ systems [45]. In addition, if the pregnancies progressed to term, then the offspring had reduced birth weight, relatively increased postnatal growth and adult-onset hypertension. Further studies have shown that the administration of glucocorticoids to pregnant rats at specific points during gestation can cause hypertension and insulin resistance in the offspring in later life, and can also lead to increased sensitivity to postnatal stress [46–48]. Postnatal stress in rat models has been shown to induce neurodevelopmental changes in the pups, and this leads to excessive responses to stress in later life. These changes may be mediated in part by effects on glucocorticoid-receptor gene expression in the brains of the offspring [49].

Further exploratory work in this area has shown that maternal dietary protein restriction in rats leads specifically to a decrease in the methylation status of glucocorticoid-receptor (GR) and peroxisomal proliferator-activated receptor α (PPARα) in the liver of the offspring after weaning [50]. These genes are of particular interest because alterations in their expression are associated with disturbances in cardiovascular and metabolic control in both animals and humans. The hypomethylation of the GR and PPARα persisted after weaning, when the direct influence of the maternal dietary restriction had ceased, suggesting stable modification to the epigenetic regulation of the expression of these transcription factors. In addition, supplementation of the restricted diet with folic acid prevented hypomethylation of GR and PPARα, and the associated increase in the expression of GR and PPARα. This observation suggests that the change in DNA methylation may reflect the impaired supply of folic acid from the mother and raises the possibility of therapeutic strategies to prevent or reduce the effects of environmental insults in early life. Other studies have demonstrated similar epigenetic changes in p53 in the kidney and angiotensin II type 1b receptor in the adrenal gland [51,52].

These studies show that the effects of maternal nutrition and behaviour appear to target the promoter regions of specific genes, rather than being associated with a global change in DNA methylation. This observation provides important clues for further work to explore epigenetic mechanisms in humans.

Epigenetic mechanisms in human disease

Epigenetic mechanisms, including DNA methylation and histone modification, are now well established in the development and progression of a variety of cancer types, including prostate, lymphoma, head and neck, breast and ovarian [53]. Data in other human diseases

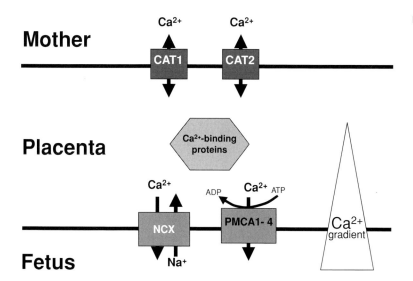

Figure 17.5 Placental calcium transport.

are limited, particularly in relation to developmental plasticity. The first example of an association between a periconceptional exposure and DNA methylation in humans was shown in Dutch subjects prenatally exposed to famine during the Dutch Hunger Winter in 1944–1945. Exposed subjects showed persistent epigenetic differences in the insulin-like growth factor 2 (IGF-2) gene six decades later compared to their unexposed same-sex siblings [54]. IGF-2 is known to be a key factor in human growth and development. This study further supports the importance of investigating how early epigenetic modification of gene expression may influence long-term health and disease.

Epigenetic regulation of placental calcium transfer

The key nutrients likely to influence fetal bone development include calcium and vitamin D, and therefore this axis provides a model for investigating the epigenetic regulation of bone mass. The human fetus requires a total of 30 g of calcium for bone development, most of which is acquired during the third trimester via active transport across the placenta (Figure 17.5), resulting in greater calcium concentration in fetal than in maternal plasma [55]. Fetal calcium needs are met primarily by increased maternal intestinal calcium absorption during pregnancy, and therefore very low maternal calcium intakes may be a risk for lower bone mass in neonates. The importance of maternal vitamin D status has already been highlighted in this chapter. The mechanism underlying the association between maternal vitamin D, umbilical cord calcium concentration and offspring bone mass is unclear, but is an area of ongoing research. 1,25(OH) vitamin D (the active form) mediates its effects by first binding to the vitamin D receptor, then by binding to the retinoic acid receptor (RXR), forming a heterodimer. This heterodimer then acts upon vitamin D response elements in target genes, initiating gene transcription by either upregulating or downregulating gene products. Vitamin D response elements are DNA sequences found in the promoter region of vitamin D-regulated genes [56]. Calcium transporters containing vitamin D response elements are therefore of particular interest.

One study has demonstrated that the expression of a placental calcium transporter (PMCA3) gene predicts neonatal whole-body bone mineral content (Figure 17.6) [57]. Modified expression of the genes encoding placental calcium transporters, by epigenetic regulation, might represent the means whereby maternal vitamin D status could influence bone mineral accrual in the neonate. Because the effects of maternal nutrition and behaviour seem to target the promoter region of specific genes rather than being associated with global changes in DNA methylation, investigating CpGs located within the promoter region of these genes, particularly those within or located near to vitamin D response elements, may provide further clues to the epigenetic regulation of bone mass. In addition, if validated, these epigenetic markers might provide risk assessment tools with which to target early lifestyle interventions to individuals at greatest future risk.

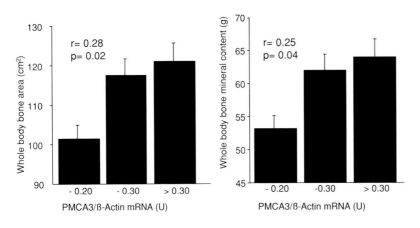

Figure 17.6 PMCA3 mRNA expression and neonatal bone mass.

Placental function and risk of fracture

The relation between placental eccentricity and the risk of hip fracture was studied in the Helsinki Cohort Study, which included 3416 men aged 50 years and over who were born in Helsinki University Hospital and who were followed forward through several decades, to ascertain the incidence of hip fracture. Table 17.1 shows the relationship between placental eccentricity and hip fracture risk in these men. There was a striking increase in hazard ratio for hip fracture (3.3; 95% CI 1.5–7.2 among men with marked placental eccentricity (defined as the difference in placental diameter for length and width) compared to those born to mothers with a round placenta. These data strongly suggest that placental capacity to deliver nutrients is an important mediating influence on intrauterine skeletal development and later risk of fracture.

Conclusion

Osteoporosis is a major cause of morbidity and mortality through its association with age-related fractures. Evidence is growing that peak bone mass is an important contributor to bone strength during later life. Many factors influence the accumulation of bone mineral during childhood and adolescence, including heredity, gender, diet, physical activity, endocrine status, and sporadic risk factors such as cigarette smoking. In addition to these modifiable factors during childhood, evidence has also accrued that fracture risk might be programmed during intrauterine life. Epigenetic processes are important mechanisms that underpin developmental plasticity, and environmental factors including maternal stress and nutritional state are known to affect the long-term epigenetic state of a number of genes during embryonic and fetal development. If validated, these epigenetic markers might provide risk assessment tools with which to target early lifestyle interventions to individuals at greatest future risk.

Table 17.1 Placental eccentricity and hip fracture risk: Helsinki Cohort Study (men 50+ years).

Placental eccentricity* (cm)	Fractures	Subjects (%)	HR (95% CI)
0/1	14	1219 (1.1%)	1.0 (baseline)
2	11	935 (1.2%)	1.1 (0.5 to 2.3)
3/4	10	894 (1.1%)	1.1 (0.5 to 2.5)
5+	12	368 (3.3%)	3.3 (1.5 to 7.2)
All	47	3416 (1.4%)	p trend = 0.009

Note: *Difference in placental diameter (length–width)

Acknowledgements

We are grateful to the Medical Research Council, the Wellcome Trust, the Arthritis Research Campaign, the National Osteoporosis Society, and the NIHR BRU in Musculoskeletal Disease and Nutrition, for support of our research programme into the developmental origins of osteoporotic fracture. The manuscript was prepared by Mrs G Strange.

References

1. **Consensus Development Conference.** Prophylaxis and treatment of osteoporosis. *Osteoporosis Int* 1991; **1**: 114–17.

2. **Cooper C.** Epidemiology of Osteoporosis. In: Favus MJ ed. *Primer on the Metabolic Bone Diseases and Disorders of Mineral Metabolism, 5th edn*. American Society for Bone and Mineral Research, 2003; 307–13.

3. **Ralston SH.** Do genetic markers aid in risk assessment? *Osteoporosis Int* 1998; **8**: S37–S42.

4. **Bateson P.** Fetal experience and good adult disease. *Int J Epidemiol* 2001; **30**: 928–34.

5. **Matkovic V, Jelic T, Wardlaw GM** et al. Timing of peak bone mass in Caucasian females and its implication for the prevention of osteoporosis. *J Clin Invest* 1994; **93**: 799–808.

6. **Seeman E.** Pathogenesis of bone fragility in women and men. *Lancet* 2002; **359**: 1841–50.

7. **Seeman E.** The growth and age related origins of bone fragility in men. *Calcif Tissue Int* 2004; **75**: 100–9.

8. **Bailey DA, McKay HA, Mirwald RL, Crocker PR, Faulkner RA.** A six year longitudinal study of the relationship with physical activity to bone mineral accrual in growing children: the University of Saskatchewan bone mineral accrual study. *J Bone Miner Res* 1999; **14**: 1672–9.

9. **Newton John HF, Morgan BD.** The loss of bone with age: osteoporosis and fractures. *Clin Orthop* 1970; **71**: 229–32.

10. **Ferrari S, Rizzoli R, Slosman D, Bonjour JP.** Familial resemblance for bone mineral mass is expressed before puberty. *J Clin Endocrinol Metab* 1998; **83**: 358–61.

11. **DeLise AM, Fischer L, Tuan RS.** Cellular interactions and signalling in cartilage development. *Osteoarthritis Cartilage* 2000; **8**: 309–34.

12. **Karaplis AC, Luz A, Glowacki J** et al. Lethal skeletal dysplasia from targeted disruption of the parathyroid hormone-related peptide gene. *Genes Dev* 1994; **8**: 277–89.

13. **Bhaumick B, Bala RM.** Differential effects of insulin-like growth factors I and II on growth, differentiation and glucoregulation in differentiating chondrocyte cells in culture. *Acta Endocrinol (Copenh)* 1991; **125**: 201–11.

14. **Sylvia VL, Del Toro F, Hardin RR** et al. Characterization of PGE(2) receptors (EP) and their role as mediators of 1alpha,25-(OH)(2)D(3) effects on growth zone chondrocytes. *J Steroid Biochem Mol Biol* 2001; **78**: 261–74.

15. **Quarto R, Campanile G, Cancedda R, Dozin B.** Modulation of commitment, proliferation, and differentiation of chondrogenic cells in defined culture medium. *Endocrinology* 1997; **138**: 4966–76.

16. **Ducy P.** Cbfa1: a molecular switch in osteoblast biology. *Dev Dyn* 2000; **219**: 461–71.

17. **Schauberger CW, Pitkin RM.** Maternal-perinatal calcium relationships. *Obstet Gynecol* 1979; **53**: 74–6.

18. **Forester F, Daffos F, Rainaut M, Bruneau M, Trivin F.** Blood chemistry of normal human fetuses at mid-trimester of pregnancy. *Paediatr Res* 1987; **21**: 579.

19. **Kovacs CS.** Skeletal physiology: fetus and neonate. In: Favus MJ (ed). *Primer on the Metabolic Bone Diseases and Disorders of Mineral Metabolism*, 5th edn. Washington DC: ASBMR; 2003: 65–71.

20. **Calvi LM** et al. Activated parathyroid hormone/parathyroid hormone-related protein receptor in osteoblastic cells differentially affects cortical and trabecular bone. *J Clin Invest* 2001; **107**: 277–86.

21. **Lanske B** et al. Ablation of the PTHrP gene or the PTH/PTHrP receptor gene leads to distinct abnormalities in bone development. *J Clin Invest* 1999; **104**: 399–407.

22. **Namgung R, Tsang RC, Li C** et al. Low total body bone mineral content and high bone resorption in Korean winter-born versus summer-born newborn infants. *J Paediatr* 1998; **132**: 285–8.

23. **Specker BL, Namgung R, Tsang RC.** Bone mineral acquisition in utero, during infancy, and throughout childhood. In: Marcus R, Feldman D, Kelsey J, eds. *Osteoporosis*, 2nd edn. New York: Academic Press, 2001; 599–620.

24. **Cooper C, Fall C, Egger P** et al. Growth in infancy and bone mass in later life. *Ann Rheum Dis* 1997; **56**: 17–21.

25. **Dennison EM, Syddall HE, Sayer AA, Gilbody HJ, Cooper C.** Birth weight and weight at 1 year are independent determinants of bone mass in the seventh decade: the Hertfordshire cohort study. *Pediatr Res* 2005; **57**: 582–6.

26. **Keen R, Egger P, Fall C** et al. Polymorphisms of the vitamin D receptor, infant growth and adult bone mass. *Calcif Tiss Int* 1997; **60**: 233–5.

27. **Cooper C, Harvey N, Javaid K, Hanson M, Dennison E.** Growth and Bone Development. *Nestle Nutr Workshop Ser Pediatr Program* 2008; **61**: 53–68.

28. **Javaid MK, Lekamwasam S, Clark J** et al. Infant growth influences proximal femoral geometry in adulthood. *J Bone Miner Res* 2006; **21**: 508–12.

29. **Oliver H, Jameson KA, Sayer AA, Cooper C, Dennison EM.** Growth in early life predicts bone

strength in late adulthood: The Hertfordshire Cohort Study. *Bone* 2007; **41**: 400–5.

30. **Cooper C**, **Eriksson JG**, **Forsen T** et al. Maternal height, childhood growth and risk of hip fracture later in life: a longitudinal study. *Osteoporosis Int* 2001; **12**: 623–9.

31. **Javaid MK**, **Godfrey KM**, **Taylor P** et al. Umbilical cord leptin predicts neonatal bone mass. *Calcif Tissue Int* 2005; **76**: 341–7.

32. **Harvey NC**, **Poole JR**, **Javaid MK** et al. Parental determinants of neonatal body composition. *J Clin Endocrinol Metab* 2007; **92**: 523–6.

33. **Javaid MK**, **Crozier SR**, **Harvey NC**. Maternal vitamin D status during pregnancy and childhood bone mass at age 9 years: a longitudinal study. *Lancet* 2006; **367**: 36–43.

34. **Harvey NC**, **Javaid MK**, **Poole JR** et al. Paternal skeletal size predicts intrauterine bone mineral accrual. *J Clin Endocrinol Metab* 2008; **93**: 1676–81.

35. **Ganpule A**, **Yajnik CS**, **Fall CH** et al. Bone mass in Indian children. Relationships to maternal nutritional status and diet during pregnancy: the Pune Maternal Nutrition Study. *J Clin Endocrinol Metab* 2006; **91**: 2994–3001.

36. **Cole ZA**, **Gale CR**, **Javaid MK** et al. Maternal dietary patterns during pregnancy and childhood bone mass: a longitudinal study. *J Bone Miner Res* 2009; **24**: 663–8.

37. **Lanham SA**, **Roberts C**, **Perry MJ**, **Cooper C**, **Oreffo RO**. Intrauterine programming of bone. Part 2: Alteration of skeletal structure. *Osteoporosis Int* 2008; **19**: 157–67.

38. **Lanham SA**, **Roberts C**, **Cooper C**, **Oreffo RO**. Intrauterine programming of bone. Part 1: Alteration of the osteogenic environment. *Osteoporosis Int* 2008; **19**: 147–56.

39. **Oreffo RO**, **Lashbrooke B**, **Roach HI**, **Clarke NM**, **Cooper C**. Maternal protein deficiency affects mesenchymal stem cell activity in the developing offspring. *Bone* 2003; **33**: 100–7.

40. **Jaenisch R**, **Bird A**. Epigentic regulation of gene expression: how the genome integrates intrinsic environmental signals. *Nature Genet* 2003; **33**: 245–54.

41. **Gicquel C**, **El-Osta A**, **Le Bouc Y**. Epigenetic regulation of fetal programming. *Best Pract Res Clin Endocrinol Metab* 2008; **22**: 1–16.

42. **Gluckman PD**, **Hanson MA**, **Beedle AS**. Non-genomic transgenerational inheritance of disease risk. *BioEssays* 2007; **29**: 145–54.

43. **Tang W**, **Ho S**. Epigenetic reprogramming and imprinting in origins of disease. *Rev Endocrinol Metab Disord* 2007; **8**: 173–82.

44. **Bird A**. DNA methylation patterns and epigenetic memory. *Genes Dev* 2001; **16**: 6–21.

45. **Kwong WY**, **Wild AE**, **Roberts P** et al. Maternal undernutrition during the preimplantation period of rat development causes blastocyst abnormalities and programming of postnatal hypertension. *Development* 2000; **127**: 4195–202.

46. **Levitt NS**, **Lindsay RS**, **Holmes MC** et al. Dexamethasone in the last week of pregnancy attenuates hippocampal glucocorticoid receptor gene expression and elevates blood pressure in the adult offspring in the rat. *Neuroendocrinology* 1996; **64**: 412–18.

47. **Nyirenda MJ**, **Lindsay RS**, **Kenyon CJ** et al. Glucocorticoid exposure in late gestation permanently programs rat hepatic phosphoenolpyruvate carboxykinase and glucocorticoid receptor expression and causes glucose intolerance in adult offspring. *J Clin Invest* 1998; **101**: 2174–81.

48. **Welberg LAM**, **Seckl JR**, **Holmes MC**. Prenatal glucocorticoid programming of brain corticosteroid receptors and corticotrophin-releasing hormone: possible implications for behaviour. *Neuroscience* 2001; **4**: 71–9.

49. **Weaver ICG**, **Cervoni N**, **Champagne FA** et al. Epigenetic programming by maternal behaviour. *Nature Neurosci* 2001; **7**: 847–54.

50. **Lillycrop KA**, **Phillips ES**, **Jackson AA** et al. Dietary protein restriction of pregnant rats induces and folic acid supplementation prevents epigenetic modification of hepatic gene expression in the offspring. *J Nutr* 2005; **135**: 1382–6.

51. **Pham TD**, **MacLennan NK**, **Chiu CT** et al. Uteroplacental insufficiency increases apoptosis and alters p53 gene methylation in the full-term IUGR rat kidney. *Am J Physiol Regul Integr Comp Physiol* 2003; **285**: R962–R970.

52. **Bogdarina I**, **Welham S**, **King PJ** et al. Epigenetic modification of the renin-angiotensin system in the fetal programming of hypertension. *Circ Res* 2007; **100**: 520–6.

53. **Gronbaek K**, **Hother C**, **Jones PA**. Epigenetic changes in cancer. *APMIS* 2007; **115**: 1039–59.

54. **Heijmans BT**, **Elmar WT**, **Stein AD** et al. Persistent epigenetic differences associated with prenatal exposure to famine in humans. *PNAS* 2008; **105**: 17046–9.

55. **Namgung R**, **Tsang RC**. Bone in the pregnant mother and newborn at birth. *Clin Chim Acta* 2003; **333**: 1–11.

56. **Kimball S**, **El-Hajj Fuleihan G**, **Vieth R**. Vitamin D: A growing perspective. *Crit Rev Clin Lab Sci* 2008; **45**: 339–414.

57. **Martin R**, **Harvey NC**, **Crozier SR** et al. Placental calcium transporter (PMCA3) gene expression predicts intrauterine bone mineral accrual. *Bone* 2007; **40**: 1203–8.

Discussion

LAMPL: I have a question about maternal oestrogen levels. How do you see these fitting in, and are there any sex differences in the offspring?

COOPER: I can answer the second more confidently. Generally speaking, where there are data on childhood trajectory of growth, the picture in the girls is one of an alteration in menarchal age. The height and weight are discordant in their deviation from the normal pattern in relation to later fracture. But where you are only talking about birth weight and weight in infancy, effects are observed in both genders. The data suggest that there is an environmental influence on bone development in both genders, but one that might be accentuated in girls as a consequence of a delay in menarchal age. It would be helpful to have measures of testosterone, SHBG and FAI in the boys, and 17β oestradiol and SHBG in the girls. As regards oestrogen status, we were talking about placental morphology and whether thickness might represent a cumulative oestrogen exposure or not, given some of the breast cancer findings. It is possible, but I personally do not have any direct evidence to support this assertion as yet.

ALWASEL: I have a question about the consistency of the LP rat model and calcium regulation. We did two experiments in Manchester University. One showed that there is severe secretion of calcium, and the other showed nothing. We did the same thing but with different sets of rats. It was confusing. The first group were excreting more calcium due to insufficient calcium retention; the other group showed that there was no difference between controls and LP rats.

COOPER: We have not been able to study urinary calcium excretion in the animal experiments. We have utilized mesenchymal stem cells derived from undernourished rat dams, and evaluated the sensitivity of osteoblast precursors to vitamin D, IGF1, and dexamethasone. Mesenchymal stem cells from the offspring of low-protein dams have blunted responses to IGF1 and vitamin D. These observations suggest a defect in bone formation, rather than accelerated bone resorption.

FERGUSON SMITH: Your PMCA expression and the genetic analysis was potentially very important. What is the hypothesis now, based on that analysis?

COOPER: It would have been helpful to evaluate PTHrP levels, but the method of sample storage did not permit this. As a consequence, we went to the second rate-limiting step in the placental calcium transport pathway, evaluating the role of plasma membrane calcium-dependent ATPases. We initially explored placental samples for *PMCA3* mRNA and found this to be related to bone mass of the baby. We then identified CG loci in the promoter region of *PMCA3*, and explored the extent to which methylatation at this site correlated with maternal and offspring attributes; it correlated with maternal height.

FERGUSON SMITH: Is the placenta expressing more *PMCA3* mRNA?

COOPER: We do not have information on the amount of PMCA3 itself.

FERGUSON SMITH: Your previous dataset showed a positive correlation between PMCA3 and bone mineral density. Your current data show increased methylation associated with a tall mother. Do you know anything about the position of these two methyl sites relative to transcription factor binding sites?

COOPER: There are known to be glucocorticoid and vitamin D response elements in these promoters.

FERGUSON SMITH: From imprinting studies we know that promoter differential methylation in placenta is not as functionally relevant as promoter differential methylation in embryonic tissues.

MOFFETT: We should not forget the uterus. Vitamin D has profound effects on uterine function. In a rat model, vitamin D is essential for decidualization. We also know that the NK cells are thought to develop in the uterus (they may migrate there as NK precursors), and one of the key transcription factors is vitamin D dependent. In addition, the macrophages are very sensitive to vitamin D in how they function: for example, whether they become immunogenic or tolerogenic depends on vitamin D. So just to go straight to the placenta risks missing out on key features of the vitamin D story.

COOPER: Are these characteristics of the uterus amenable to non-invasive imaging?

SMITH: Endometrial thickness can be measured by ultrasound, but this doesn't tell you much.

THORNBURG: Why is this programming effect? Why is it that there is something permanent about calcium deposition and the way bones are made that can't be recovered during postnatal life with good nutrition and adequate levels of vitamin D?

COOPER: That's a good question. The observations suggest that although linear, growth of the skeleton can recover to some degree after transient insult but doesn't do so fully. If you make the hit early enough, you don't get back to the preordained genetic trajectory.

THORNBURG: Does it have to do with the bone skeleton architecture that has to calcify?

COOPER: We don't know, but if you were to make an inference about what is going on, it is that you get your envelope as large as you can make it towards your genetic ceiling, but you are struggling to lay down an appropriately thick cortex and appropriately connected trabecular architecture. These, coupled with the shape abnormality, seem to be the developmentally programmed characteristics.

ROBERT BOYD: The first bit doesn't quite fit with Harris' lines, which are over-calcification at times of impaired growth.

COOPER: If you look at linear growth catch-up, once you get to insult in childhood, there is some evidence that this catch-up doesn't quite get you back to the centile that you started on.

BARKER: I'd like to add a historical rider. Fifty years ago Elsie Widdowson was thinking about why, after rickets during infancy, the shape of the pelvis is never restored, despite growth during childhood and adolescence. This is what triggered her interest in programming.

Chapter 18

Final general discussion

ROBERT BOYD: I'd like us to finish with a discussion vis-a-vis some general points, beginning with Michelle Lampl who has volunteered to consider the fetus as a driver.

LAMPL: I have noticed that the fetus has not been at the centre of our discussions. The perspective that I find interesting is the cross-talk between the fetus, the mother and the placenta. Taking the perspective that the placenta is fetal tissue is important. It is the maternal genome and the paternal genome as it comes together in the fetus, which is behind the development of the placenta. I'd be interested in hearing more about that interaction as we know it. Going back to the fetus, the mother, and placenta as a system, we have had a great deal of conversation that is clinically oriented, and I appreciate that. In our conversations, the fetus appears to be on the receiving end of programming, a passenger at risk from a number of insults. By contrast, the fetus itself is a genetic individual who has a growth pattern that is being expressed. How is it that the fetal growth programme and the growth of the fetus interact with the environment that we have all been talking about? There is more to fetal growth than as a stress on the mother. Fetal growth is a process of facultative adjustment to its environment that is important in our considerations of the fetal origins of chronic disease. I hope that going forward we can have some collaboration on this topic. In longitudinal studies on fetal growth, a number of participants here have contributed ideas and have access to data. It will be useful going forward to consider how it is that the placenta is the translation point of the observations for maternal size, for example, and the fetus itself. We have talked often about critical time periods that the placenta presents to the fetus. I'd be interested in more information on how that interaction is responding to the fetus and the fetal growth process itself.

BURTON: We never think about the dynamics of the umbilical circulation in the growth of the placenta. We talk in terms of VEGF and growth factors, but we never think about sheer stress, strain and these sorts of factors. The placenta is seeing the cardiac output from the fetus.

LAMPL: There are interesting observations from longitudinal data that identify dynamic changes in fetal growth alterations in blood flow between the placenta and the fetus. It is interesting to consider the potential of that circulatory loop as an opportunity for compensatory responses involved in the kinds of stresses that we have talked about, and perhaps being part of the programming at the metabolic level.

RICHARD BOYD: I want to comment on 'safety factors'. This is a plug for someone else's science, and the person is Jared Diamond. Towards the end of his experimental research career as a physiologist he wrote a very good article on what he called 'safety factors'. He gave an example talking about how you build an elevator. Your engineer has to decide how much rope or string to hang this up on, and has to build in a safety factor. For engineers building lifts this safety factor will be substantial. In placental biology, in relation to imprinting, where are the safety factors? We have two organisms talking to each other. It might be quite clever not to have much of a safety factor for some events, in order for example to allow that pregnancy to be lost, another pregnancy ensuing when things get better. There is thus inevitably a need to have quantitative thinking about each step in the sequence.

BARKER: I'd like to mention ethnic issues. In Caroline's Indian studies she sees small placentas and small babies. In Saleh's studies in Saudi Arabia he sees the same size placentas but babies that are similar to those in the west. In Helsinki there are bigger placentas but the same sized babies as in Saudi Arabia. The only other ethnic data we have heard about is among people living at altitude. I was left confused about this: I know that Dino's data from La Paz say that there is a big placenta in relation to birth weight, whereas Lorna's don't. The general point, however, is do we know enough about placentas around the world?

The Placenta and Human Developmental Programming, ed. Graham J. Burton, David J. P. Barker, Ashley Moffett and Kent Thornburg. Published by Cambridge University Press. © Cambridge University Press 2011.

JACKSON: Are you talking about ethnic differences or geographical differences? Given a healthy environment, children grow remarkably similarly wherever they are in the world.

MOFFETT: As an immunologist, one of the greatest differences in human evolution as people moved out of Africa were in the two sets of genes, the KIR and HLA genes. These are outliers in terms of their extreme polymorphism. We performed a computer-based project to ask whether immune system genes have any characteristics that are unique compared with other genes. They don't, except for these two outliers, the KIR and HLA. We must take them into account when looking at the different types of pregnancy in different parts of the world. There is drift, but there is also huge selection pressure on them.

LOKE: When we talk about what drives the growth of the placenta, there are two levels where the maternal and paternal genomes can contribute. One is in the development of the placenta itself. This is when genomic imprinting comes in. But the other level, which people tend to forget, is the contribution of the uterine environment to the development of the placenta. This second level is also very important. The paternal genome will drive the invasion and the maternal genome will resist the invasion. To me, with respect to the placenta, size doesn't seem to matter. It might be better to be small and perfectly formed.

POSTON: The reason I don't work on the placenta is that it is phenomenally difficult to work on transport. You have to understand flux, total amount, blood flow, the expression of the transporters and the membrane potential. Each of these contributes to the net transport of the nutrient. I still feel that we don't really know how much of a given nutrient actually gets across. You can measure arterial/venous differences, but in terms of the net transport it is extremely difficult to do properly. This is one major problem about placentology.

HUNT: One thinks about trends and concepts when listening to this terrifically well informed group of scientists speaking about things they know a lot about. It reminds me of different things that I have learned over the course of the years that might be of interest. One has to do with the immune cells in the decidua. Ashley Moffett is the world's authority on the NK cell, which is an exciting cell. I have worked on the macrophage, which is very different. It has many different activities and receptors, but the interesting thing about the macrophage is that it comes in as a benign monocyte, and is environmentally programmed to do whatever needs to be done. It stays with the uterus during the course of pregnancy. We still don't know all of the jobs that are performed by this versatile cell. The other thought I had was about the placenta. This is again related to immunology. In the placenta there are many markers that are traditionally associated with the immune response, but in the placenta they do entirely different things. As you look at studies that involve cytokines and chemokines, keep in mind that when they are in the placenta their functions may be opposite to that expected.

SMITH: It is almost as if there are two parallel descriptions of things with the human data. The bit where it is most interesting is where you can go to the human data with a specific prediction and test it. There is a tendency in the human side of things to be constantly describing in the absence of a clear hypothesis. The thing I'll take from this meeting is that uterine artery Doppler is a good way of predicting whether people will get early-onset severe pre-eclampsia but what is the biological correlate? I thought I knew, but now I don't think I do.

MAGNUS: I am fascinated by the fact that much of this field has come about through epidemiological studies, such as ecological studies and historical cohort studies. Now there is all this basic biology. I think we should go back to the epidemiology. We now have some large pregnancy cohorts where we have high quality biological material. You should come to us with specific hypotheses and we can look at the stability of changes over time.

JANSSEN: If you think about the placenta and programming, there are two problems. One is semantic: as a tissue the placenta doesn't have an adult or postnatal life. Placental programming doesn't really fit the strict definition of programming. More importantly, with few exceptions we can only ascribe indirect roles to the placenta in programming. For example, placenta dysfunction leads to growth restriction, which leads to programming. Of course, if we came up with an intervention that could alleviate growth restriction, we could decrease the incidence of cardiovascular disease in future generations, but this is still indirect. As placentologists, we might start thinking about direct effects of placental function on the fetus. The placenta is very good at regulating maternal metabolism by secreting a large number of factors. It is plausible that the placenta could, for example, directly influence fetal skeletal muscle insulin resistance.

MOFFETT: One thing we haven't discussed is the difficulty in humans in experimenting on trophoblast. First, there's the great difficulty of using primary tissue because of the ethics required to obtain first trimester samples. It's hard to isolate and it doesn't grow in culture. The cells need to be characterized as villous or extravillous trophoblast. Also, human trophoblast cell lines are still too poorly characterized to be used reliably.

PIJNENBORG: I agree with Ashley Moffett: there is a need for good *in vitro* models of invasive trophoblast. This should also be looked at on the comparative level. It would be interesting to know why the shallow invaders have less invasive activity. How does this compare among different species? The other issue that strikes me is our ignorance of the uterus and decidualization effects. The most important point to start on is the luteal phase of the menstrual cycle, because this is when the early decidual changes occur. These are important for implantation. I would like to see decidualisation in a broader context than just endometrium only, because of the involvement of the inner myometrial layer underneath. There is good evidence that this layer is also changed in early pregnancy. Perhaps better knowledge of the uterine environment, decidualization and the myometrium could also teach us something about why some placentas have different shapes than others.

BURTON: I have a plea concerning programming: there should be emphasis on the first trimester and we should also see it in the evolutionary context. The early placenta is using some very primitive metabolic pathways, and we should keep an open mind and not impose what we see in the adult on early placental development.

BAGBY: I'm thinking of two kinds of growth – height versus accrual of mass – as being differently regulated. In the zebra fish fin, morphogenic growth gets it to its basic shape, and this is not nutrient sensitive, whereas growing from that point is highly nutrient sensitive. We haven't talked about this. We see an asymmetry of height compared with weight. The kidney does this during development when there is nutritional restriction: it maintains its length much better than its width.

FALL: David Barker's data shows the power of the measuring tape. I have a couple of cohorts that I am associated with where we have these sorts of data. I'll be going back to these.

SIBLEY: I have tried to put together a simple model of placental adaptations, compensations and plasticity. Spiral artery conversion is key here. There are various types of conversion (good, medium, poor) which will affect early placental development. There might be awful spiral artery conversion, which means that the placenta is so damaged that you get a failed adaptation, which leads to IUGR. If it is OK, then you have maternal factors feeding into the early placental development, maternal factors feeding into the early placental and spiral artery conversion, and then you get fetal growth feeding in from middle-second to late-second trimester, putting demand on the system. If the placenta is OK you might not need any adaptation, and you get flow-through to a normal birth. More than likely there will be some extra fetal demand, and this could alter the placental adaptation compensations in a number of different ways. These include blood flow, morphology changes, nutrient utilization by the placenta itself will be important, and transporter adaptation. The work Abbey Fowden and we have done on the mouse suggests that all these things can happen at different times in gestation and independently. One placenta, though it might be the same size as another, has quite different ways of feeding the baby, depending on those adaptations. We get variable baby composition. Homeostatic set points could be altered, which could lead then to later programming.

FOWDEN: What I'd like to say follows on from this. There must be some combination of morphological functional markers that we can develop for a placenta at term that tells us about the history of that placenta and/or the intrauterine environment. This would allow us to better predict how the baby will develop in later life.

FLEMING: My prejudice is that I think it is the early developmental steps that will be most critical in programming. For example, on the placentation front for the human, it will be very interesting if we can learn more about the yolk-sac-mediated pathway and its role in histiotrophic nutrition. Speaking as a developmental biologist, and stepping away from the placenta for a moment, there are many areas that have yet to be explored in the programming field. As Lewis Wolpert once said, 'gastrulation is the most important thing that happens in our lives'. The spatial patterning of gastrulation coupled with the changes in cells from an undifferentiated state to a differentiated state – the formation of the three germ layers – will be critical in understanding programming. But like all good things in life, it is an intractable area to try to get at with no ideal *in vitro* models currently available. Perhaps this will be a future role for stem cell biology and *in vitro* models to investigate this area. The oocyte is another big area that we don't really understand, providing a link from one generation to the next.

SFERRUZZI-PERRI: In terms of Colin's thoughts about spiral artery conversion, trophoblasts invade the spiral arterioles to convert them, but they are also secreting abundant hormones which act on the mother and her capacity to adapt or prepare for pregnancy. There are association studies that have been done, but there is a lot more to be gained from better studies, both in terms of placental secretion of hormones into the maternal circulation and their impact on the mother, but also the other way: placental secretion into the fetal circulation. Hormones secreted into the fetal circulation (either initially from the mother or synthesised by the placenta) have the capacity to act on the fetal liver and its secretion of hormones, including the IGFs, which in turn, regulate and influence development of particular fetal organs, which may be programmed for later life.

JACKSON: I suppose we have to accept there isn't infinite resource, but the question is whether there is limited resource. The implication is that there is, for one reason or another. In that case there have to be choices and trade-offs. Presumably the placenta's challenge is to optimise the outcome for mother and fetus. It makes choices in this respect. What is the least investment for the maximum return? My problem is that I don't feel that we have even started to have a conceptual framework to capture these relationships. There have to be a limited range of critical factors that make a big difference and have multiple pleiotropic effects. It was suggested here that oxygen is the big limiting concentration: if it is, then we ought to pay more attention to it.

TYCKO: I have been trying to think about intervention. Intervention could occur early during gestation, or later in childhood, or even in adulthood. Even though there are anatomical differences between mouse and human placenta, since obstetricians are reluctant to intervene, and human intervention studies post-natally are very expensive, we should hammer away a bit more at this question of producing meaningful mouse models of fetal growth restriction. The genes are conserved, and the expression of the critical genes in the placenta is conserved. It is just that certain details seem to have changed between the two species. Mouse trophoblast does have invasive properties, but it is not quite the same invasion.

ALWASEL: Most fetal programming work has been done in laboratories and centres in Europe and the USA. Any conclusions that can be drawn from these studies will be incomplete until we hear from centres in other parts of the world. For example, we know nothing about the Chinese placenta and its role in fetal programming.

FERGUSON SMITH: We have heard a lot about the physiology of the normal and compromised placenta. Picking up from some of the things that Ben Tycko said, we do want to think about intervention and therapies. One powerful approach is to look forward to understanding these mechanisms and pathways in more detail at the molecular level. Certainly, the rapidity of the response of the conceptus to a compromised environment indicates that epigenetic mechanisms will likely be important in this process. I don't think we can understand how epigenetics might be modulating the compromised behaviour of genes until we understand how these genes are regulated normally. We should not forget however that there are many ways to regulate genes; epigenetic mechanisms represent just one part of this.

COOPER: I think there is room within the normal range encountered, for more placental epidemiology. I am not left with a clear picture of the diversity of placental morphology and function that one encounters across the population. This would help me to make inferences about what might be at the edges of the distribution, and what might be close to the middle of it. Also, we now have for human studies fantastic human tissue resources, very well phenotyped with characterisation of genetic variation. What we are doing is playing around at the edges of that in terms of the placental experiments. I'd love to have an established way of looking at placental function that we might apply in those human tissues.

Chapter 19

The placenta and developmental programming

Some reflections

Robert Boyd and Richard Boyd

The concept of developmental programming of future health has had a galvanizing impact on thinking, both by those concerned with the aetiology of disease in adult life and its impact on future global human health, and by those interested in developmental biological mechanisms. Evidence for 'fetal origins' of future ill-health through such programming comes from both epidemiological and animal data. The papers and discussions reported in this book consider from the perspective of a range of disciplines the specific interface between placental structure and function and later health outcomes. This is an important perspective, as several observations indicate that the placenta may play a necessary, perhaps even sufficient, role in programming. The meeting was a fascinating occasion, both for the data presented and for the ensuing discussion.

The mark of a successful seminar is the richness of reflection, criticism and mental argument that continues thereafter. By this token, we have found ourselves debating several angles that emerged – some discussed at the meeting, some not. Here we take the opportunity to share them in the hope of encouraging further discussion and perhaps, if any take hold, the production of future data.

How key is the placenta to the ability of the developing organism to program?

Programming was conceived through consideration of epidemiological relationships between placental and fetal weights and later health. The placenta plays a key role in fetal nutrition, excretion, endocrinology, intermediary metabolism and respiratory function during the maximal period of fetal growth and development. It has a major role in immunological cross-talk between mother and fetus. It is therefore inevitably central to programming, and many new and interesting findings relevant to a possible placental role in programming were indeed presented at the seminar. However, it is also worth briefly considering the theoretical framework in order not to fail to think about other possibilities.

Much remains very uncertain. Placental differences found between differently programmed individuals may not themselves be causal: they might be secondary to programming within the embryo/fetus, or they might be parallel but separate changes induced in both extraembryonic and embryonic tissues by the same environmental pressure.

Furthermore, the list of non-placental candidates for roles in programming is long. The gonadal period in either parent, the process of fertilization and the early life of the conceptus in the oviduct and *in utero* before placentation, especially of those elements that will become the future embryo, are all parts of development during which programming might plausibly take place in response to the maternal environment. An important topical example raised was the association of different techniques in assisted reproduction with the programming of different long-term outcomes.

During the placental period of fetal life the proportionate role of the placenta in programming also remains uncertain. Possible extraplacental routes of maternofetal exchange could also play a role in enabling the fetus to respond to maternal environmental experience. There might even be a more direct interface between the fetus and the wider environment via the vaginal flora.

Programming continues after birth. There is strong evidence of long-term programming taking place in the neonatal period in the marked inverse relationship between neonatal microbial exposure and

---

The Placenta and Human Developmental Programming, ed. Graham J. Burton, David J. P. Barker, Ashley Moffett and Kent Thornburg. Published by Cambridge University Press. © Cambridge University Press 2011.

later airways hyper-reactivity [1]. Nutritionally, early postnatal caloric restriction in the rhesus monkey has a dramatic impact on resistance to illness and mortality 20 years later [2]. Milk/colostrum provided by the mother can program at least neonatal immune behaviour: might this be long term? Social mechanisms of neonatal programming are seen in the relationship between infant learning and later mothering behaviour in non-human primates.

Bearing in mind the importance of climate and food-gathering needs in survival and long-term health, it is conceivable that environmental physical signals could play a role in programming mediated directly to the fetus or embryo: ambient temperature or maternal physical exercise are potential examples.

Predicting the future: what might be sensed, and how?

In our view the identification of sensing mechanism(s) will be central to a full understanding of programming, because (as discussed below) to be biologically meaningful, programming capacity must have evolved under selective pressure to benefit a population group. Programming must therefore use previous group experience to sense and integrate those environmental signals that have historically been most predictive of the future environment for members of that group. Interesting data were presented on such potentially important environmental triggers of programming as the availability of oxygen (pregnancy at altitude) and nutritional adequacy (general diet and the dietary availability of possibly rate-limiting components). Other microbiological, or more speculatively toxicological or social, factors possibly predictive of the future were not considered on this occasion.

To respond in an integrated fashion to environmental cues a central mechanistic requirement is that they be sensed directly or indirectly by the embryo/fetus. Some important possibilities are highlighted in this book. Some of the recent more conceptual advances should also be considered. For example, where in the conceptus, placental or otherwise, might conserved enzymes that sense ('senzymes', [3]) or transporters that signal ('transceptors', [4]) be found?

Most sensing mechanisms ultimately depend on receptor–ligand interations and, if this is true of the sensing underpinning programming, there are likely to be interesting lessons from the pharmacological/toxicological epidemiology of programming.

Is clinical evidence under-regarded?

It is sometimes commented that overt clinical abnormalities of pregnancy and their sequelae are categorically different from programming during normal development. In our view there is a need to bring together more effectively clinical evidence with population-wide epidemiological evidence. A hallmark of programming tends to be the expectation of a 'silent period' between insult and consequential biological cost; is this expectation soundly conceived? Thalidomide's damaging effects are apparent immediately after birth, but those of stilboestrol not for decades; yet are they categorically different? Similarly, it has been known for over a century that intrauterine distress or birth injury may be associated with very severe 'programming', i.e. long-term neurological deficit, as with cerebral palsy. This may not be apparent in the early neonatal period. More recent work shows that, in a population cohort, less optimal neurological development after birth correlates with low fetal growth (e.g. [5]). It appears to us unsound to categorize such initially invisible programming of later neurological impairment (associated as this is with increased mortality) as entirely different from more overt programming of future deficits that are apparent from birth or soon after.

One important conclusion from clinical studies is the existence of short or very short windows of timing in relation to later outcome following rubella *in utero* or exposure to thalidomide. It seems plausible that more subtle 'programmers' will be subject to similar constraints.

There is extensive evidence that normally grown substantially preterm fetuses have a life course very different from their intrauterine contemporaries; thus they are programmed differently following early expulsion from the uterus. Do we take sufficient account of the extensive literature on preterm birth in developing the programming paradigm as a whole, and indeed fail to sufficiently integrate clinical experience?

Developmental programming and the origin of species: two different paradigms?

There is a conceptual challenge in the fetal origins paradigm that needs to be more explicit: the potential conflict between individual benefit and species benefit in considering future outcomes. The roots of the developmental programming paradigm lie in medical epidemiology and thus in the impact of earlier environmental experiences on individual future lives, albeit considered as groups. The roots of Darwin's paradigm lie in consideration of the impact of environments on the outcome for populations. Even at the simplest level these different perspectives are obvious. Is adolescent human pregnancy to be considered as a clinical abnormality or as normal biology?

The extent to which strategies for the group and for the individual may be in conflict or may overlap is central to programming. Is the afferent limb of a mechanism sensing environment which may program an individual to thrive in its predicted future environment (e.g. moderate nutritional challenge leading to a more thrifty phenotype) the same when it leads only to reproductive benefit to the DNA carried by its parents (e.g. severe nutritional challenge leading to abortion)? This may be the bread and butter of Darwinian teaching, but perhaps needs to be factored in more clearly when considering programming.

An evolutionary perspective is also relevant to the work on fetal origins in recent years leading to the important concept that if environmental circumstance changes faster than the evolution of mechanisms to predict the future, then a programmed response may be dysfunctional or adverse.

Conclusion

To quote an earlier placentologist: 'for any satisfying explanation of the relation of the unborn child to its mother the darkness of intrauterine workmanship must first be made visible and the inscrutability replaced by biological answers to rational questions' [6]. Programming provides ever more challenging questions – and, as this book shows, some of the answers are beginning to emerge with regard to establishing the role of the placenta in this critical process. And for this we must be grateful to Charlie Loke, without whose foresight and generosity the meeting reported here would not have taken place.

References

1. **Eder W**, **Ege MJ**, **von Mutius E**. The asthma epidemic. *N Engl J Med* 2006; **355**: 2226–35.
2. **Colman RJ**, **Anderson RM**, **Johnson SC** *et al*. Caloric restriction delays disease onset and mortality in rhesus monkeys. *Science* 2009; **325**: 201–4.
3. **Frommer WB**. Common sense. *Science* 2010; **327**: 275–6.
4. **Hundal HS**, **Taylor PM**. Amino acid transceptors: gate keepers of nutrient exchange and regulators of nutrient signaling. *Am J Physiol Endocrinol Metab* 2009; **296**: E603–13.
5. **Van Batenburg-Eddes T**, **de Groot L**, **Steegers EA** *et al*. Fetal programming of infant neuromotor development: the generation R study. *Pediatr Res* 2010; **67**: 132–7.
6. **Boyd JD**. Some aspects of the relationship between mother and child. *Ulster Med Journal* 1959; **28**: 35–46.

Index

Note: Page numbers in italics – Tables; Page numbers in bold – Figures; Page numbers with 'n' – Notes

ACE, see angiotensin converting enzyme
ACTH, see adrenocorticotropic hormone
activin A, 193
   pre-eclampsia, 193
   see also pregnancy associated placental protein
adaptive regulation, 149
   see also amino acid transport
adolescencent pregnancy, 25–6
adrenocorticotropic hormone (ACTH), 19
advanced villous maturation (AVM), 203
   placental symmetry, 204
AGA, see appropriate for gestational age
AIMs, see ancestry-informative gene markers
AKT/mTOR signalling, **167**
allostasis, 19
alpha-fetoprotein (AFP), 192, 199
amino acid transport
   factors involved, 155
   sodium-dependent transporter, 148, 149, 153
   sodium-dependent transporter stimulators, 149
   sodium-independent transporter, 148, 149
amino acid transporters
   and fetal size, 150
   and IUGR, 150
   large babies placentas, 154
   placenta receptors, 151
   regulation, 151
   sodium-dependent transporter, 160
   trophoblast regulators, 150

amniotic cavity, 50, 55
amniotic fluid
   analysis, 55
   antioxidant molecules, *51*
amniotic sac, 48
anaemia, meternal, 56
ancestry-informative gene markers (AIMs), 136
androgenetic embryo, 75
Angelman syndrome, 91
   and ART, 91
angiotensin converting enzyme (ACE), 37
antechamber, 162
   functions, 162
antigenic paternal genome, 103
antioxidant molecules, *51*
appropriate for gestational age (AGA), 115
   see also small for gestational age
aquaporins, 21
areal density, 217
arginine
   cardiovascular adaptations, 27
   conditionally essential amino acids, 26
   inter-organ co-operativity, 27
   see also glycine
arginine vasopressin (AVP), 21–2
   activation, 21
   on liver, 21
   nitrogen balance, 21
   on placenta, 21
   role, 21
ART, see assisted reproductive technologies
arterial O_2 concentration (CaO_2), 130–1
   altitude impact, 127
   determinants, *130*

arterial O_2 saturation (SaO_2)
   altitude impact, 130
   see also arterial O_2 concentration
assisted reproductive technologies (ART), 35
   and Angelman syndrome, 91
   and Beckwith-Wiedemann syndrome, 91
   in vitro, 23
   long term effects, 25
   nutritional impact, 23
   pre-implantation nutrition, 24
   supplementation, 23
AVM, see advanced villous maturation
AVP, see arginine vasopressin

basal plasma membrane (BM), 148
BAT, see brown adipose tissue
Beckwith-Wiedemann syndrome, 91
β-cells proliferation, 85–6
bilaminar embryonic disk, 50
   see also amniotic cavity; yolk sac
binucleate cells (BNC), 178
   placental lactogen level, 180
birth weight
   adult bone mass, 218–19
   adult bone strength, **220**
   adult longevity, 175
   altitude and resident time, **128**
   altitude impact, **127**, 126–9
   altitude *vs.* surname analyses, **135**, 135–6
   blood pressure, 7, *7*
   and diseases, 22, 114, 175
   infant survival,
   ischemic heart disease mortality, **201**
   as marker, 5
   mTOR signalling, 154
   placental weight, 115
   risk factors, 1, 5, 20, 201

socio-economic status, 128
triglycerides, 159
blood
   factors affecting flow, 132
   placental impact on pressure, 8–9
   pregnancy impact on volume, 131
blood pressure, mean systolic, *7*
BM, *see* basal plasma membrane
BNC, *see* binucleate cells
bone
   calcium regulation, 227
   density and PMCA3, 227
   density, geometry and strength, 219
   in utero growth, 217–18
   infancy weight and femoral geometry, **219**
   mass, 216
   mass and birth weight, 218–19
   mass and maternal vitamin D, **221**
   mass and PMCA3expression, **224**
   maternal nutrition, 218
   strength and birth weight, **220**
   strength and infancy weight, **220**
brain, 194
   DLK1 neurogenesis, 91
   epigenetic programming, 91
   perfusion and SGA fetuses, **118**
brown adipose tissue (BAT), 85
buffering, 46

C^{14}-methyl amino-isobutyric acid (MeAIB)
   across placenta, **180**
   small *Igf2P0* null placenta, 181
   system A, 148
cancer
   pre-eclampsia, 14
   tumor suppressor gene, 64
      see also lung cancer; placental area
cardiometabolic syndrome, 25
cardiomyocytes
   embryonic, 209
   factors affecting, 3, 10
   fetal heart, 209, 210
   heart growth, 208, 211
   heart vulnerability, 208
cardiovascular disease
   offspring hypertension, 15
   placenta, 14
   placental growth, 7
cardiovascular system
   endothelium development, 208

heart features, 208
stress effect, 208
vasculature, 208
caruncles, 210
CF, *see* coelomic fluid
chondrocyte proliferation, 218
chorion
   endometrial interaction, 172
   frondosum, 163
chorionic cavity, *see* exocoelomic cavity
chorionic fluid, *see* coelomic fluid
chronic diseases
   heart failure, 10
   lung disease, 146
   5-L-oxoprolinuria, 27
   pattern,
   plasticity, 35
coelomic cavity, 55
coelomic fluid (CF), 50
   antioxidant molecules, *51*
compensatory placental growth, 5, 9, 10, 46
   hypertension, 11
   inadequacy effect, 9
   pre-eclampsia, 6
compensatory processes, 46
conceptus
   androgenetic, 74
   in compromised environment,
   healthy development, 18
   maternal nutritional state, 18
   nutrient suppy factors, 18
   nutrients, 18
   parthenogenetic/gynogenetic, 74
      see also syncytiotrophoblast
coronary artery disease prevalence, 215
coronary heart disease
   placental impact, 9
   SMR, **206**
cytotrophoblast, 49
   layer, 102
   shell functions, 48

decidua
   B cells, 106
   basalis, 105
   erosion, 93
   function, 109
   leukocytes, 105, 106
   macrophages, 106

   NK cells, 106
   T cells, 106
decidual glands, *see* endometrial glands
decidualisation, 46, 98, 105
   definition, 105
   in human, 96
   trophoblast-cell invasion, 105
developmental delay, 87
'Developmental Origins of Health and Disease' (DOHaD), 35, 57
   Cited1 model, 66
   Esx1 model, 66
   Igf2P0$^{+/-}$ *model*, 65
   imprinted genes, 65
   mouse IUGR models, 67
   oocyte maturation, 36
   Phlda2 KO and Tg mice, **63**
   Phlda2 over-expression model, 66
   plasticity, 35
developmental plasticity, 222
developmental programming
   after birth, 233–4
   cardiovascular disease, 208
   and clinical evidences, 234
   and Darwin's paradigm, 235
   environmental signals, 234
   extra-placental routes, 233
   fracture risk, 224
   implications, 168
   imprinted genes, 74
   key factors, 233, 235
   mechanisms, 9
   non-placental candidates, 233
   placental differences, 233
   sex impact on, 186
   see also developmental plasticity
dexamethasone
   leptin expression, 180
   PGE2 production, 181
   umbilical lactate uptake, 179
   vasculature, 186
diet, poor, 25
   preimplantation, 39–41
differentially methylated regions (DMRs), 77
   secondary/somatic, 77
DMRs, *see* differentially methylated regions
DNA methylation, 77, 222
   as epigenetic processes, 22
   folic acid impact, 222
   functions, 37
   gene regulation, 34
   in humans, 223

Index

DNA methylation (cont.)
  osteoporosis epigenetics, 221
  in preimplantation development, 37
  see also histone modification; imprinted genes; non-coding RNA
DNA methyltransferases (Dnmt), 222
Dnmt, see DNA methyltransferases
DOHaD, see 'Developmental Origins of Health and Disease'
Down's syndrome
  AFP level, 199
  detection issues, 194
dual energy x-ray absorptiometry (DXA), 217
dually-perfused circuit, 134
DXA, see dual energy x-ray absorptiometry

ECC, see exocoelomic cavity
'echoic cystic lesions', see 'placental lakes'
Emb-LPD model, 38
embryo
  active, 24
  androgenetic, 75
  blood flow pattern effect, 208
  culturing effect, 91
  gender based responses, 45
  interaction with environment, 40
  maternal diet impact, 39–41
  nutrient supply, 55
  over-growth, 81
  parthenogenetic/gynogenetic, 74
  preimplantation in human, 41
  quiet, 24
  transfer, 46
emphysema mortality, 146
endochondral ossification, 217
endocytosis, gestational
  LPD impact, 41
  yolk sac role, 39
endometrial glands
  conceptus survival, 56
  first trimester, 48
  functions, 48, 49
  second trimester, 48
  secretions, 48
  "uterine milk", 49, 56
endometrium, 105
  chorion interaction, 172
  NK cells, 172

uNK cells, 106
  see also decidualisation
endoplasmic reticulum stress, 166–7
epidermal growth factor, 49
epigenesis, 22–3
  developmental plasticity, 224
  epigenetic mark, 75
  human disease mechanism, 222–3
  see also DNA methylation
epigenetic regulation
  gene methylation, 147
  placental calcium transfer, 223
epigenetic reprogramming
  at embryo, 78
  environmental modulation, 79–80
  after fertilization, 78
  in utero undernourishment, 79
  in vitro embryo culture, 79
  low protein diet, 79
  pre-implantation, 78–80
  at primordial germ cell, 80–1
  uterine environment effect, 78
  at zygote, 78
epigenetics, 221
ER stress, 173
  ER homeostasis, 137
  on placental function, 168
  placental transporter activity, 168
essential amino acids, conditionally, 19
  functions, 26
  metabolic interactions, 27
"eutherian" mammals, see 'placental mammals'
EVT, see extravillous trophoblast cells
exocoelomic cavity (ECC), 50, 54
  coelomic fluid, 50
  materno-embryonic transfer activity, 50
extra-embryonic lineages, 39, 40
extra-embryonic tissues
  in androgenetic conceptus, 75
  imprinted genes, 81
  in parthenogenetic/gynogenetic conceptus, 74
extravillous trophoblast cells (EVT), 47, 102
  HLA-B + A expression,
  interstitial invasion, 93
  mRNA for PAPP-A, 191
  spiral artery conversion, 163
  types, 102

  see also placentation; syncytiotrophoblast; trophoblastic plugs

fAd, see full-length adiponectin
fat diet model, high, 154
femoral geometry and infancy weight, **219**
fetal cortisol
  concentration effects, **176**
  developmental programming, 186
  glucose uptake, 179
  MeAIB uptake, 179
  placental amino acid delivery, 179
fetal growth, 115
  abnormality indicator, 114
  altitude-associated reductions scans, 137
  amino acid, 147
  amino acid transporters in altered, 149
  feto-placental O_2 consumption, 139–41
  imprinted placental genes' role, 155
  kidney's role,
Fetal Growth Restriction (FGR), 114
  gene deletion effect, 177
  impact on gender, 205
  mTOR activity, 154
  protection against, 144
  trophoblast invasion, 103
fetal heart
  developmental regulators, 210–11
  sensing hemodynamic force, 209
fetal heart rate (FHR), 119
fetal nutrition
  demand anticipation, 18
  diet variation impact, 20
  factors affecting supply, 152
  hypertension, 11
  impact on development, 24
  maternal diet, 21
  nutrient prerequisites, 18
  undernutrition, 22
fetal programming, 5
  amino acid, 147
  amino acid transport role, 154
  chronic diseases, 5
  decidualisation, 105
  early steps' importance,
  gastrulation,
  geographic prevalence,
  HLA-C, 107
  KIR, 107

Index

fetus
  maternal immune response in, 102
  NK cells receptor, 106–7
  u NK cells, 106
  see also developmental programming

fetus
  adaptation, 125
  adolescent pregnancy, 33
  adult diseases, 57
  Cdkn1c expression, 61
  Cited1 deficient placenta, 64
  developmental plasticity, 74
  developmental programming, 1, 4, **167**
  DNA in maternal serum, 200
  epigenetic alterations, 74
  Esx1 deficient placenta, 64
  fatty liver, 207
  feed back, 125
  gestational diabetes on weight, 15
  Grb10 expression, 62
  head circumference *vs.* AIMs
  hypoxia, 116–17
  Igf2r expression, 62
  nutrient sensor, 3
  nutritional impact, 8, 9, 10, 11, 22
  oxygen requirement, 55
  Peg1/Mest expression, 62
  Peg10 role, 63
  Peg3 deletion, 63
  placenta size, 14, 71
  placental mitochondrial DNA *vs.* umbilical pO_2, **118**
  Plagl1 deletion, 63
  reduced aminoacid, 15
  resource limitation for, smoking mother, 26
  thyroid hormone production, 55
  under-oxygenation, 144
  Zac1 growth promoter, 64

FHR, *see* fetal heart rate

fibroblasts, 45
  see also trophectoderm

full-length adiponectin (fAd), 149

gastrulation,

gene expression
  DNA methylation, 37
  fetal undernutrition, 22
  IGF2 expression, 64
  in inflammation, 146
  knockout, 61
  maternal nutrition, 22
  methylation, 34
  over-expression, **63**, 61

Phlda2 expression, 61
suppression, **63**

gene transcription
  altered phenotype, 23
  as epigenetic processes, 22
  silencing, 22

gene, natural selection signatures, *137*

genomic imprinting, **75**, 57, 75
  DNA methylation, 77
  histone modification, 77–8
  intra-genomic conflict model, 60
  non-coding RNA, 78
  in placenta, 58–60
  see also imprinted genes

gestation
  age and UA blood flow, 132
  gestational sac, **50**
  length and adult diseases, 100

glucocorticoid receptors (GR), 20, 181
  distribution, 186
  expression, 187
  functions, 20
  integrated responses, 33
  maternal grooming, 20

glucocorticoids, 20–1
  action mechanism, 181
  amino acid metabolism, 179
  in cell differentiation, 176
  feto-placental exposure, 176–7
  functions, 181–2
  hormone metabolism, 181
  intrauterine environmental signal, 175
  intrauterine programming, 175
  in late gestation, 175
  in maternal and fetal circulation, **182**
  maturational signals, 175
  placental and fetal weights, *178*
  placental development, 177–9
  placental efficiency, 177
  progesterone/oestrogen, 180

glucose
  disposal site, 84
  umbilical uptake, 179

glycine, 27
  across placenta, 33
  conditionally essential amino acids, 26
  "pacemaker of metabolism", 27
  see also arginine

glycodelin, 49

GM-CSF, *see* granulocyte-macrophage colony-stimulating factor

GR, *see* glucocorticoid receptors

granulocyte-macrophage colony-stimulating factor (GM-CSF), 37

GRIT, *see* Growth Restriction Intervention Trial

growth, 17
  blastocyst expansion, 17
  factor gene, 57
  implantation, 18
  velocity of body length, 217
  velocity of bone, 217
  vulnerability period, 17

Growth Restriction Intervention Trial (GRIT), 119

haemochorial placentation challenges, 162

hazard ratio
  for hip fracture, 224
  pre-eclampsia, **6**

heart, 208
  blood vessel development, 208
  cell detection, 214
  developmental features, 208
  diseases, 208
  endothelium development, 208
  growth, 209–10
  stress effect, 208

heart attack, *see* myocardial infarction

hemodynamic force, 209

hemodynamic signal, 208–9

hip fracture
  and placental eccentricity, *224*
  placental efficiency, 224

histiotroph, *see* "uterine milk"

histone modification, 77–8, 222
  in cancer, 222
  heritability of, 78
  in programmed tissue, 67
  and transcription, 23
  see also DNA methylation; imprinted genes; non-coding RNA

HLA, *see* human leucocyte antigen

'hose-pipe' effect, 167

HPA, *see* hypothalamo-pituitary-adrenal

human chorionic gonadotrophin (hCG), 192

human evolution, immunological perspective,

human leucocyte antigen (HLA),
    104–5
  against HIV,
    HLA-C, 107
    HLA-C and reproductive success,
      107–8
  in human evolution,
hypertension
  fetal nutrion, 11
  mothers' height, 9
  odds ratio, **11**
  offspring cardio vascular risk, 202
  placental area, 8, 9, **8**
  sex difference and placental weight,
    11
hypothalamo-pituitary-adrenal (HPA),
    17
  Avp role, 21
  axis prepartum activation, 176
hypoxemia
  cardiac cell numbers, 210
  IUGR, 210
hypoxia
  cardiac cell enlargement, 145
  intermittent, 140
  reperfusion and spiral artery
    conversion, 171
  reperfusion injury, 125
  uterine vessels' growth, 134
  vascular responses, 145

I/R injury, *see* ischemia/reperfusion
    injury
IAPs, *see* intra-cisternal A particles
ICRs, *see* imprinting control regions
IGF, *see* insulin like growth factors
IGFBP, *see* insulin-like growth factor
    binding protein
Ig-SF receptors, *see* Ig-superfamily
    receptors
Ig-superfamily receptors (Ig-SF
    receptors), 106
impedance, 209
implantation
  blastocyst orientation, 164
  human, 161
  placental site, 189
  uterine cornuae, 165
  variations, 161
imprinted genes, 74, 151
  on adipose tissue, 84–5
  adult diseases, 87
  arrangement, 75

β-cells proliferation, 85–6
  chromosomal locations, **58**
  Cited1 model, 66
  DNA methylation, 57, 60
  DOHAD models, 65
  dosage sensitive, 87
  Esx1 model, 66
  functions, 57, 81–7
  homeostasis control, **82**
  Igf2 knockout, 83
  Igf2 placental-specific model
    possibility, 67
  Igf2 role, 87
  $Igf2P0^{+/-}$ *model*, 65
  IUGR, 64–5
  monoallelic restriction, 87
  multiple, 60–4
  neonatal diabetes, 86
  neurological disorders, 91
  parental imprinting, 57
  Phlda2, **62**
  Phlda2 marker, 71
  Phlda2 over-expression model, 66
  in placenta, 58–60
  polymorphism, 91
  psychiatric diseases, 86
  role of, 74, 75
imprinting, 74, 136
  and ART, 41
  brain's scope, 91
  defect, 72
  and histone modification, 60,
    77–8
  intra-genomic conflict model, 60
  IUGR, 65
  loss of, 61, 62, 85, 87
  monoallelic expression, 57
  mutations, 71
  in placenta, 59
  placentation, 81
  see also genomic imprinting
imprinting control regions (ICRs), 75
  12 Dlk1-Dio3 domain, **77**
  categories, 77
  DNA methylation reprogramming,
    80
in utero
  drug effect, 174
  stress impact, 86
in vitro culture (IVC), 36
  GM-CSF, 37
in vitro maturation (IVM), 36
  long term effects, 36
inhibin A and pre-eclampsia, 193
insulin like growth factors (IGF), 191
  functions, 191

insulin-like growth factor binding
    protein (IGFBP), 191
insults, 3
  impact on generations, 199
intergenerational communication
    consequence, 20
  epigenetic factor, 20, 22
  glucocorticoid's role, 20
intermittent perfusion, 165
intervention studies,
intervillous circulation, 47–8
intervillous space blood pressure, 162
intra-cisternal A particles (IAPs), 80
intra-genomic conflict model, 60, 61
intramembranous ossification, 217
intrauterine environmental conditions,
    175
intrauterine fetal death, 119
intrauterine growth restriction
    (IUGR), 64
  acidemia, 117
  amino acid level, 149
  amino acid transporters, 149
  amino acid uptake, 119
  blood flow, 124
  BM calcium pump, 150
  brain damage susceptibility, 119
  clinical relevance, 114–15
  etiology, *115*
  fatty acid level, 159
  fetal adaptation, 125
  fetal signal, 159
  head circumference, 124
  hypoxia, 117
  Igf2 knockout mice, 124
  imprinting defect, 72
  monitoring, 119–20
  placental mitochondrial DNA *vs.*
    umbilical pO_2, 118
  placental phenotype, **117**
  reduced *Igf2* expression, 83
  reduced nutrient uptake, 116
  seveity classification, 120
  thyroid impact, 211
  truglyceride level, 159
  see also Fetal Growth Restriction
ischaemia reperfusion
  episodes, 165
  hypoxia, 172
ischemia/reperfusion injury (I/R
    injury), 140
IUGR, *see* intrauterine growth
    restriction

IVC, *see in vitro* culture
IVM, *see in vitro* maturation

junctional zone, 99

Killer Immunoglobulin receptors (KIR), 106, 107
  AA and pregnancy,
  against HIV,
  in human evolution,
  reproductive success, 107–8
KIR, *see* Killer Immunoglobulin receptors
knockout (KO), 57, 61
  HOXA13mice, 209
  Igf2, 83
  liver-specific *Igf1*, 158
  TNF receptor impact, 73
KO, *see* knockout

large offspring syndrome (LOS), 36
LDL receptor functions, 39
LOS, *see* large offspring syndrome
low protein diet (LPD)
  effects, 158, 222
  model, 153–4
  osteoporosis, 221
lung cancer
  amino acid level, 10
  placental impact, 10
  short at birth, 10

macrophages, 45
  see also trophectoderm
Major Histocompatibility Complex (MHC), 106
  antigens, 104
mammalian target of rapamycin (mTOR), 147
  complexes, 150
  protein translation, 150
  signalling, 3, **153**, 154
maternal
  BMI and neonatal fatty acid, 207
  care and plasticity, 35
  and embryonic transfer activity, 50
  fetal blood flow, 116
  fetal boundary formation, 102
  fetal exchange, 48, 57
  fetal interface, **49**
  grooming effects, 20
  nurturing behaviour and *Peg3*, 73

nutrition and placental sensitivity, 207
obesity, 33
obesity consequences, 35, 207
oestrogen on offspring, 227
perfusion, 93
placental-fetal phenotypes, 9, 10
plasma with antioxidant molecules, 51
pre-conceptional nutrition, 206–7
maternal immune response
  decidualisation, 105
  HLA, 104–5
  HLA-C, 107
  KIR, 107
  NK cells receptor, 106–7
  u NK cells, 106
  uterine mucosal lymphocytes, 105–6
maturational signals, 175
metabolic axis control, 86–7
MHC, *see* Major Histocompatibility Complex
microvillous plasma membrane (MVM), 148
mid-gestational lethality, 81
miscarriages
  pheochromocytoma, 173
  reasons, 55
  trophoblast invasion, 100
mitochondrial DNA (mtDNA), 129
mtDNA, *see* mitochondrial DNA
mTOR, *see* mammalian target of rapamycin
muscle pump, *see* fetal heart
MVM, *see* microvillous plasma membrane
myocardial infarction, 208

neonatal diabetes, 86
nitrogen balance, 21
NK cells receptor, 106–7
non-coding RNA, 78
  see also DNA methylation; histone modification; imprinted genes
non-imprinted gene IUGR models, 67
non-shivering thermogenesis (NST), 85
NR, *see* nuclear receptors
NST, *see* non-shivering thermogenesis

nuclear receptors (NR), 119
nutrients
  essential, 19
  factors affecting supply, 152
  protected pathways, 19
  as stressors, 19
nutrition, 26
  adolescent pregnancy, 34
  conditionally essential amino acids, 26
  demand in pregnancy, 26
  endogenous formation, 26
  follicle development, 36
  gender selection, 24
  genome activation, 45
  histiotrophic nutrition, 48
  integrated responses, 33
  IUGR fetus, 119
  low protein diet, 32
  maternal-fetal blood flow, 116
  metabolic interaction, 23
  metabolic stress consequences, 25
  periconceptional undernutrition, 38
  placental transport, 179–81
  poor maternal diet, 39–41
  preconceptual, 32
  preimplantation development, 38
  prepregnancy obesity, 25
  undernourishment, 36
  uterine – umbilical blood flow, 116
  villi surface, 115

obesity
  effects on pregnancy, 25
  factors, 33
  maternal, 33, 35, 207
  pregnancy, 154
  prepregnancy, 25
odds ratio
  men, **11**
  women, **11**
oocyte donation and pre-eclampsia,
ossification, 217
osteoblasts, 217
osteoid matrix, 217
osteoporosis, 216, 224
  developmental origins, **216**, 221
  epigenetic mechanisms, 221–2
over-expression, 61, 71
oxidative stress
  apoptotic cascade activation, 164
  causes, 165
  placental, 166
  on placental function, 168
  placental shape, 172

oxidative stress (cont.)
  vascular regression, 214
  vasoconstriction, 169
  villi regression, 169, 173
oxygen
  altitude impact, 126–7
  delivery normalization 143n.4
  fetal hypoxia, 116–17
  high altitude fetus, 138
  hypoxia–reperfusion injury, 125
  maternal-fetal blood flow, 116
  supply chain, 126
  umbilical pO_2, 116
oxygen consumption (VO_2), 138
  feto-placental oxygenation, *139*, 139–41
  at high altitude, 139
oxygen saturation SaO_2
  high altitude neonate, 138
  subacute infantile mountain sickness, 145

P/F ratio, *see* placental/fetal weight ratio
PAMP, *see* Pathogen-associated Molecular pattern
pancreas, 85
  Igf2 role, 85
  neonatal diabetes cause, 86
  Neuronatin role, 86
PAPP-A, *see* pregnancy associated placental protein-A; pregnancy associated plasma protein-A
parental imprinting, *see* genomic imprinting
parthenogenetic/gynogenetic embryo, 74
Pathogen-associated Molecular pattern (PAMP), 104
PCU, *see* periconceptional undernutrition
PE, *see* primary endoderm
peak bone mass, 217
  heriditary, 216
  and strength, 224
periconceptional undernutrition (PCU), 38
  effects, 38–9
perinatal mortality
  Doppler flow velocimetry, 190
  IUGR, 120
  Peg3 loss and, 87
  placental maturity, 189

perinatal stressors, 201
peripheral computed tomography (pQCT), 219
PGCs, *see* primordial germ cells
PI, *see* pulsatility index
placenta, 60, 103, 211
  abnormal placentation, 6, 14
  adaptiveness, 3, 5
  adult diseases, 115
  altitude impact, 140, 168
  amino acid transport, 118–19, 123, 147–8, 149–52, 158
  amino acid transporters, 148–9
  anatomical study, 47
  arterio-venous fistula, 140
  basal plate, **93**
  bloodflow geometry, 99
  calcium acquisition, 218
  chorion frondosum, 163
  circulation, 2–3, 47–8, 116, 168
  circumvallate, 174
  compensatory growth, 7, 9
  cross-sections, **94**
  damage, 116
  demand anticipation, 27
  development and glucocorticoids, 177–9
  diabetic, 73
  dual perfusion, 117
  eccentricity and hip fracture, *224*
  efficiency, 7, 71
  endocrine function, 84, 180–1
  endometrium and, 2
  energy consumption, 60
  exocoelomic cavity, 50
  factors affecting development, 1, **2**
  on fetal growth, 19
  as fetal tissue,
  feto-maternal integration, 159
  fetus feed back, 125
  function assessment, 3
  functions, 1, 3, 17, 18, 20, 47, 54, 116, 233
  gender diference, 10, 205
  glucocorticoid exposure, 176–7
  glucose, 10, **179**, 118, 158
  GR expression, 20
  in heart development, 211
  historical record, 18
  hormone, 149, 151
  human fibroblasts, 45
  human IUGR, 73
  human macrophages, 45
  hypertension, 1, 8–9
  IGF1regulator, 159
  imaging, 19
  imbalanced free radicals, 47
  immune barrier, 147

  implantation, 116
  insufficiency, 159
  insulin signalling, 149–50
  insults, 3, 16
  intermittent perfusion, 165
  invasion, 100, 101
  maternal factors, 2, 3, 5, 7, 19
  and maternal immune system, 103
  metabolic control, 147
  mitochondrial DNA *vs.* umbilical pO_2, **118**
  morphology, 177–9
  mTOR, 207
  nutrient demand, 18
  nutrient sensor, 3, 5, **152**, 118, 152–3, 155, 207, 210
  nutrient transport, 18, 100, 179–81, 202
  effect on offspring health, 207
  oxidative stress, 214
  Phlda2 protein, 72, 73
  placentomegaly, 61, 62
  polarisation, 6, 15
  prednisolone's effect, 186
  pre-eclampsia, 6
  primitive metabolic pathways, and programming,
  pro-inflammatory cytokines, 207
  'pseudo-labyrinthine', 92
  receptors, 151
  rodent models controversy, 100, 101
  secondary yolk sac, 51
  shape, 5–6
  size, 5, 7, **6**, 56, 72, 73
  structure, **59**
  syncytiotrophoblast, 47
  system A activity study, 215
  thicker, 167
  trophoblast invasion, 2
  VEGFR-1, 192
  venous equilibrator, 116, 140
  volume, 19
  weight, 1, 14
placenta creta, 103
placenta previa, 189
  see also implantation; placentation
placenta, human
  abnormalities, 47
  development, 49–50
  spiral artery, 94
placental
  adaptation, 83
  barrier, **148**
  calcium transport, **223**, 223, 227
  calcium transporter gene, 223
  homeostatic pathways, 3
  hyperplasia, 81
  impaired development effects, 161

inadequacy, 202
infarcts, 15
lesion, 189
maturity, 189
morphology, 177
nutrient transport, 150
nutrient transport regulation, **151**
oxidative stress, 166
oxygen delivery, 144
phenotype, small, **167**
symmetry, 204
uptake regulation, 119
vasculature and heart growth, 209
volume, 188–9

placental area
  amino acid level, 10
  blood pressure indicator, 9
  coronary heart disease, 9
  hypertension, 8, 9, **8**
  lung cancer, 10
  mother's height, 8

placental development
  abnormal maternal blood flow, 165–7
  failure consequence, 61
  Igf2 gene, 181

placental efficiency, 7, 99
  biological constraints, 99
  Dutch famine, 8
  glucocorticoid's role, 177
  hip fracture, 224
  maternal body composition, 15
  Ramadan eating pattern, 7–8

placental growth, 115
  Igf2P0, 83
  imprinted genes, 57
  maternal nutrition, 19, 26, 205
  maternal anaemia, 56
  multiple imprinted genes, 60–4
  Phlda2, **62**
  Phlda2/Ipl expression, 83
  restriction, 177
  sex difference, 10–11

placental growth factor (PlGF)
  pre-eclampsia, 192, 193
  SGA, 193

placental insufficiency, 114, 115–19, 120
  animal models, 210
  etiology, 115, 209
  genetic model possibility, 67
  histological features, **204**
  human, 203–6
  IUGR, 81, 114–15
  placental weight, 115–16
  reverse flow, 209

placental lacunae, 189
  see also trophoblast invasion

'placental lakes', 167

'placental mammals', 48, 63

placental programming
  chronic disease, 11, 12
  sex difference in, 11

placental shape
  abnormality effect, 161, 168
  determinants, 161
  impact on gender, **205**
  human, 161
  oxidative stress, 172, 173
  pre-eclampsia, 202–3
  stable form, 168
  variations, 161
  see also placental surface

placental size
  blood pressure, 9
  cardiovascular risk, 201
  compensatory growth, 211
  mother's birth weight, 15
  offspring disease/hypertension, 211
  pre-eclampsia, 6
  uterine blood flow, 7

placental surface, 1
  blood pressure, 9
  chronic heart failure, 10
  disease predictor, xi
  function, 5
  hypertension, 9
  malnutrition indicator, 9
  pre-eclampsia, 6
  transfer capacity, 115

placental thickness, 8, 14
  gestational age, 188
  see also pre-eclampsia

placental weight, 3, 7
  abnormalities, 115–16
  adult diseases, 202
  gender diference, 10
  and glucocorticoids, 177
  hypertension, 5, 7, 8, 9, 11, 116
  hypotension, 7
  mean systolic blood pressure, 7
  see also birth weight; placental thickness

placental/fetal weight ratio (P/F ratio), 115

placentation, 172
  abnormality, 101
  AFP, 192
  angiogenesis associated proteins, 192–3
  assessment, 188–9, 194–5

  biochemical markers, 191–4
  decidua basalis, 105
  defect indicator, 192
  driving forces,
  failure cause, 81
  hCG, 192
  HLA-C antibody,
  imprinting role, 81
  as indicator, 188
  KIR:HLA-C interaction, 107–8
  optimal development, 96
  PAPP-A, 191–2
  placenta derived proteins, 193–4
  preconceptual nutrition, 32
  regulatory T cells, 105
  trophoblast invasion, 189
  upright gait, 108
  see also haemochorial placentation challenges; trophoblast invasion

placentomegaly, 61, 62

placentomes, 179, 210

PlGF, *see* placental growth factor

polymorphic genes in reproduction, 108

POP assessment, 195

postpartum depression and maternal nutrition,

pQCT, *see* peripheral computed tomography

Prader-Willi syndrome, 91

pre-eclampsia, 6
  abnormal flow-mediated dilatation, 199
  ADAM12, 193
  adult diseases, 198
  AFP-PAPP-A indicator, 200
  angiogenesis associated proteins, 192–3
  anorexic mothers, 124
  cancer, 14
  Cdkn1c deficiency, 61
  "dangerous" males,
  effects, 116
  genetic cause, 103
  hazard ratios, **6**
  hCG, 192
  immune recognition, 103
  impaired invasion, 92, 95, 96, 97
  IUGR, 96
  marker's scope, 200
  maternal cardiovascular risk, 124
  maternal endothelial damage, 124
  maternal low birth weight, 124
  offspring cardio vascular risk, 202
  oocyte donation,

pre-eclampsia (*cont.*)
   oxidative stress, 133
   PAPP-A, 191, 193
   placental oxidative stress, 166
   placental shape, 202–3
   placental volume, 188
   PlGF, 193
   pre-pregnancy Doppler, 199
   pulsatility index, 190
   risks, 103
   triglycerides level, 159
   trophoblast invasion, 102, 103, 124
   UA Doppler, 199, 200
   UPI *vs.* normal placenta, **206**
   utero-placental haemodynamics, 173
   see also placental size
pregnancy
   adolescent, 25–6, 33, 34
   arginine's role, 27
   AVP sensitivity, 21
   blood volume, 131
   complications, 162, 188, 194
   decidual erosion, 93
   dexamethasone role, 47–8
   glucocorticoids, 20, 177
   GR promoters, 23
   high metabolic demand, 26
   HLA, 104–5
   HPA axis stress response, 20
   immunology of, 103–4, 108
   intergenerational communication, 20
   IUGR, 114
   leukocytes, 105
   low protein diet, 23
   macrophages, 106
   maternal diet impact, 24
   maternal obesity, 154
   micronutrient supplements, 33
   mothers' preparation, 109
   normal features, 18
   nutritional metabolic stress, 25
   overnourishment, 26
   5-L-oxoprolinuria, 27
   placental blood flow, 163, 168
   and safety factor,
   smoking, 26
   as stress, 19
   UA blood flow, **133**, 131, 140
   UA diameter, 131
   vasoactive substances, 133–4
pregnancy associated placental protein-A (PAPP-A), 3
   femur length, 194
pregnancy associated plasma protein-A (PAPP-A)
   see also activin A

preimplantation
   development, **37**
   Emb-LPD model, 38
   IVC effects, 36–7
   LOS, 36
   maternal nutrition, 38
   mono-allelic gene expression, 37
   post-fertilisation cleavage, 36
   undernourishment, 36
prenatal diagnosis limitation, 55
prepregnancy obesity, 25
primary endoderm (PE), 39
primordial germ cells (PGCs), 80
   imprint reacquisition, 80
principle component (PC)
   analysis, 129
   SNP, **129**
protein synthesis inhibition, **167**
puberty delay etiology, 87
pulsatility index (PI), 119
   flow resistance, 190
   pre-eclampsia prediction, 190
   SGA infant, 190

reactive oxygen species (ROS), 140
reproduction
   histo-incompatibility, 103–4
   HLA-C and KIR, 107–8
   KIR AA effect, 108
   maternal receptor significance, 108
   polymorphic genes, 107–8
ROS, *see* reactive oxygen species

SaO$_2$, *see* arterial O$_2$ saturation
SCOPE assessment, 195
secondary yolk sac (SYS), 50
   coelomic fluid, 54
   formation, 51
   functions, 51–2
   serum proteins, 51
sEng, *see* soluble endoglin
sFlt-1, *see* soluble fms-like tyrosine kinase 1
SGA, *see* small for gestational age
short mothers
   amino acid level, 10
   amino acid synthesis, 8
   hypertension, 8
   lung cancer, 10

systolic pressure, 9
   see also lung cancer; tall mothers
Signal Transducer and Activator of Transcription [3] (STAT3), 149
'silent period', 234
single nucleotide polymorphism (SNP), 129
   principle components, **129**
skeletal development, 217–18
   factors affecting, 219–21
skeletal recovery, 228
small for gestational age (SGA), 64, 114, 190
   brain perfusion, **118**
   hCG, 192
   placental volume, 188
   see also intrauterine growth restriction
SMR, *see* standardised mortality ratios
SNP, *see* single nucleotide polymorphism
soluble endoglin (sEng)
   pre-eclampsia, 192, 193
   SGA, 193
soluble fms-like tyrosine kinase 1(sFlt-1), 192, 193
Southampton Women's Study (SWS), 195
spiral arteries, 173
   alternative nutrition source, 163
   anatomy, 162
   antechamber, 162
   bipedalism, 97
   blood flow, 162
   cross-sections, **94**
   erosion, 99
   fate of, 174
   functions, 161
   haemochorial placentation, 162
   'hose-pipe' effect, 167
   invasion, 95
   myometrial sections, **95**
   placental efficiency, 99
   placentation, 6
   and pregnancy, 162
   research scope, 14
   velocity, 173
   villus-free cavity, 162
spiral artery conversion, 162, 163, 165, 172
   deficiency, **168**, 163, 165, 167
   impact on placenta,

standardised mortality ratios (SMR), **206**
STAT3, *see* Signal Transducer and Activator of Transcription [3]
stem cell differentiation, mesenchymal, 218
stillbirth
　AFP and hCG, 199
　gender impact, 198
　hCG, 192
　risk analysis, 190
　trophoblast invasion, 103
stress
　allostasis, 19
　diet variations, 20
　GR expression, 20
　hormonal responses, 19
　in periconceptual period, 20
　regulatory genes, 23
　response, 20
stressors, 17
subacute infantile mountain sickness, 145
superficial fibrinoid layer, 100
surname analysis, **135**
SWS, *see* Southampton Women's Study
syncytiotrophoblast, 47, 49
　glucose transporters, 55
SYS, *see* secondary yolk sac
system A, *see* amino acid transport
system L, *see* amino acid transport

tall mothers
　hypertension, 8
　placental area, 9
　systolic pressure, 9
　see also short mothers
TAM, *see* time-averaged mean velocity
TE, *see* trophectoderm
TGF-β, *see* transforming growth factor β
"Thrifty phenotype", 74
thyroid hormones regulation, 87
time-averaged mean velocity (TAM), 131, 131n.3
TNF-α, *see* Tumor Necrosis Factor-alpha
tocopherol transfer protein, 49
toxaemia and cardiovascular risk, 15

'trabeculae', 92
transforming growth factor β (TGF-β), 193
"tri-mark", 78
trophectoderm (TE), 39
　blastocoel, 45
　blastocyst differentiation, 39
　endocytosis, 39–40
　maturation, 39
　nutrient sensing, 40
　see also fibroblasts; macrophages
trophoblast
　active transport, 55
　Cited1 expression, 64
　degeneration, 164
　extravillous, 163
　HLA, 104–5
　HLA-C, 107
　hormones,
　properties, 104
　sample limitation,
　sub-populations, 102
　transporters, 54
trophoblast invasion, 2, 92, 93, 99, 100, 140, 162, 189
　abnormal, 102–3
　altitude impact, 145
　blood flow, 124
　decidual NK cells, 109
　deficiency effect, 165
　depth advantages, 95–6
　effect of shallow, 165
　end points, 100
　endovascular, 94–5
　factors affecting, 165
　function of, 102
　junctional zone, 99
　maternal alleles role, 100
　miscarriages, 100
　pattern, 163
　pre-eclampsia, 92, 102
　prematurity, 103
　spiral arteries, **96**, 165
trophoblastic plugs, 48
Tumor Necrosis Factor-alpha (TNF-α), 149

UA, *see* uterine artery
ubiquitous phenomenon, 217
umbilical
　artery Doppler, 190–1, 199
　circulation, 54
　cord centricity, 14

umbilicoplacental embolisation, 210
　cardiomyocytes activity
　see also placental insufficiency
uniparental disomies (UPDs), 57
uNK, *see* uterine natural killer cells
UPDs, *see* uniparental disomies
UPI, *see* uteroplacental insufficiency
"uterine", 124n.1
uterine – umbilical blood flow, 116
uterine artery (UA), 126, 199
　Doppler, 199
　enlargement, 134
　growth, 134
　in twinning, 127
　uterine definition, 124n.1
　O_2 delivery and VO_2, *139*
uterine artery (UA) blood flow, 131–3, 173
　abnormal velocity, 167
　AIMs, 136
　altitude impact, 139–40
　determinants at altitudes, *132*
　Doppler velocimetry, 189–90
　fetal growth, 138–9
　feto-placental development, 126
　vs. feto-placental oxygenation, 139–41
　genome scans, 136–7
　I/R injury, 140
　intermittent hypoxia, 140
　maternal blood viscosity, 162
　non-pregnancy *vs.* pregnancy, **133**
　normalization, 143n.4
　vasoactive substances, 133–5
uterine blood flow, 131
　high altitude populations, 3
　impact, 26, 152
　IUGR fetus, 124
　in labour, 166
　placental resistance, 116
　pre-eclampsia prediction, 173
　regulation, 137
　see also placental efficiency; placental size
uterine cornuae, 165
uterine glands, *see* endometrial glands
"uterine milk", 48, 54
　epidermal growth factors, 49
　production, 49
　see also endometrial glands
uterine mucosal lymphocytes, 105–6
uterine natural killer cells (uNK cells), 92, 106, 107

uterine O$_2$ delivery, 130
  altitude impact, 130–5
  fetal growth, 139–41
uterine vasculature changes, 162
uteroplacental circulation, Doppler, 189–91
uteroplacental insufficiency (UPI), **204**, 203

variable risk, 20
vascular deaths, 208
vascular endothelial growth factor (VEGF), 134
vascular endothelial growth factor receptor 1 (VEGFR-1), 192
vascular resistance, 134
vasoactive substances at altitudes, 133
VDR, *see* vitamin D receptor
VEGF, *see* vascular endothelial growth factor (VEGF)
VEGFR-1, *see* vascular endothelial growth factor receptor [1]
villi
  development, 49–50
  hypovascularisation, 178
  regression, **164**, 163–5
  'trabeculae', 92
villous trophoblast, 102

visceral yolk sac endoderm (VYSE), 40, **40**
vitamin D
  bone mass, **221**
  and uterine function, 227
vitamin D receptor (VDR), 219
VO$_2$, *see* oxygen consumption
VYSE, *see* visceral yolk sac endoderm

yolk sac
  functons, 55
  GLUT1 transporter protein, **51**
  mesenchymal cells, 55

zinc-finger protein, 84

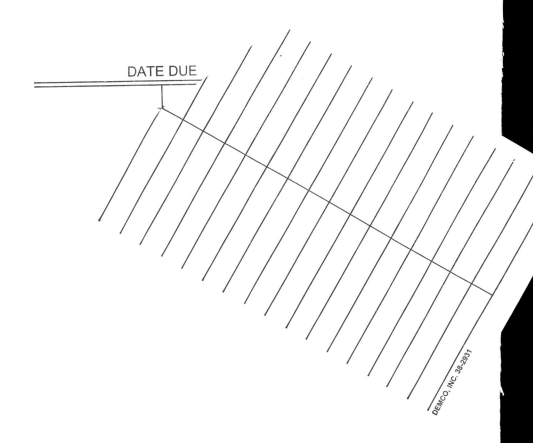